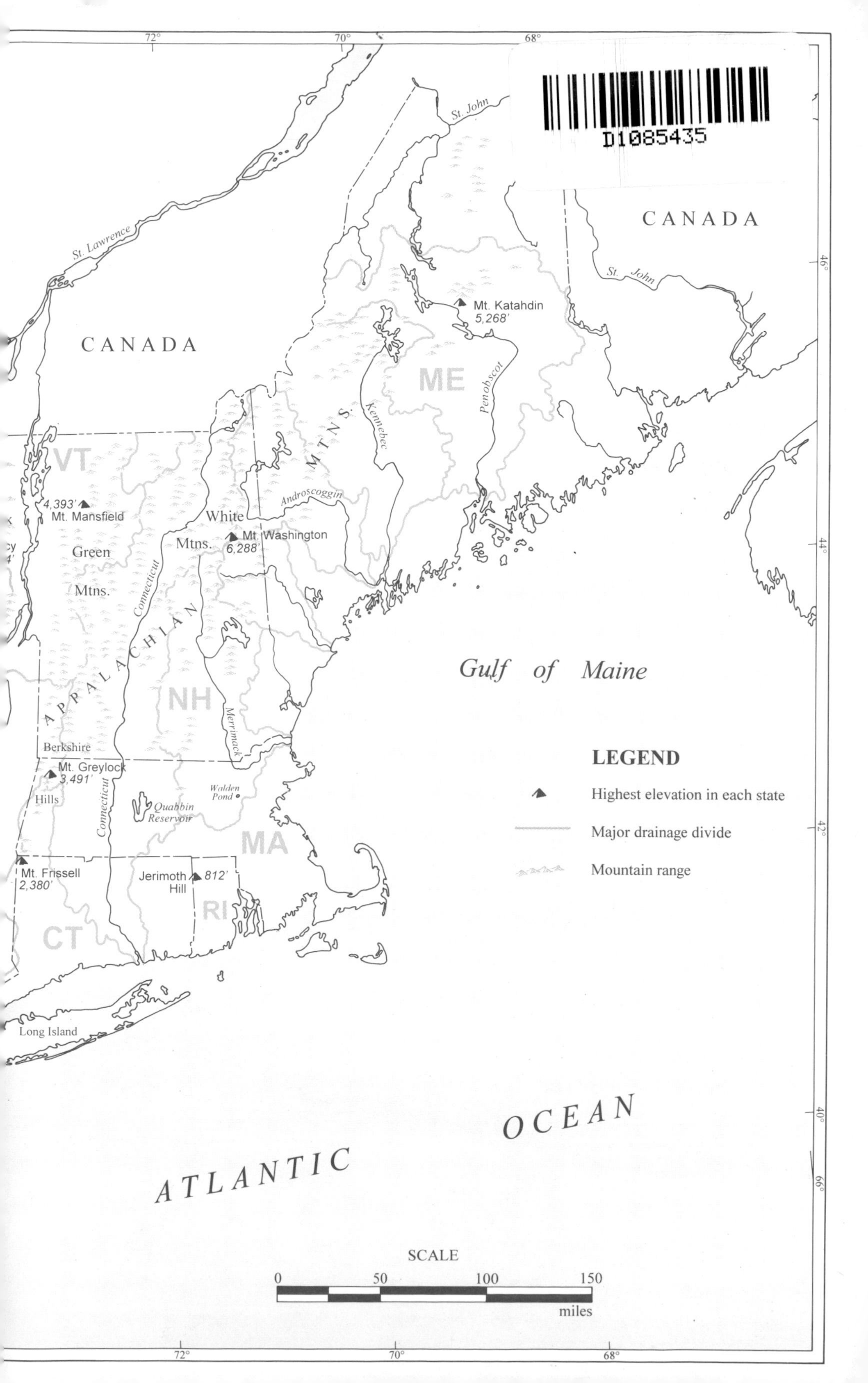

D1085435
CANADA
CANADA
St. Lawrence
St. John
St. John
Mt. Katahdin
5,268'
ME
Kennebec
Penobscot
MTNS.
Androscoggin
VT
4,393'
Mt. Mansfield
Green
Mtns.
White
Mtns.
Mt. Washington
6,288'
Connecticut
APPALACHIAN
NH
Merrimack
Berkshire
Mt. Greylock
3,491'
Hills
Connecticut
Quabbin
Reservoir
Walden
Pond
MA
Mt. Frissell
2,380'
Jerimoth
Hill
812'
RI
CT
Long Island
Gulf of Maine
LEGEND
Highest elevation in each state
Major drainage divide
Mountain range
ATLANTIC OCEAN
SCALE
0
50
100
150
miles
72°
70°
68°
46°
44°
42°
40°
66°

The Northeast's Changing Forest

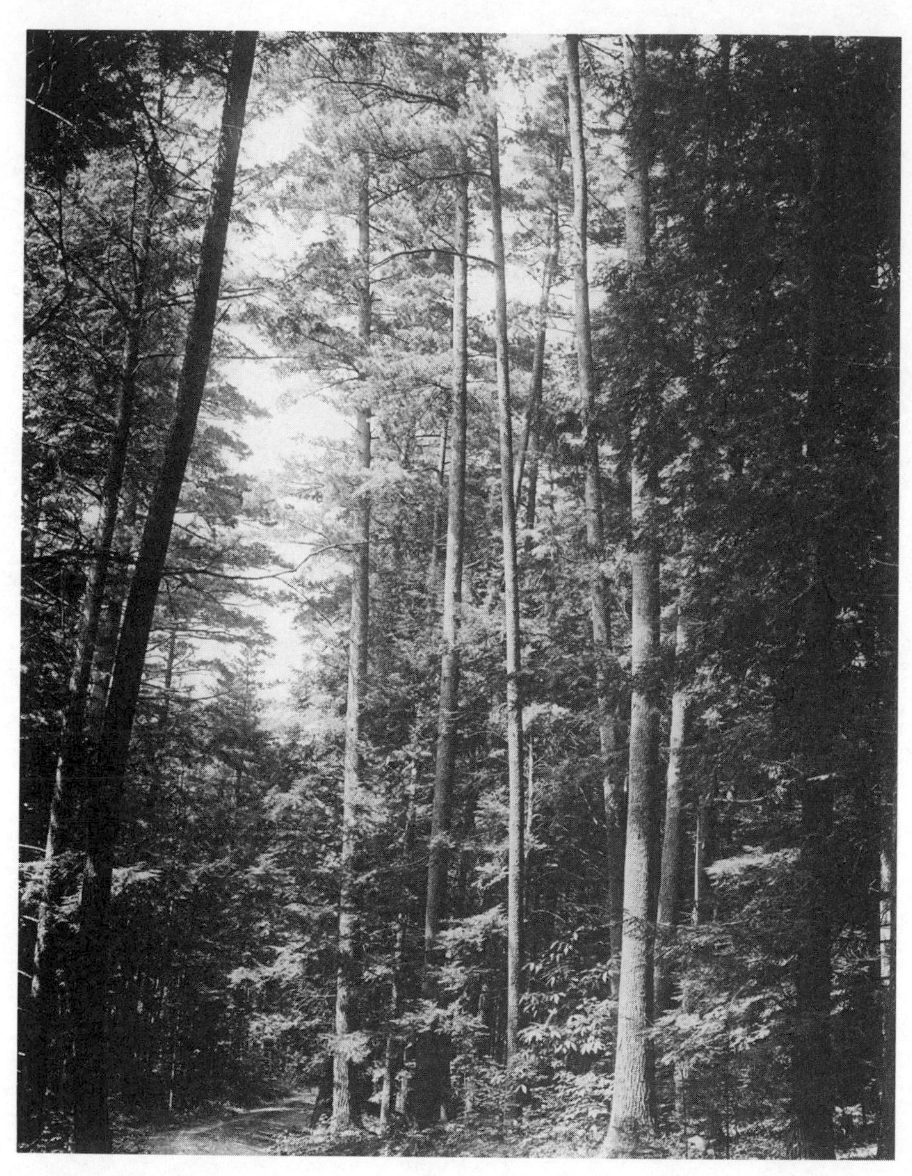

Virgin white pine stand, Colebrook, Connecticut, 1911.

The Northeast's Changing Forest

LLOYD C. IRLAND

DISTRIBUTED BY HARVARD UNIVERSITY PRESS
FOR HARVARD FOREST
Petersham, Massachusetts
1999

Also by Lloyd C. Irland

Wilderness Economics and Policy
Wildlands and Woodlots: The Story of New England's Forests
Ethics in Forestry

This book is printed on acid-free paper, and its binding materials have been chosen for strength and durability.

Library of Congress Cataloging-in-Publication Data

Irland, Lloyd C.
The Northeast's changing forests / Lloyd C. Irland.
p. cm.
Includes bibliographical references.
ISBN (invalid) 0–87462–680–X
1. Forests and forestry—Northeastern States. 2. Forest management—Northeastern States. I. Title.
SD144.A127I75 1999
634.9'0974—dc21 99-27087
CIP

Text design by Joyce C. Weston

Printed in the United States of America

To Dr. Ernest Morton Gould, Jr.
1918–1988
Longtime forest economist at
the Harvard Forest

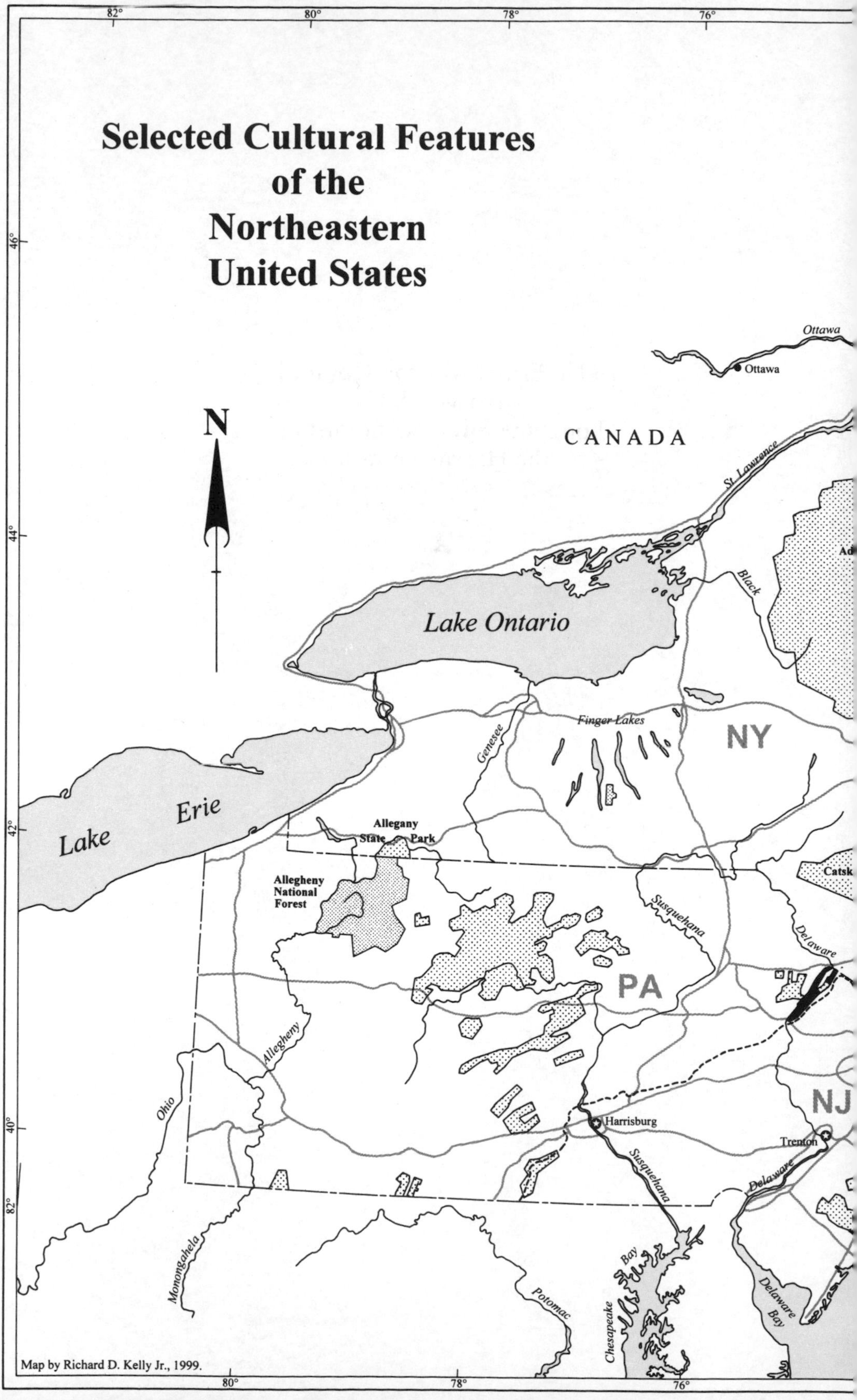

Selected Cultural Features of the Northeastern United States

82° 80° 78° 76°

46° 44° 42° 40° 82°

N

CANADA

Ottawa
Ottawa
St. Lawrence
Black
Ad
Lake Ontario
Finger Lakes
NY
Genesee
Lake Erie
Allegany State Park
Allegheny National Forest
Catsk
Susquehana
Delaware
PA
Allegheny
NJ
Ohio
Harrisburg
Trenton
Susquehana
Delaware
Monongahela
Potomac
Chesapeake Bay
Delaware Bay

Map by Richard D. Kelly Jr., 1999.

80° 78° 76°

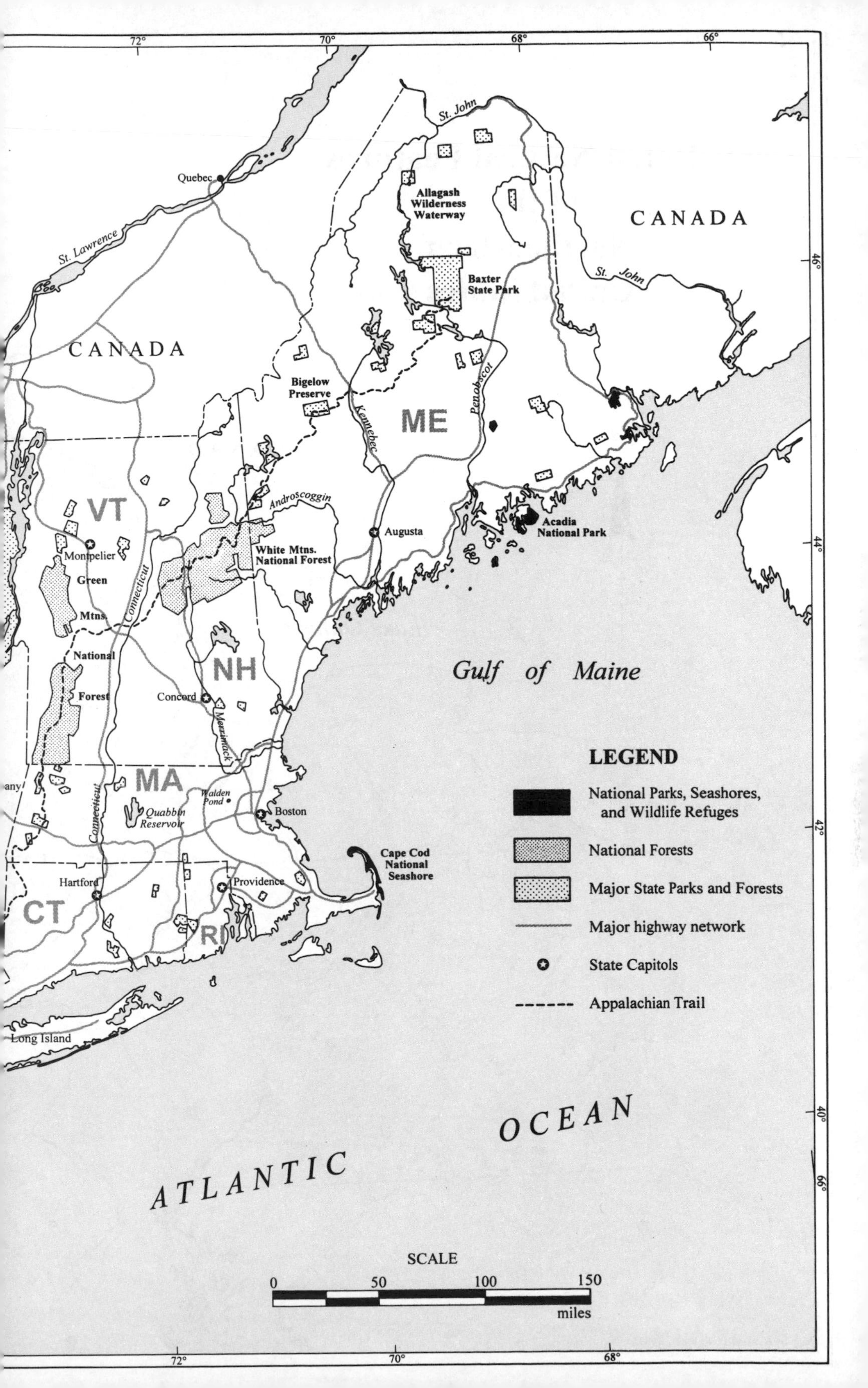
72°
70°
68°
66°
Quebec
St. Lawrence
St. John
CANADA
Allagash
Wilderness
Waterway
Baxter
State Park
CANADA
46°
Bigelow
Preserve
ME
Kennebec
Penobscot
VT
Androscoggin
Augusta
Acadia
National Park
Montpelier
Green
Mtns.
National
Forest
White Mtns.
National Forest
44°
Connecticut
NH
Concord
Merrimack
Gulf of Maine
LEGEND
National Parks, Seashores, and Wildlife Refuges
National Forests
Major State Parks and Forests
Major highway network
State Capitols
Appalachian Trail
MA
Walden Pond
Quabbin Reservoir
Connecticut
Boston
42°
Cape Cod
National
Seashore
Hartford
Providence
CT
RI
Long Island
40°
OCEAN
ATLANTIC
SCALE
0
50
100
150
miles
72°
70°
68°

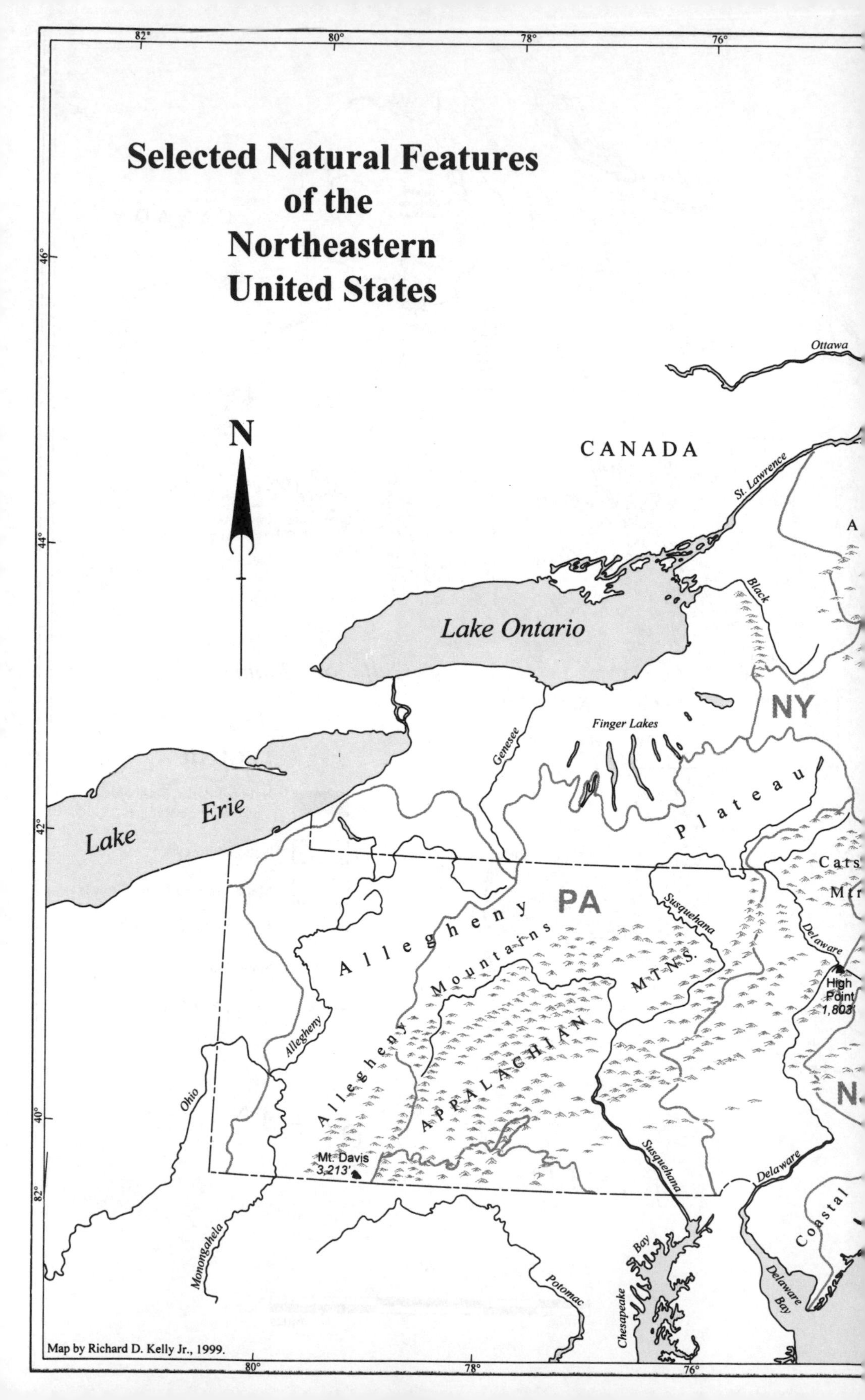
Selected Natural Features of the Northeastern United States
82°
80°
78°
76°
46°
44°
42°
40°
N
CANADA
Ottawa
St. Lawrence
Lake Ontario
Black
NY
Finger Lakes
Genesee
Lake
Erie
Plateau
Cats
Mtn
PA
Allegheny
Mountains
Susquehana
Delaware
High Point 1,803'
APPALACHIAN
MTNS.
Allegheny
Ohio
Mt. Davis 3,213'
Susquehana
Delaware
Coastal
Monongahela
Potomac
Chesapeake Bay
Delaware Bay
Map by Richard D. Kelly Jr., 1999.

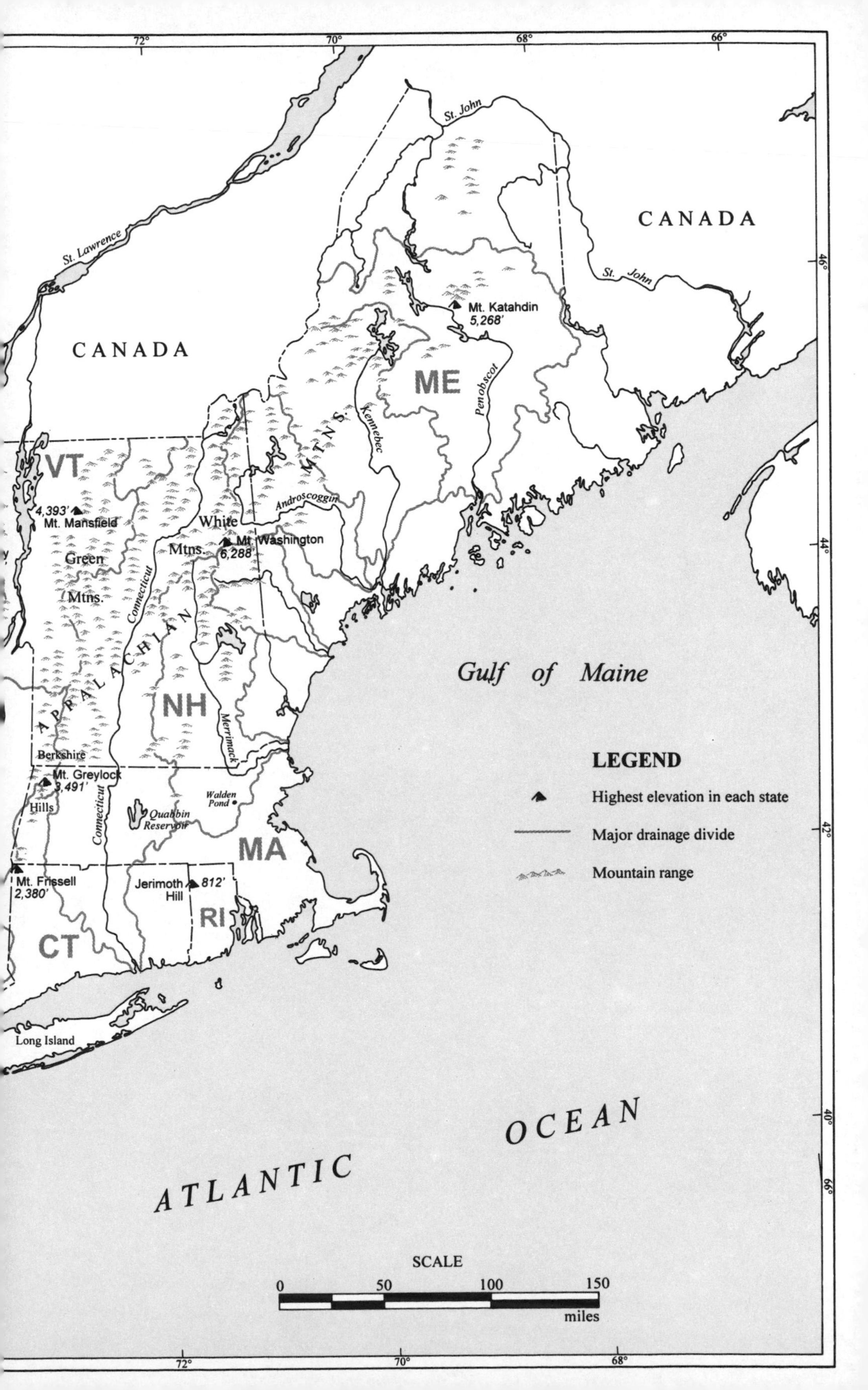
72°
70°
68°
66°
St. John
CANADA
St. Lawrence
46°
St. John
Mt. Katahdin
5,268'
CANADA
ME
Penobscot
Kennebec
MTNS.
VT
4,393'
Mt. Mansfield
White
Androscoggin
Mt. Washington
6,288'
Mtns.
44°
Green
Mtns.
Connecticut
APPALACHIAN
Gulf of Maine
NH
Merrimack
Berkshire
Mt. Greylock
3,491'
Hills
LEGEND
Highest elevation in each state
Walden Pond
Quabbin Reservoir
Connecticut
Major drainage divide
42°
MA
Mountain range
Mt. Frissell
2,380'
Jerimoth Hill
812'
RI
CT
Long Island
40°
OCEAN
ATLANTIC
66°
SCALE
0
50
100
150
miles
72°
70°
68°

Contents

Figures

Tables

Abbreviations

AFPA	American Forest and Paper Association
CONEG	Conference of Northeastern Governors
DNR	Department of Natural Resources
ERS	Economic Research Service
FAO	Food and Agriculture Organization (United Nations)
GPO	Government Printing Office
HCRS	Heritage Conservation and Recreation Service (unit of NPS which no longer exists)
INTFRES	Intermountain Forest and Range Experiment Station
LURC	Land Use Regulation Commission (Maine)
NCFES	North Central Forest Experiment Station
NEFES	Northeastern Forest Experiment Station
NPS	National Park Service
PNWFRES	Pacific Northwest Forest and Range Experiment Station
RMFRES	Rocky Mountain Forest and Range Experiment Station
SAF	Society of American Foresters
SOFES	Southern Forest Experiment Station
SPF	State and Private Forestry (part of USDA-FS)
TWS	The Wildlife Society
USDA-FS	U.S. Department of Agriculture, Forest Service
USDI	U.S. Department of the Interior

Preface

Over the past century the landscape and forests of the Northeastern U.S. have undergone an ecological transformation. As a consequence of major changes in agriculture, industry, and social conditions, large areas of former farmland have been abandoned and extensive new areas of forest have developed, one of the most remarkable environmental stories in U.S. history. This new forest has brought a natural appearance to much of the land. It has created new habitat for many woodland plant and animal species that were rare a century ago. However, this forest resurgence has also generated many new social and policy issues. Commercial logging on industrial and public lands is being challenged by wilderness advocates in the Northern Forest. The sprawl of suburbia has placed countless homes on small lots in woodland areas around most metropolitan areas, making forest management more difficult and forcing homeowners to confront the region's growing wildlife population through firsthand experience.

Whether one's interest lies in the appreciation and preservation of nature, conservation of open space and its diverse values, or coherent management of a vast resource, the major challenge facing everyone interested in forest lands is that of information and its synthesis. How do we collect and make sense of the data necessary in order to make wise decisions about the future? One of my prime sources of both information and insight has been Lloyd Irland's delightful *Wildlands and Woodlots: The Story of New England's Forest*. My copy has been well-used by myself and countless students and colleagues to whom I loaned it after it went out of print in the mid-1980s. Thus, I was immensely pleased to learn that Lloyd was writing a new book that widened the geographic and topical scope of his previous volume while, at the same time, addressing new literature, data, and issues.

While this work is not without a point of view, it avoids polarized viewpoints. Its breadth of approach in considering history, land use, and economic forces makes it useful to a wide audience.

Given Harvard Forest's nearly century-long history of research and education in forest ecology and conservation, it is appropriate that we join with Lloyd in publishing this new volume. My hope is that the book will help to educate a wide range of students, practitioners, scientists, and citizens, and help us all as we seek to understand, appreciate, and manage our forest landscape.

David R. Foster
Director, Harvard Forest

Foreword

I AM GRATEFUL to the Harvard Forest and to Harvard University Press for their interest in issuing this book, which grows out of my earlier work, *Wildlands and Woodlots*. A new book would be warranted by the mere passage of time since 1981 when *Wildlands* was finished. There are new and more detailed forest surveys of the region; new industrial trends, such as paper recycling, are coming to the fore; and the regional economy has seen significant changes. Changing wood supplies in the former USSR, the tropics, and western North America are likely to give new importance to the potential wood production of the Northeast's forests. And finally, my widening knowledge of the region, gained through work as a consultant, has enabled me to incorporate New York, Pennsylvania, and New Jersey in a work of widened geographic scope. Certainly public interest in the forest, from a variety of perspectives, has increased dramatically. Media articles now commonly address issues and use concepts that were only in the lab scientists' vocabulary in 1982.

Yet an even more important development compels attention. This is the extraordinary change in the intellectual tools with which we think about the forest. In this work, I refer to these as "metaphors"; they could more formally be termed "paradigms." These metaphors are more fully developed in Chapter 1. Some are based on deeper scientific understanding, as is the metaphor of the Global Commons' effects on the forest. Others, the Town Meeting View, attempt to formulate a basic feature of our society's ways of forming policies about land and forests. I hope that incorporating these metaphors into the exposition will enrich the book and increase its relevance to a wider range of readers.

It is a story of the Northeast's forests, of how they have changed, and of what they can mean to the region's future. The attitudes about the forest expressed by the impersonal market and by public policy have altered, but the importance of forests in the region's life has never changed.

This book is full of numbers, most of them estimates. Before the 1950s little serious sampling of the forest was done; few aerial photos were available. Foresters offered their best judgments on forest area, volume cut and grown. Census enumerators did their best, but there are inevitable gaps in the data. In this book, I refrain from detailed comment on the accuracy of the figures. I draw frequent historical comparisons, using employment, production, and wage figures, but recognize that data from before 1950 must be used with special care. Changing definitions, different standards of accuracy, and the frequent guesswork of early estimates must be remembered.

This book is a general essay and not a report of an extensive and costly research project. I usually offer my estimates in very round terms, so that readers will not be tempted to rely overmuch on their precision. Many of the things I define (the suburban forest, the area regrown after farm abandonment) are difficult to measure precisely, and basic data do not exist. While I do not think that slightly different numbers would change the story, I hope that these rough estimates will challenge others to refine them by detailed local and regional studies.

This is a book about forests, by a forester, written for nonforesters. It tries to identify those themes about the Northeast's great forest that may interest the general public. Stories of technical forestry do not fit here, nor do details of silviculture, taxation, pest and fire control, forest land use planning, forest research, and administrative policy. These subjects have been my daily concerns as a forestry administrator, civil servant, and consultant, but they interest, I suspect, few others.

Other topics have been omitted, not because they are unimportant, but because they are more distant from my main focus. I do not discuss the region's wildlife resource and its management, though the region's sportsmen have been important conservation leaders. There is nothing on presettlement forest history and on the role of science and education, and little on federal policies and the federal land managing agencies. Nor do I consider the literary, artistic, and cultural history and values of the region's forests.

I would like to acknowledge the assistance given by all of my former teachers, who equipped me for this work. More particularly, I owe thanks to friends and colleagues who helped me years ago with *Wildlands and Woodlots*: Ernie Gould of the Harvard Forest, who patiently read several drafts, testing and challenging ideas; Neal Kingsley and Steve Boyce of the United States Department of Agriculture Forest Service; David Field of the School of Forestry, University of Maine at Orono; Philip Conkling of Cushing, Maine; Al Worrell and Clark Binkley of the Yale School of Forestry and Environmental Studies; Tom Rumpf, Dick Arbour, and Ellen Baum of the Maine Forest Service; and Tim Glidden and Chuck Hewett,

then of the Resource Policy Center, Thayer School of Engineering, Dartmouth College and Carl Reidel of the University of Vermont. Many of my associates in the Maine Department of Conservation, the region's forest products companies, and New England's forestry community have taught me things without even realizing it.

The current work has been made possible by the thinking and the work of many scientists, foresters, and observers of the northeastern land and forest scene. I would especially like to acknowledge review comments on all or portions of earlier drafts by Jack Lutz, Ted Howard, Mitch Lansky, Pat Flood, Bill Leak, Malcolm Hunter, Jr., D. M. Smith, Bill Leak, Ralph Nyland, Craig ten Broeck, Ellen Baum, Dave Kittredge, and Paul Sendak.

Research help, computer graphics, and document processing by Rondi Doiron and Amy Zwicker made the work possible. Susan Hayes managed the editing, printing, and production process. Dick Kelly prepared the endpaper maps. Johanna helped me check the index. A check of proofs by John O'Keefe of the Harvard Forest went beyond the call of duty.

Many people helped by supplying photos: Chris Ayres supplied several of the photos and skillfully prepared prints from borrowed negatives for others; the Maine Department of Conservation; the Maine State Library; Paula Broydrick of the International Paper Company; Paul McCann of Great Northern Paper Company; Keith Ruff and Ray Kozen of Georgia-Pacific Corporation; Bill Leak, USDA Forest Service, Durham, New Hampshire; and Charles Gill, USDA Forest Service, Milwaukee, Wisconsin; Lester DeCoster, American Forest Institute; John Lawrence, Rhode Island Division of Forest Environment; Tom Siccama, Yale School of Forestry; Bill Gove, Vermont Division of Forests, Parks, and Recreation; Ellie Horwitz, Massachusetts Division of Fish and Wildlife; Greg Gerdel, Vermont Agency of Development and Community Affairs; Wilfred E. Richard, Outdoor Ventures North, Inc.; Marshall Wiebe, Maine Department of Conservation; Roger Lee Merchant; David R. Foster, Harvard Forest; Steve Jones, currently of the Alabama Cooperative Extension Service; Yuen-Gi Yee, USDA Forest Service; S. F. Manning, Virginia Museum of Fine Arts; the Concord Free Library; Gary Randorf and The Adirondack Council; Cheryl Oakes, Forest History Society; Susan Stout and Barbara McGuinness, USDA Forest Service; the Pinelands Commission; the Delaware Water Gap National Recreation Area; the U.S. Fish and Wildlife Service; the Hahn Machinery Company; Steve Lawser of the Wood Components Manufacturers Association; the U.S. Forest Products Laboratory; the Maine Department of Inland Fisheries and Wildlife; the Maine State Archives; the U.S. Geological Survey; the New York City Department of Environmental Protection; the National Gallery of Art; Robert Seymour; Fred Hathaway, Norbord Industries, Inc.; Monadnock Forest Products; Bud Blumenstock, Maine Cooperative Extension; Roger Lee Merchant; and Julie Lalo of the Western Pennsylvania Conservancy. Priscilla Bickford, Maine Department of Conservation librarian, obtained for me many of the sources used in this book as did the staff of the Maine State Library.

For encouragement, discussion, and tolerance while this book was being written, as well as for a helpful reading of an early draft, I am indebted to my wife, Connie.

The Northeast's Changing Forest

1 Forests in Northeastern Life

THE FOREST has set the stage for much of life in the Northeast. Today it covers about seven-tenths of the region's rural landscape. Forests provided colonists with raw materials for ships, with masts, boards, and staves for exports, with fuel, and with game; with raw materials for their distinctive architecture and their principal industries. The region is full of Sawmill Roads, Papermill Roads, Clapboard Hills, and Mill Rivers.

It is useful to define "Northeast" as the six states of New England plus New York, Pennsylvania, and New Jersey (the "Mid-Atlantic" states). There are other ways, but this defines, as we shall see, a sufficiently uniform region botanically, historically, and economically to be a useful basis for our inquiry. Stretching from the Mason-Dixon line to Northern Maine's Canadian boundary, and from Lake Erie to Chesapeake Bay and the North Atlantic, this landscape bore the impacts of the young nation's earliest achievements and excesses. It has spent the better part of a century recovering from them.[1]

This chapter reviews the changing importance and meaning of the region's forests. Its task is to describe the region's five forests — distinctive forest regions used as a basis for the book's analysis. It then sets out a set of seven metaphors representing major themes in thinking about forests.

Forests: Growing Importance, Changing Meanings

For much of this century, forests were of little concern to legislators and the public, except on occasions such as the creation of National Forests or the Adirondack Park. Forest policy was a concern of specialists, citizens, and legislators of rural areas, and the directly affected groups and industries. But during the 1970s, a series of dramatic changes swept over the region. A new investment boom in paper and sawmills boosted wood use in the northern woods, at the same time that a spruce budworm outbreak threatened the softwood timber supply. Rising land prices prompted legislative clashes over forest taxation. A wave of land speculation hit rural areas, converting woodlots to subdivisions. Controversy erupted over the proper management of state forests, over wilderness areas in the National Forests, and over management of the region's municipal water supply watersheds. Two oil supply crises spurred an increasing harvest of fuelwood in the later 1970s that raised once again the specter of overcutting in some areas. In the 1980s, another regional land boom drew wide attention. Booming export markets began to draw noticeably on the region's hardwood logs and lumber for the first time in generations. Concerned people explored municipal regulations, public subsidies, and a spate of education programs to cope with this increase in cut, unequaled since the 1890s. Even as these concerns were being examined and debated, a regionwide economic boom spawned new pressures on suburban and rural land, and by the late 1980s, the "Northern Forest Lands" of Northern New York and New England became a regional conservation issue.

The forest values of aesthetics, watershed protection, wildlife, recreation, and timber production are all increasing in importance to the region's residents. All of these values are threatened by careless cutting, land posting, suburban sprawl, speculative rural subdividing, and landowner

Old growth oak, Massachusetts, represents the forest values threatened today. *Courtesy of Bill Leak, USDA-FS.*

indifference to sound land stewardship. These trends no longer concern only foresters, paper companies, and government planners. In addition, the potential changes in the global climate and the chemistry of rain and snowfall pose new uncertainties and challenges for future forest composition and productivity. At the same time as society's needs for tax revenues, jobs, and raw materials are increasing, we are ever more aware of the need to manage the forest landscape as an ecosystem rather than as simply a source of logs and a place to camp. These issues concern all of us.

The Northeast is today about 70 percent forested. The shrinkage of cropland area has released enough land so that, despite the inroads of suburbia, there are still about 70 million acres of forest (Table 1.1). The Northeast contains significantly more forest than it did in 1909, having gained forest acreage equivalent in size to the entire present forest of Maine. In contrast, the nation as a whole lost 60 million acres since then. Today's forest problems and opportunities emerge from these past cycles of land use change. The forests of today were largely produced by natural regeneration following human disturbances. Cutting of recent decades has harvested this cost-free timber. But another cycle of free forest regeneration cannot be taken for granted.

The forest sets the visual tone for most of the region. It is difficult to find a place outside of a central city where a woodlot or wooded hillside is not in view (Figure 1.1). The pine-lined lakeshores, wooded hills, and stone walls rambling through the woods are essential ingredients of the region's scenery and quality of life. This quality of life is in turn a key attraction for the region's bustling tourist trade. The institutions related to ownership and management of the forest are rooted in colonial times, and in the history of a region dominated by private landownership. Only 12 percent of the region's forest is publicly owned. The agencies, nonprofit groups, corporations, and laws that are the fabric of the region's forestry institutions emerged in the conservation movement of 1880–1910, during the Great Depression, and in the 1950s and 1960s.

Today, about 270,000 jobs depend directly on the region's forest industry, and additional employment in turn relies on them. Many of these jobs depend directly on timber produced in the region. In the 1950–1970 period, many residents placed a higher value on the forest as a green backdrop for subdivisions than as a source of useful products. During the 1970s fuelwood boom, this attitude began to shift toward more sympathy for using the forest's wood for products. By the 1990s, however, a strong public sentiment, motivated by rising concern for natural forests and habitats worldwide, began shifting in the former direction once again. As noted below, new paradigms about the value of natural ecosystems began to be more widely accepted.

TABLE 1.1. Forest Land Area in the Northeast, 1600-1997 (thousand acres)

	1600 Original	1909	1952	1997	Loss 1600–1909	Increase 1909–1997
Maine	18,180	14,900	17,088	17,711	3,280	2,811
New Hampshire	5,490	3,500	4,848	4,955	1,990	1,455
Vermont	5,550	2,500	3,730	4,607	3,050	2,107
No. New England	29,220	20,900	25,666	27,273	8,320	6,373
Massachusetts	4,630	2,000	3,288	3,264	2,630	1,264
Rhode Island	650	250	434	409	400	159
Connecticut	2,930	1,600	1,990	1,863	1,330	263
So. New England	8,210	3,850	5,712	5,536	4,360	1,686
New York	27,450	12,000	14,450	18,581	15,450	6,581
New Jersey	4,330	2,000	1,958	1,991	2,330	(9)
Pennsylvania	27,260	9,200	15,205	16,905	18,060	7,705
Mid-Atlantic	59,040	23,200	31,613	37,477	35,840	14,277
Northeast	96,470	47,950	62,991	70,286	48,520	22,336
Nation	856,690	544,250	664,194	736,681*	312,440	(54,695)
Northeast as % of U.S.	11.3%	8.8%	9.5%	9.5%	15.53%	-40.84%

Source: D. S. Powell, et al., *Forest resources of the U.S.,* 1992, USDA-FS Rocky Mountain Forest and Range Experiment Station, General Technical Report, RM-234. 1993; and R. S. Kellogg, *The timber supply of the U.S.,* USDA-FS Circ. 166, 1909; USDA-FS, Timber Resources Review, 1958; and USDA-FS RPA Website.

*data for 1992.

Waves of Land Use Change

The early colonists encountered a forest abounding in useful trees and game.[2] By 1700, the region's population was only 195,000, 130,000 in New England and 65,000 in the Mid-Atlantic states. In 1760, Philadel-

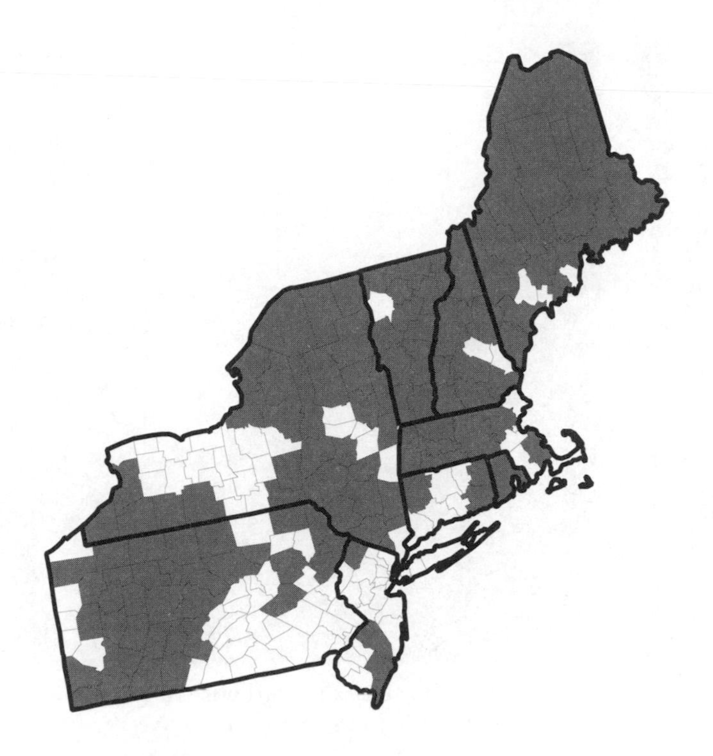

Figure 1.1 Timberland Acreage by County in the Northeastern U.S.
Shaded counties each contain 219,327 acres or more of timberland.
Source: Copied from USDA-FS, Timber volume distribution maps for Eastern U.S., Gen. Tech. Report. WO-60.

phia was the leading city, and New York State's population was sixth among the colonies.[3] Most of the region remained an untouched and little explored wilderness. Loggers cut only the largest pines near the streams. At times, such as the 1730s and 1780s, speculative booms in wildland townships occurred, though little of the land traded was promptly (or ever) cleared and settled. After 1790, it took 30 years for the region's population to double to 4.4 million persons, and farms were appearing even in remote corners of the region's hills. At the peak of landclearing, in 1880, only 14.5 million people lived in the region's mostly rural society. The years 1880 to the 1950s saw the slow, steady decline of farming. During the same time, the lumber industry reached an all-time high production level, and a new paper industry, based on wood pulp, emerged.

From 1920 to the 1960s, the automobile changed the face of the Northeast. As more and more farms "went back" to forest, cottages and camps sprang up along rural and northwoods lakesides. The seacoast was dotted with recreational developments. From the late 1940s to the 1980s,

Forest landscape of Allegheny Plateau, Northwestern Pennsylvania.
Courtesy of Stephen B. Jones, Penn State University.

the landscape again changed dramatically. While total forest acreage remained stable, farm acreage plummeted. Suburban sprawl on a massive scale engulfed millions of acres of farm, orchard, and forest in an unprecedented wave of land use change. In the early 1970s, and again in the mid-1980s, explosions in recreational land speculation converted large areas to partially built, partially sold, and occasionally bankrupt subdivisions. The automobile was at it again. The Interstate Highway System had rendered remote corners of the region easily accessible to suburbanites eager to escape from the traffic jams of their new communities.

Today, 52 million people live in the Northeast, 20 percent of the nation's population. Their landscape has only a fraction of the farmland that it had a century ago — they eat food grown elsewhere. Farmland varies as a proportion of the rural land (see also Chapter 6):

Northern New England	9%
Southern New England	13%
Mid-Atlantic	34%

More than 10 million acres are devoted to urban, suburban, and industrial uses. The area of untouched virgin forest is nominal, as is the acreage of artificial forest plantations. Woodgrowing is, by acreage, the dominant

land use. A far larger area of land is managed for growing commercial wood crops than is devoted to farming, housing, office buildings, and factories.

Patterns of wood use have changed significantly over the centuries. In colonial times, fuelwood was the dominant use. Its abundance, by allowing warmer homes, has even been credited with fostering a higher level of health than existed in Europe in those times. Despite its commercial importance, I doubt that the colonial export trade in lumber, masts, staves, and other products equaled the wood consumption of fuelwood and locally used lumber. In many areas, large quantities of wood were simply burned to clear the land for farming. Throughout the nineteenth century, fuelwood and charcoal for industry remained key forest uses, except in the north, where lumbering dominated and people were few. It is difficult today to visualize the role wood held in 1900 as a raw material for boxcars, wagons, machinery, construction, and containers.

Wood pulp paper arrived in 1880. The early cradles of the American paper industry were in the Philadelphia area and on the lower Connecticut River in Massachusetts. However, not until the mid-twentieth century did pulpwood use exceed fuelwood. After its peak in 1909, lumber output fell, until the regional recovery in the 1970s and 1980s. Wood use for household woodstoves has been on the decline once again, as oil prices moderated after 1985. In local areas, wholetree chips, mill wastes, and landclearing and urban woodwastes fuel boilers generating steam and electricity.

Despite its forest-covered landscape (Figure 1.2), the region continues to be a major net importer of lumber, newsprint, panel products, and a host of products made from wood. It also depends heavily on imported oil and food, fertilizer, and feed grains. The region's Economic Interdepen-

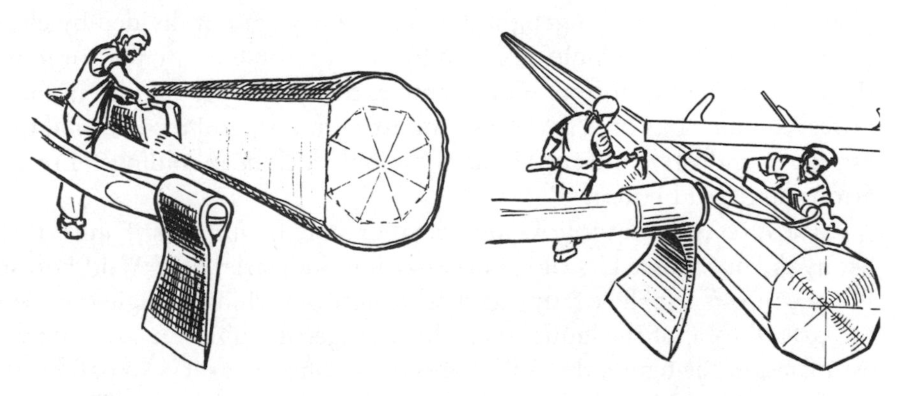

Hewing white pine for the mast trade, an early strategic export from Northeastern forests. *Courtesy of S. F. Manning.*

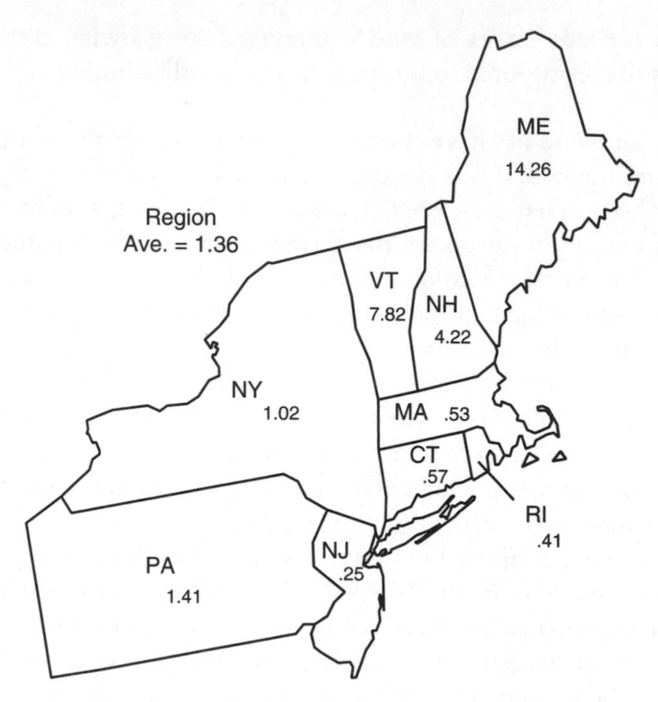

Figure 1.2 Acres of Forest Land Per Capita, 1997
Source: Based on USDA-FS website, and the U.S. Bureau of the Census.

dence in resource products endures as a major theme in its economic history.

The Northeast's Five Forests

This book is the story of the region's five forests — industrial, recreational, suburban, wild, and rural (Table 1.2). These forests are defined by distinctive economic and geographic traits. They are not divided by clear boundaries but intermingle. Some individual ownerships are part of more than one of the five forests. For example, the White Mountain National Forest contains bits of Wild Forest, industrial forest, and recreational forest. Forest industry lands in Maine commonly include inholdings of recreational and Wild Forest.

The forests do not follow municipal or state boundaries (Figure 1.3). Many towns include bits of suburban, rural, industrial, and Wild Forest. But the forests usually occupy general regions, in which a single type sets the overall tone of the landscape. The acreages in Table 1.2 are my own estimates. In the figure, the Wild Forest is not shown, as it is so patchy. Its largest units are Maine's Baxter Park and New York's Forever Wild lands in the Adirondacks.

TABLE 1.2. The Northeast's Five Forests and Nonforest Land, 1997

	Acres (millions)	Percent of All Land	Percent of Forest Land
Forest Land			
Industrial	14	13%	20%
Recreational	7	7%	10%
Suburban	10	10%	15%
Wild	5	4%	6%
Rural	34	33%	49%
Total Forest	70	67%	100%
Nonforest Land			
Cropland	14	13%	
Pasture	8	8%	
Developed	10	10%	
Total Nonforest	34	33%	
Total Land Area	104	100%	

Source: Author estimates, based on USDA-FS and other data.

About two-thirds of the region's forest is in the industrial and the rural forests. These provide the timber for the region's mills and fireplaces, and a wide range of hunting, fishing, watershed protection, and aesthetic benefits. The remaining forests are devoted to more specialized uses, but still provide diverse benefits.

A sixth forest, and a very important one, is not discussed in this book. This is the urban forest, consisting of forested parks and parkways in the cities and suburbs. These forests exist for their amenity values, and the financial values that real estate markets attribute to them are extremely high. The management problems of this forest are distinctive and have produced a new academic subdiscipline and a series of bureaucratic programs — urban forestry.[4]

The Northeastern Region

Though it is not a distinctive botanical or economic unit, the Northeast recommends itself as the scope of this work for several reasons. The New England states, comprising Maine, New Hampshire, Vermont, Massachusetts, Rhode Island, and Connecticut, share a common heritage of town

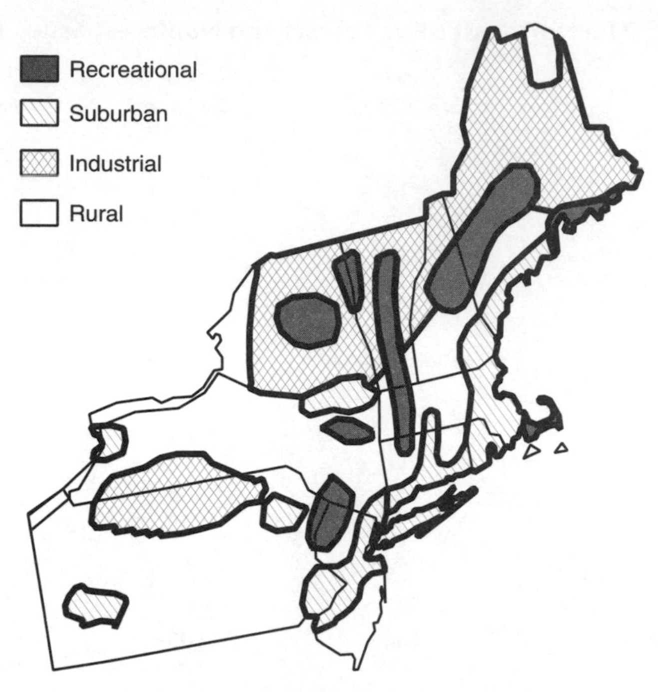

Figure 1.3 Forest Regions of the Northeast (Schematic)

Fields, woods, brushland, and houselots present a panorama of land use change. *Photos by Chris Ayres.*

government, a unique pattern of industrial forest ownership, and a strong sense of themselves as a region. The Mid-Atlantic states are closely connected to New England economically and socially and share similar forest conditions.

This region is highly urbanized, with New Jersey being the only state with no nonmetropolitan population (Table 1.3). Yet Vermont is the third least urbanized state after Idaho and Montana, and extensive rural hinterlands spread from Maine to North Central Pennsylvania.

Is the story of Northeast's forests of interest to persons concerned with forest conservation elsewhere? I believe it is. This region has been ahead of the rest of the nation in many ways in its forestry trends and concerns. Farmland abandonment, returning land to forest, occurred early in the nineteenth century. The Northeast was the first region to "cut out," to exhaust its mature pine and spruce for lumber. It then made a transition — pioneering in the nation — to a pulp economy based on a renewable resource.

During the same period, private initiative prompted government conservation efforts in land acquisition for national, state, and local forests. The region is home to major early land reservations and to the earliest forestry schools. During the 1960s and 1970s, the region led in developing town and state land conservation groups, such as land trusts and conservation commissions. These groups and a few wealthy individuals have created or fostered the creation of the bulk of the region's Wild Forest.

This is a region of private forests. Its forest future depends almost entirely on what private owners and land markets do. Until recently, its

TABLE 1.3. Percent of Population in Nonmetropolitan Areas, 1994

United States	20.2%
Northeast	10.6
Maine	64.1
New Hampshire	40.4
Vermont	72.8
Massachusetts	3.9
Rhode Island	6.2
Connecticut	4.3
New York	8.3
New Jersey	0.0
Pennsylvania	15.3

Source: Statistical Abstract of the U.S.

Maine log drive, Penobscot River, 1920s. This was the dominant mode of log transportation in New England for three centuries. *Courtesy of Maine State Library, Myron H. Avery Collection.*

wood manufacturing economy was scarcely influenced by federal timber policy, over which loud debates constantly rage in Oregon. Now that western federal timber harvests are sharply declining, however, a new economic environment is being created for timber cutting, for longterm forest management, and for manufacturing wood products. Private lands will be the principal source of growth in future wood production and other forest services. The increase in demand creates the potential for more exploitive cutting, even as it provides the economic base for improved longterm management.

In the 1990s, government agencies faced severe budget stringency. During this decade, the ability of government to meet land conservation and recreation needs declined. A notable improvement has occurred as this is written, but it is too early to tell if it will last.

Metaphors about the Forest: Concerns and Policies for the Coming Century

Public concerns with the region's forest have evolved slowly. These concerns can usefully be summarized by seven metaphors that reappear throughout our discussion (Table 1.4).

TABLE 1.4. Metaphors about Northeastern Forests

Metaphor	Representative Meanings
Town Meeting	Localism in policymaking Resistance to regional and federal initiatives Tension between public/private rights
Eagle's Eye View	Conservation biology Landscape management Loss of forest area Public Access
Global Commons	Climate change affecting forests Atmospheric deposition effects Forests affecting climate change? Role of biomass energy
Worm's Eye View	Nutrient pools Nutrient cycling Biodiversity
Economic Interdependence	Need for rural economic development Global economy — threats and opportunities City/countryside interdependence Self-sufficiency in materials and energy
Cultural Heritage	Importance of wildness to regional sense of place Cultural significance of resource industries and occupations Artistic expressions
Bequest	Individual concerns with descendants' heritage Society's view of resource condition as a Bequest Sustainability

During colonial times, the forests were seen as indispensable sources of raw material. Local public authorities regulated everything from roadside trees to grazing to promotion of mills, as well as the pricing, measurement, and export of products. During the nineteenth century, however, public regulations vanished, and the forests were seen as significant assets by individual citizens but only as wastelands by public policy. What few laws about forests existed were not implemented by administrative organizations, and forest policy was dormant. During the period, certain areas came to be used for recreation, hunting, and fishing. Traditional rights of public use of the land were respected, or even guaranteed, though recreation use of the forest was limited. The Colonial Ordinances of 1640, which still apply in Maine and Massachusetts, provided that citizens have the right to walk over undeveloped land to reach Great Ponds (lakes larger than 10 acres). Thoreau rarely met other casual strollers in the Concord woods.

As suburbs spread, community leaders began to work for preserving patches of forest through public and nonprofit groups. Their efforts were based on interest in preserving forests from despoliation or on preserving future timber supply. The great legacies of the early Conservation period were the Adirondack Park, major state land systems, a number of forestry schools, and a few research stations. The rest of the forest landscape was largely ignored by public policy, though its use for timber continued.

The principal area of public and government concern with forests from the 1950s to the early 1970s was the value of the forest as a "green backdrop," a wooded setting for houses or parks. Open space was the perceived public need. Many suburban residents saw harvesting timber products as a threat to higher forest values. The green backdrop view is often more concerned with simply preventing development than it is with the positive values of wildland.

These stories illustrate the *Town Meeting* metaphor, shorthand for the various political factors and processes by which policies are shaped for forests. The Town Meeting metaphor hints at the intense localism, a polity that is suspicious of control from Harrisburg or Augusta, much less Washington or multistate regional entities.

Today, public interest has revived in a host of problems that would have been recognized by colonial town officials. Planners attempt to identify and conserve prime farmland for future crop production. They work to preserve forests for watershed protection and to conserve aquifer recharge areas. Faced with increasing timber cutting, more and more towns are considering or adopting zoning or other restrictions on logging. The fuelwood market has revived not only an active concern for the ability of the forest to provide timber but also for the old colonial custom of timber stealing. The king's timber agents would feel at home with today's trespass situation and would feel just as helpless as they did in their own time. These tensions express the enduring theme of conflict over private rights versus public responsibilities in land use.

Several concerns, however, are new. First is the concern over continued public access to the forest. This concern is a product of modern social conflicts, of higher use pressures, and of the suburban attitude toward land — as a commodity for private use and not as a community asset. To stem the loss of public access, a team of researchers from Cornell University has urged that the Northern Forests of the region be viewed and managed as a Commons.[5]

Second is the concern, familiar from earlier times, over the future forest land base. The loss of productive forest land through recreational and suburban subdividing, the fragmentation of parcels as land passes between generations, and inundation by land-hungry, inefficient development, have turned the tide after a century of increases in woodland area. Likely future trends raise the possibility of dramatic reductions in forest area. Management and access problems will be magnified by the difficulties of managing tiny parcels for timber, wildlife, and recreation.

In the backyard of major metropolitan areas, land prices and property taxes are high. They can often push owners into cutting when they would prefer not to. The towns and counties are at the focal point of debates over what, if anything, should be done about this. These concerns are especially acute in the suburban forests, valuable for open space, wildlife, aquifer and watershed protection, and amenity. The suburban forest remaining in private ownership could be substantially destroyed in the next half century. As this is written, regional real estate markets are in depression following the excesses of the late 1980s. But there will be other booms. If projected population growth continues, millions of additional forest acres will be urbanized by the year 2020, and millions more will become vacant lots, tiny patches of unused land surrounded by shopping malls.

Loss of forest area and the land use changes occurring in the suburban, recreational, and rural forests highlight the *Eagle's Eye View.* From the Eagle's perspective, the change in the overall forest landscape is what counts. For the past century, forests have spread over large areas of open land, shifting habitats for many species. In the future, these regrown habitats will become more fragmented and isolated from one another by development, roads, or timber cutting. Some scientists are concerned that the various food, cover, nesting space, and other needs of birds and game animals will be met less and less effectively. The emerging science of conservation biology attempts to study these relationships, and to develop guidelines for management practices that will maintain the habitat value of entire landscapes.[6]

Another new concern is the potential effect on the region's forests of the changing *Global Commons.* Acids and other chemicals from precipitation, the possible changes in rainfall, temperatures, and radiation

Global Commons and Northeastern Forests: Report of the National Acid Precipitation Assessment Program

Sugar Maple Condition

In the late 1980s, concern about suspected deteriorating sugar maple condition in the Northeastern United States and Eastern Canada was expressed by diverse public groups and private individuals. Acidic deposition was often implicated. In response, a joint program to monitor and evaluate sugar maple condition, the North American Sugar Maple Decline Project, was organized by formal agreement between the United States and Canada.

Forestry experts from Canada and the United States representing seven states, four provinces, and both federal governments developed and conducted a program to annually evaluate sugar maple trees at 171 identified locations from Wisconsin to Ontario to Massachusetts to Nova Scotia. Half of the locations were active sugarbushes, and the remainder were unmanaged and recently (10-plus years) undisturbed forests.

Recently released analyses of the data from the first four years of measurement indicate that over 90 percent of the 11,000 sugar maple trees examined each year continue to be healthy. Many of the trees not rated as healthy had major damage to their main stems or roots. No differences in tree condition or proportion of healthy trees were found between sugarbush stands and the unmanaged forests. Furthermore, over the observation period (1988 to 1991), the trees examined were exhibiting improvements in their crown health each succeeding year. Trees with poor or unhealthy crowns usually were diagnosed to be suffering from insect defoliation and drought. No correlations were found between the distribution of wet sulfate deposition and trees with unhealthy crowns.

In summary, the maple project data to date show an apparent improvement in sugar maple tree health since 1988 and detect no visible effects from sulfate deposition.

Red Spruce Decline in the Northeast

Work on red spruce during the first phase of NAPAP identified an increasing susceptibility to winter injury for trees growing at the highest elevations on mountains in the Northeastern United States. Atmospheric monitoring in these mountains revealed that both the rain and the cloud water were quite acidic. Overall, the amounts of acidic pollutants deposited at these high-elevation sites are the highest in North America, although deposition is higher in parts of central Europe. Observations revealed that the red spruce trees were often exposed to very acidic clouds. The nearer to the top of the mountains, the more frequently the trees were exposed.

Scientists were concerned that the acidic cloud water and precipitation might be reducing the cold tolerance of the red spruce in the Northeast. Injury might be caused by freezing of the current year's foliage. Such injury seriously affects the vigor of trees, reduces growth, and causes physical damage to the crown, as well as death of trees. However, until recently, acidic deposition could not be confidently identified as a cause of change in forest condition.

In recent field studies, mature trees growing at a high-elevation site in the Northeast were protected from the acidic rain and cloud water, while neighboring trees continued to receive them. As a result of these experiments, scientists have now demonstrated that ambient concentrations of the pollutants in the rain and cloud water are associated with reducing the midwinter cold tolerance by 4 to 10 degrees C compared with trees growing at the same location but protected from the acidic cloud water and rain. The acidic deposition is associated mainly with the sulfur and nitrogen content of clouds, air, and precipitation; however, scientists are not certain which of the components of the acidic rain and cloud water are the principal causal agents, or what the damage mechanisms are.

This demonstration of the reduction in the cold tolerance of red spruce helps to explain the damage occurring to these Northeastern mountain forests. Furthermore, it provides more information on the role that acidic deposition plays in the complex processes that result in the visible decline, or death, of red spruce trees.

Source: National Acid Precipitation Assessment Program, 1993. 1992 Report to Congress. June. p. 67–68.

regimes, and in global carbon dioxide concentrations are full of implications for the future composition, health, and productivity of the forest.[7] On many points, a scientific consensus has not emerged as to the likely severity of the problems or the policy implications. But in addition to our traditional earthbound concerns, the effects of a changing atmosphere are now firmly on our agenda.

Our base for understanding the impacts of the Global Commons on the forest was first built by scientists who developed the *Worm's Eye View* of the forest. This metaphor expresses our understanding of the nutrient pools and cycles of forest ecosystems; in a way, it is a chemical engineering perspective. The Worm's Eye View tells us how forest growth and cutting, rainfall, litterfall, decomposition, nutrient chemistry and leaching, food webs, and other processes are tied together in biological and chemical cycles. This view also helped develop a sophisticated understanding of how natural forests are maintained through different disturbance regimes. The Worm's Eye View also expresses the importance of *maintaining biodiversity,* the natural diversity of species, including those existing in the soil, forest floor, the dead wood, and the treetop ecosystems.[8]

As the demands on the forest escalate, competing supplies of wood are becoming more costly, and regional economic and population growth are slowing down. The *Economic Interdependence* metaphor expresses the globalizing economy, and the interdependence of city, countryside, and wildlands. George Bellows' 1916 painting *Shipyard Society* vividly expresses this. Not only did the shipyards draw workers and materials from

Researchers study the chemistry of rainfall to determine its possible effect on Northeastern forests. *Courtesy of Yuen-Gi Yee, USDA-FS.*

nearby communities, but the ships themselves, once in service, connected these communities to the world. The complex economic forces that determine forest ownership and use have not diminished in force. The old tensions over nonresident landownership go back to Colonial times. They persist in new forms, and they mirror the dependence of rural areas on urbanites for investment capital, for markets, and for tourist visitors.[9]

That the forests are vital to this region's sense of place and identity is appreciated in many ways. It seems to take threats and worries, however, to bring this latent cultural value to the surface. The role of the forest backdrop to the region's history is an essential *Cultural Heritage* that concerned citizens want to preserve.[10]

The Cultural Heritage metaphor leads us naturally to the seventh and last one. This is the idea of the *Forest as a Bequest,* a heritage of landscape, of biodiversity, and of resources and economic opportunities for future generations. The concept of sustainability has become an increasingly accepted way of describing this concept.[11]

Shipbuilding illustrates historic dependence on raw materials and trade of the Northeastern economy. *George Bellows (1916), Courtesy of Virginia Museum of Fine Arts. The Adolph D. and Wilkins C. Williams Fund.*

For a century, forests have been recovering the region's landscape. It is unfamiliar, and uncomfortable, to be talking of the forests disappearing again in a new cycle of land use change, and being threatened by changes coming from the sky. The balance of this book is designed to provide a better basic understanding of this forest, its importance, and trends affecting its future. The discussion is organized around the region's five forests and several general categories. Topics such as forest taxation, landowner assistance programs, and land use control take quite different forms in each of the five forests. Since this book does not aim at an encyclopedic treatment, I take a selective approach to these issues.

Chapters 2 and 3 describe the Northeast's forests in biological and ecological terms. The succeeding five chapters examine the region's five forests, using the metaphors as tools for discussing the issues. The final chapters examine factors that cut across the five forests: landownership, the region's timber budget, the economic significance of forests, and the major policy issues.

Allegheny National Forest: Emerging Issues and Public Concerns

Most of these issues were previously identified, yet are still relevant. All of these issues may significantly affect the Forest Service's ability to implement the Allegheny Forest Land Management Plan and could result in future amendments or revisions.

Biological Issues

Riparian Area Management: A Riparian Team is analyzing riparian area management on the Forest. Utilizing a landscape ecology approach to management and delineation of riparian areas, they will provide recommendations to the Forest's Leadership Team. Areas of emphasis include appropriate management in headwater stream areas and on sensitive soils. The analysis will be completed in 1994. The team has won a Regional award for their work in watershed management.

Deer Herd Management: During the development of the Forest Plan, the overpopulation of deer on the Forest was well recognized by both the Forest Service and the Pennsylvania Game Commission. The Game Commission was increasing antlerless license allocations at a steady rate. Research had revealed that 70 percent of all clearcuts that had failed to regenerate into new forest stand could be directly attributed to excessive deer browsing. Game Commission biologists and other researchers had documented the impacts of high deer populations on turkeys and other wildlife species.

The intent of managing habitat on the National Forest is to provide habitat to maintain viable populations of all species. This requires managing for a variety of habitats, managing for unique habitats and managing for specific features that may be needed by a species. Habitat for forest interior species as well as forest edge species must be provided in a balance that maintains ecological integrity. We believe that the Forest Plan provides sound guidance for not only managing deer, but all native wildlife species on the Forest. . . .

Landscape Corridors: The Forest Plan specified amounts of acreage in each Management Area that should be designated to provide a late successional vegetation component for the Forest. However, it did not provide direction as to where these should be located or identify appropriate vegetative types. Comments from the public suggested that the Forest develop a landscape approach for designation rather than allocating areas in a piecemeal fashion.

A Forest team initiated a landscape corridor analysis in 1992; conducted an extensive public involvement effort; and has produced a landscape map for the Forest to use in its site-specific analyses. The proposed corridors will connect the large parcels of contiguous late successional vegetation on the Forest. These include the Allegheny National Recreation Area, the Tionesta Scenic and Natural Area, and the Hickory Creek Wilderness.

The team is currently considering management objectives for these areas and will complete a report on the entire analysis in 1994.

This led to an agreement between the Game Commission and the Forest Service. The Commission agreed to continue to bring the deer herd population down, striving to reach a goal of about 19 deer per square mile, Forest-wide. The Forest Service agreed to provide more early succession vegetation, mainly through timber harvest. By increasing the food supply through the creation of early successional vegetation, and by reducing the population through increased antlerless allocations, a balance should be reached where deer are in equilibrium with their habitat.

Allowable Sale Quantity (ASQ): Because of continuing concerns about the Allowable Sale Quantity specified in the Forest Plan, a preliminary review of monitoring results was initiated in 1991. Several factors were identified that have seriously impacted the availability of timber on the Forest. These include: insect, disease and wind damage; a lack of regeneration opportunities; a lack of accurate inventory information; management area mapping errors (more riparian areas exist than were originally mapped); and FORPLAN spatial allocation limitations.

A Forest team has been evaluating these factors and has produced three reports for the public (copies were mailed to the public in February 1993). To acquire more accurate inventory information, a 6,000-plot survey was conducted in 1992 in Management Area 3. The results of this study and an analysis of all the relevant factors will be completed and available to the public in 1994.

Forest Health: A variety of insects and diseases are affecting tree health locally. Pear thrips, forest tent caterpillars, gypsy moth, cherry scallop shell moth, fall cankerworm, elm spanworm, linden looper, beech bark disease complex, maple decline, and ash dieback are of particular concern. Damage from most of these was observed in 1993, and an increasing number of trees are showing symptoms of decline from repeated impacts by these species. The most significant impact in 1993 was from elm spanworm.

In 1993, close to 261,000 acres of National Forest land were moderately to severely defoliated by elm spanworm. This marked the third year of such defoliation on 7,500 acres, and the second year on 51,800 acres. Tree mortality is expected to develop in 1994 from the effects of this defoliation; however, the amount is difficult to predict. Spray treatment (with Bacillus thuringiensis or B.t.) proposed for 1994 would limit additional defoliation stress on the surviving trees, and help limit the number of those that may die in subsequent years. The total impact elm spanworm defoliation has had and will have in years to come remains largely unknown. It depends on environmental conditions and the success of proposed spray treatments. If defoliation continues, permanent effects on wildlife habitat, vegetation diversity, recreation, and timber harvest volumes could be severe.

Tree Seedling Development in Upland Hardwood and Northern Hardwood Forest Types: Data collected in 1992 in Management Area 3 indicates that tree seedlings of a variety of species are not becoming established beneath the overstory tree canopy of these forest types. Selective deer browsing, dense interfering plants, and erratic seed production all play a role in limiting seedling development. If this situation is not corrected, it will have devastating effects on forest structure and tree species composition. Over the long term, it raises serious questions about tree composition and sustainability in Management Areas where the Forest Plan direction permits little human intervention to control forest and ground vegetation structure, composition, and development. Trees that die will not be replaced by similar species or, in many cases, by any species of vigorously-growing tree seedlings that are capable of growing up to replace the trees that have died.

Source: Allegheny National Forest, 1994. Monitoring and Evaluation Report FY 1993. p. 5–7.

Tall spruce in mid-slope mixedwood forest, mountains of Western Maine.
Photo by author.

2 Forest Landscapes of the Northeast

THE NORTHEASTERN forests are formed from the interweaving of the great boreal forest flora to the north and the temperate hardwood forest flora to the south.[1] From the north come the paper birch, fir, cedar and spruces, and the bog flora that is of generally uniform appearance throughout the region. From the south come the oaks, Atlantic white cedar, hickories and tulip poplar. White pine, hemlock, and black cherry are species characteristic of this transition belt and not of the far Canadian North or the Southern United States. This broad zone of transition gives the region its distinctive vegetational contrasts, which are especially vivid in the fall color season (Figure 2.1; Table 2.1). The forests confronted by the earliest white settlers had undoubtedly been shaped by burning, farming, and other activities of the native tribes. These influences undoubtedly varied in character and intensity from place to place. One author argues that the huge Indian population declines caused by introduced diseases reduced these influences and resulted in significant ecological changes in the forest.[2]

The region's forests can be grouped into five distinctive ecological zones:

- Forests of the mountains;
- The Northern New England lowlands;
- The lowlands from Southern New England to New Jersey and Southeastern Pennsylvania;
- The Coastal strip from Cape Cod to Cape May; and
- The Ridge and Valley and Allegheny Plateau regions.

This chapter offers a broad Eagle's Eye View of the region's present forests.

The distinctions are not so much in the presence or absence of groups of species as in their relative abundance and the way in which the species are associated.[3] White pine, for example, occurs across the entire region.

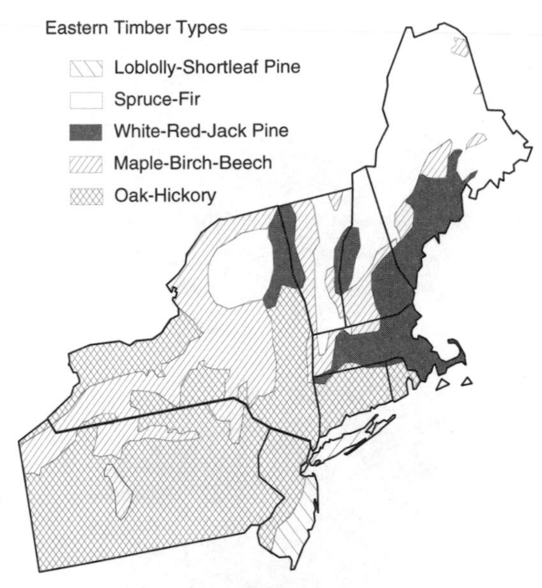

Figure 2.1 Present Distribution of Major Forest Types

The forest types described in Table 2.1 are useful for national comparisons, but I will use more locally familiar terms in the balance of this chapter. The northern hardwoods type of maple beech, birch and associates

Table 2.1. Major Types of Commercial Forest Land in the Northeast, 1997.

Type	Acres (1,000)	Percent
White/Red/Jack Pine	6,257	8.9%
Spruce-Fir	8,600	12.2%
Loblolly/Short Leaf Pine	1,065	1.5%
Oak/Pine	1,737	2.5%
Oak/Hickory	14,715	20.9%
Oak/Gum/Cypress	239	0.3%
Elm/Ash/Cottonwood	2,330	3.3%
Maple/Beech/Birch	31,033	44.2%
Aspen/Birch	4,041	5.7%
Total	70,286	100.0%

Source: USDA-FS RPA Website.

Open grown, multi-stemmed white pine overlooks a rural hillside. *Photo by author.*

accounts for roughly half of the forest area of this region. Second in importance is the spruce-fir type of the north, followed by the pines.

Forests of the Mountains: New York and Northern New England

The mountains from Northern New England to the Adirondacks support the most dramatic forest scenes in the region. Their special character derives from the rocky, cold, humid, windblown mountainsides where they grow. Their grandeur stems not from the size of the trees but from their setting, high on the peaks, overlooking the region's low hills and lakes.[4]

Walking up any of a dozen different peaks, from the Adirondacks to the Green Mountains to Katahdin in Maine, the hiker passes a characteristic pattern of forest. First, at the roadhead in the valley, are lush mixed hardwood forests of maple, yellow birch, beech, and a scattering of oak, ash, and paper birch, with a few small spruce and fir. In these stands grow scattered large pines and hemlocks. On sandy areas are stands of pine, and in the wettest spots surrounding bog lakes, spruce, fir, and tamarack. Farther up the trail, old burns support paper birch, aspen, and pin cherry. Higher on the rocky slope, the burns seem to come back to dense, even aged stands of white or red spruce, with little understory vegetation.

In steep ravines spared from past cutting giant pines, hemlocks, yellow birches, and maples may be found. The ancient yellow birch are often flat-topped and have dark, deeply wrinkled bark that resembles yellow birch not at all.

Up the trail, spruce and fir increase. Occasional dense stands occur. If they have been undisturbed long enough, the overstory will be breaking up from wind-snapped tops or falling trees, opening the forest floor to sunlight and allowing a dense carpet of tiny seedlings to grow freely. As the trail steepens, there is an occasional glimpse at the broad gray peak, and the view opens outward over the valley. The trees are now all spruce and fir, densely packed. They stand in a soft, spongy mat of moss that covers rocks. When you step on it, it gently oozes water after a rain. Tiny birches mix in with the fir and spruce. Soon the trees are head-high or shorter, and the hiker is free of the dark forest. The vegetation takes on a matted, dense, wind-formed aspect, called "Krummholz" by botanists. Upward from the Krummholz, the tiny trees fade into grass, sedge, and moss, rocks, and tiny alpine shrubs and flowers. The dwarf forest will penetrate to higher elevations on south slopes and in sheltered locations.

These mountain forests take shape in response to gradients in growing season, temperature, soil, aspect, and moisture. The vegetational changes with elevation are similar to those found in traveling north to the treeline in the Arctic. The shrub and grassy vegetation above the treeline in both situations — arctic and alpine — share the name of tundra.

In the forests of these mountains, scientists have been probing the possible effects of acidic deposition and other atmospheric pollutants on the forest. In these high elevation areas, precipitation is high and fogs are common. These conditions tend to expose vegetation to large amounts of whatever contaminants and acids the air is carrying. On these peaks, the Global Commons meets the forest, returning industrial and automotive effluents to earth. Scientists continue to debate whether declines of

View from Mount Marcy, New York. *Courtesy of Gary Randorf, Adirondack Council.*

individual species at high elevations are due to pollutant stress or to other causes.[5] Recent literature continues the debate.[6]

Forests of the Lowlands: Northern New England and New York

A traveler through these lowlands passes diverse forests, shaped by soils, drainage, cutting history, insects, wind, and human rebuilding of the landscape.

A drive from Maine's coast to the interior of the "Big Woods" and then westward across the region south of the Adirondacks as far as Western New York will pass through most of the forest types of this region. On the coastal points and islands are forests of spruce and fir, mostly on thin soils (Figure 2.2). Their growth is slow, and the forest floor is made up of moss. Coastal windstorms occasionally flatten patches of these forests, allowing regeneration. Where deep soils occur in sheltered locations, mixtures of birch, ash, oak, and other hardwoods are found. Some distance inland, pines appear and large oaks occasionally dominate the stands.[7]

In farming areas, roadside fields, recently grazed, will be dotted with ground juniper and an occasional wispy pine or birch. In wet pastures, tamarack will invade from its haunts in wet spots and along streams. Gi-

Red spruce in a subalpine zone in the White Mountains, New Hampshire. *Courtesy of Bill Leak, USDA-FS.*

ant, broad-crowned maples and pines along the roads and old fences recall the former landscape of tree-lined fences amid fields. A hardwood forest of maple, beech, yellow birch, cedar, hemlock, pine, and occasional fir and spruce covers most of the landscape (Figure 2.3). Passing through a sandy region, the pines appear more prominently. A large roadside bog displays the pattern of pale green tamaracks slowly invading the margins, followed by spruce and pine, all tapering up in height to meet the surrounding hardwood forest.

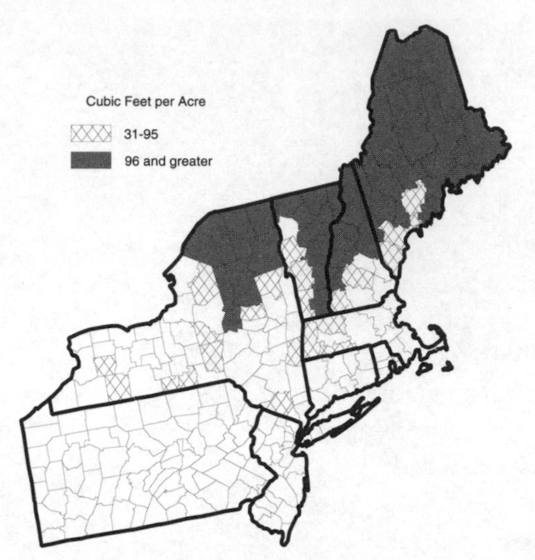

Figure 2.2 Spruce and Balsam Fir Growing Stock Volume Per Acre
(Growing stock includes live trees 5 inches in diameter or larger that are suitable for use as pulpwood or other products)
Source: USDA-FS, Timber volume distribution maps for the Eastern U.S., Gen. Tech. Rept. W0- 60 .

To the north, the road crosses bouldery streams and skirts large lakes, lined with pines. The terrain becomes gently rolling, and a distant mountain appears from time to time. Log trucks appear more frequently, and

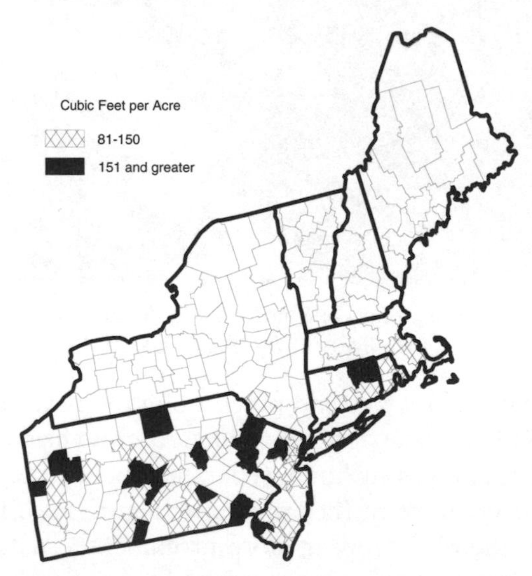

Figure 2.3 Hard Maple Growing Stock Volume Per Acre
Source: USDA-FS, Timber volume distribution maps for the Eastern U.S., Gen. Tech. Rept. W0- 60 .

now and then a slash-littered landing appears, with a skidder and muddy pickup parked by a long woodpile. You pass through a milltown of old tenements clumped around a mill with a huge pulpwood pile. A giant Catholic church dominates the end of Main Street. A row of roadside motels disappears behind you, and you move onto a broad hardwood ridge dominated by beech, maple, and paper birch. You notice that something has cleaned the leaves off the birch. You pass a large clearcut of recent vintage, with dark green spruce and fir trees poking through raspberries and pin cherry.

Coming down off the ridge, you enter a dark "fir thicket" — two miles of densely stocked trees, few larger than six inches in diameter. The dense overstory has shaded out all but the occasional understory plant, and few spruce or fir seedlings are visible. The trees have tiny spindly tops from being overcrowded. The stand is full of dead stems, standing and sprawled over the mossy ground. These fir flats are found on the low slopes of the ridges and the flat areas along streams. The old timber cruisers called these areas "black growth." These spruce-fir stands are the prime source of pulpwood for the region's paper mills. These dense stands have often been created by outbreaks of spruce budworm, a defoliating insect that periodically kills most of the mature trees and starts a new forest cycle. In the late 1970s, visual evidence of the recent budworm outbreak was widespread in the thin foliage and occasional dead tops in these

White Birch, Keene, New Hampshire. *Courtesy of Gary Randorf, Adirondack Council.*

stands. In early June, the foliage is covered with tiny green caterpillars; in early July, the damaged needles gave the forest canopy a rust-brown appearance over millions of acres. In mid-July, hordes of small brown budworm moths collected around lights at night. By the late 1980s, the insect had virtually vanished from the region.

The road rises up a moist slope. Here, the spruce-fir and hardwoods mix in equal numbers to form mixedwood stands. In such situations, rapid wood growth is typical, because the soils are deeper and well-drained. Hemlock is a common component of these and nearby hardwood stands. From 1850–1900 in New York, Maine, and Pennsylvania the tanneries cut the hemlock, leaving the trunks to rot in the woods after removing the bark for tanning chemicals.

The northwoods landscape is sprinkled with tall pines that emerge above the general forest canopy (Figure 2.4). Early surveyors and timber cruisers used to climb these trees to view the surrounding area. Many of them were not cut in the logging days — they were too large to move, too far from water, and often rotten. The dense, extensive stands of pine are found in Southern Maine, New Hampshire, Massachusetts, and parts of the Hudson Valley but are uncommon in Northern Maine and the Adirondacks.

A gentle hillside now displays a stand of large spruce, mixed with

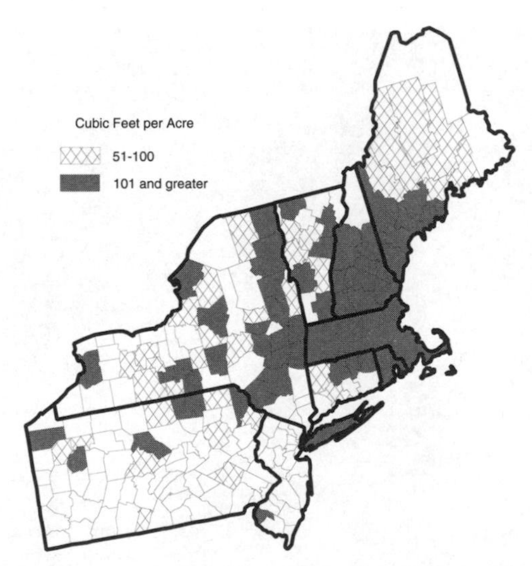

Figure 2.4 Eastern White Pine and Red Pine Growing Stock Volume Per Acre
Note: Volumes on Long Island are pitch pine.
Source: USDA-FS, Timber volume distribution maps for the Eastern U.S., Gen. Tech. Rept. W0- 60 .

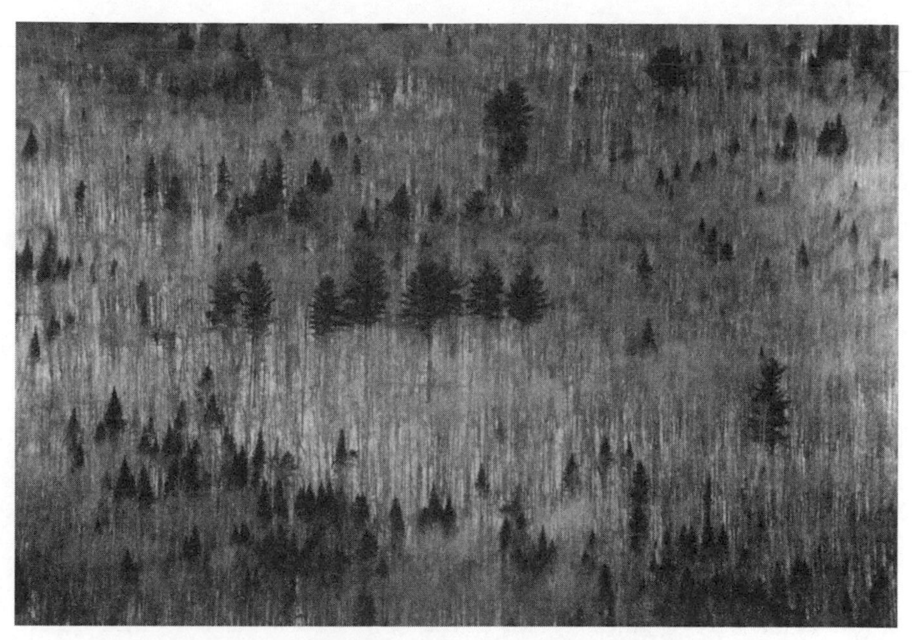

White pines often reach their best development in hardwood and mixed wood stands. *Photo by Roger Lee Merchant.*

Second-growth northern hardwoods, Bartlett Experiment Station, White Mountains, New Hampshire. *Courtesy of Bill Leak, USDA-FS.*

small fir, spruce, paper birch, and maple. These large spruce display healthy deep crowns. They are old enough to have survived the massive budworm outbreak of 1912–1920 and they now tower over younger trees. They will grow vigorously until age 200 or more. Because of their large size, straight grain, and strength they are prized for lumber. Such trees formed the basis for the region's softwood lumber industry from the 1840s to World War I.

Coastal Forests

From Cape Cod to the tip of Cape May stretches a narrow belt of an unusual forest. The pitch pine and scrub oaks, often gnarled and thin-crowned in appearance, grow on the sandy and worn soils of this coast.[8] Early settlers found these areas inhospitable to farming, and agriculture was largely a sideline to fishing and other industries. In New Jersey this belt widens considerably to encompass the Pine Barrens. Bits of this forest are preserved in the small state, national, and local parks that offer urbanites a few square feet of sand to enjoy in the summer, with a view of the ocean.

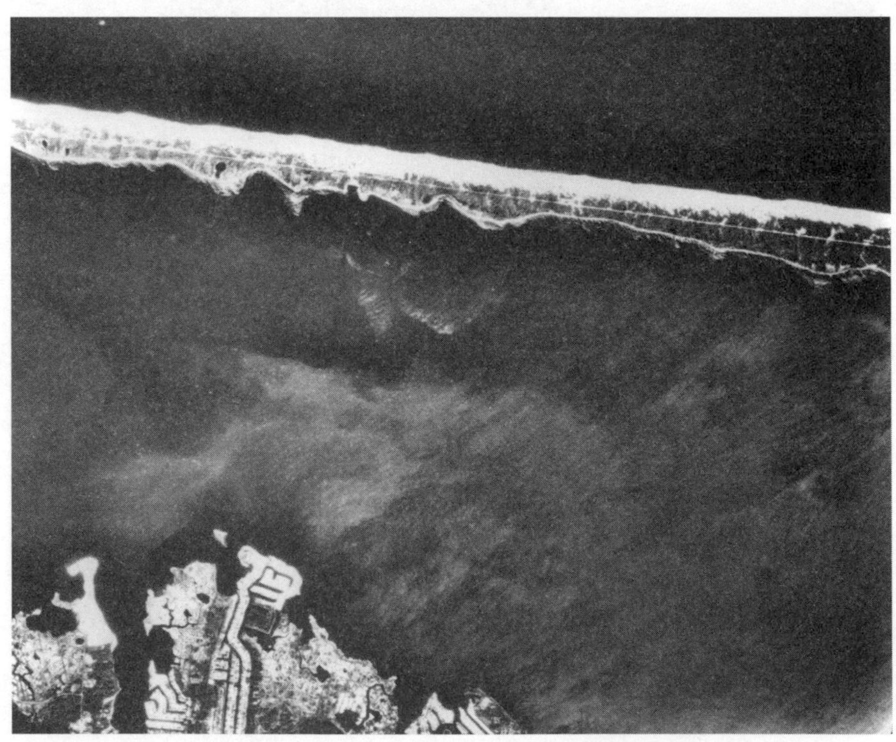

New Jersey dune forest, Island Beach State Park.
Courtesy of U.S. Geological Survey.

Cedar logs ready for market, New Jersey, 1897. *Courtesy of USDA-FS.*

A major inland extension of this coastal forest forms the New Jersey Pine Barrens, now protected in the Pinelands National Reserve (Figure 2.5). In colonial times, the pitchpines were large enough to yield a board two feet wide; the wood was valued for its durability. Sporadic efforts to use pitchpine as the basis for a naval stoves industry (hence "pitch" pine) were abandoned when the southern pine tar and turpentine industry began. In the cedar swamps of the region, cedars, even logs dug from the peat, were used for shingles, organ pipes, and water pipes. Observers spoke of "mining" the cedar. Benjamin Franklin advocated replanting the cedars.

The Pine Barrens occupy a flat, sandy region. The forests have been adapted to frequent fires. A seventeenth century explorer noted the smell of smoke offshore at sea, before the land was even visible. One fire in 1902 burned 75,000 acres; in 1930, 267,000 acres burned; in 1963, 202,000 acres burned with fires averaging 96 acres. A well publicized fire in spring 1995 burned 19,225 acres in Ocean County. Some landowners learned at an early time to ignite deliberate fires frequently to reduce fuel accumulations and reduce risks of catastrophic fires; this is still being done.

Lying between the shipyards and large populations of New York and Philadelphia, the forests of New Jersey were quickly cutover for lumber, timbers, and fuel. Yet the meager soils preserved a measure of wildness in the area up to the present. The Pinelands ecosystem covers 30 percent of

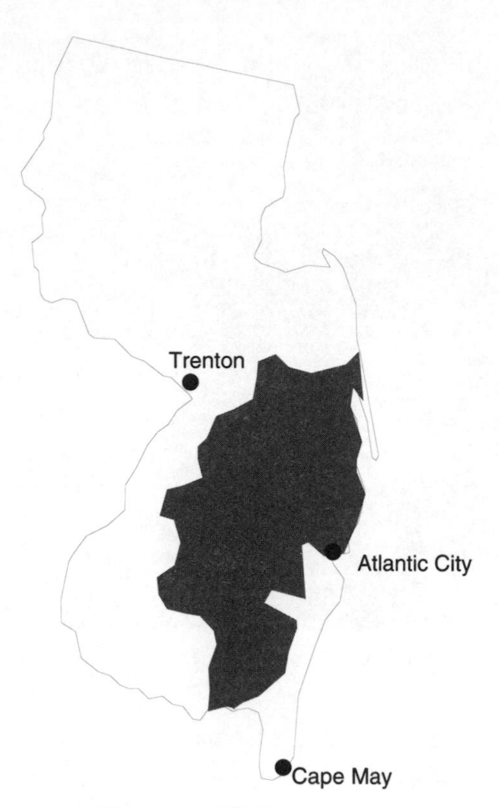

Figure 2.5 The Pine Barrens

the state, or 1.4 million acres. The Pinelands National Reserve encompasses 1.1 million acres, and a State Commission has jurisdiction over 934,000 acres. A "core" of 340,000 acres includes several state land units and is highly protected.[9]

Forests of the Lowlands: Southern New England to Southeastern Pennsylvania

The glaciers left behind a wet, ill-drained landscape over much of the Northeast. They left thin, "ledgy" soils and wide areas of sand plain in Southern New Hampshire, Southeastern Massachusetts, Rhode Island, and New Jersey. Walden Pond, surrounded by oaks and pines, is an example of these sandy landscapes (Figure 2.6).[10]

Chestnuts once dominated this landscape, with oak beneath them. Occasional hillsides of maples and yellow birches are found. The occasional extensive stands of pines contrast with the scattered pine of the far north, as does the prominence of oak. Fir is absent. Wet spots with poor

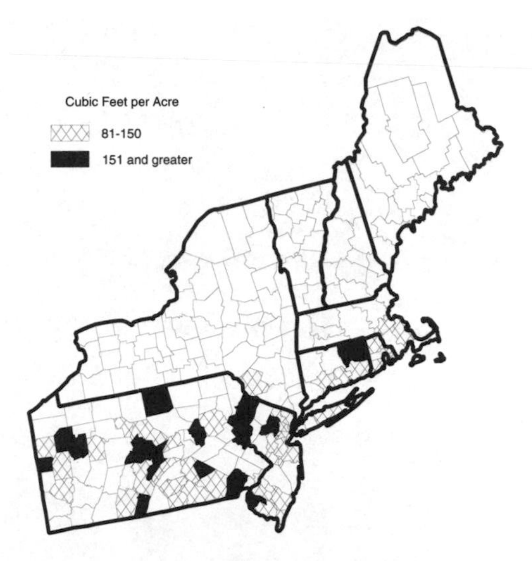

Figure 2.6 Select White Oak Growing Stock Volume Per Acre
Source: USDA-FS, Timber volume distribution maps for the Eastern U.S., Gen. Tech. Rept. W0- 60 .

drainage are often known as red maple swamps or may contain hemlock, ash, and elm. A few acid bogs resemble those farther north, with mosses, black spruce, and tamarack. A few bogs scattered from coastal New Jersey to Southern Maine contain Atlantic white cedar. Along the Delaware and into Pennsylvania, the northern hardwoods increase.

Abandoned farmlands make up much of the forest acreage of this region. A woodland of young oaks with mature and dead red cedar scattered through it is surely an old pasture, colonized by cedars, unappetizing to cattle, and later filling in with oaks.

Most of area's woodlots contain sprout growth; nineteenth century forest vegetation maps refer to this area as the sprout hardwood region. The sprouts come from heavy, repeated cutting. In many areas, heavy cutting on short rotations for iron furnaces and for charcoal kept the forest young for generations. Many trees bear basal fire scars from light fires. The centuries of repeated cutting, fire, grazing, and neglect have left many of these woods burdened with cull trees of poor form or quality.

Forests of the Ridges and Valleys and the Allegheny Plateau

This region properly includes the ranges of hills along the Delaware, and the ridges and valleys of Pennsylvania and New York to the Ohio Valley. These forests consist largely of hardwoods, though the region at one time supported extensive pine stands and hemlock forests, which failed to

Pitchpine on the sandy soils left by glaciers, Rhode Island. *Courtesy of Rhode Island Division of Forest Environment.*

regenerate following cutting and heavy burning. In parts of New York and the Allegheny Plateau, valuable stands with a high percentage of fast-growing and valuable black cherry were created by disturbances, primarily clearcutting and fire (Figure 2.7). The sawlogs from this species appear

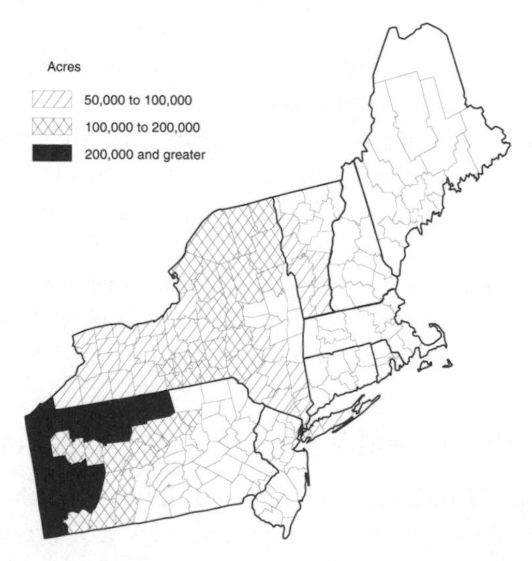

Figure 2.7 Distribution of Black Cherry Acreage by USFS Survey Units
Source: USDA-FS Resource Bulletins.

in expensive furniture and are often exported at high prices. Across the middle of Pennsylvania lies a boundary forming the southerly extent of the northern hardwoods type. In this northern hardwoods belt, forests dominate the landscape, as any traveller along Interstate 80 can see. The southerly half of the state lies in the oak-hickory belt, containing species more characteristic of the states to the south. In the Ridge and Valley Province, the forests occupy the long ridges that separate the productive, neatly kept farms of the valleys.[11]

Planted Forests

Trees have been planted for timber, erosion control, and aesthetics. Old fields and waste places have been planted with white pine, Norway and white spruces, red pine, and Scotch pine. Occasionally, venturesome foresters or landowners have planted larches, domestic and exotic, and other less common species.[12] Hardwood planting has been infrequent, except in attempts to revegetate coal mine spoils in Pennsylvania in which, for example, black locust serves well.

In local areas, Christmas tree plantations are prominent features of the landscape. These plantings have generally been established on former farmland, and have maintained early stand development stages of habitat as surrounding pastures grew up to forest.

In some instances, white pine planted in old fields has disappointed foresters and landowners. The soils, degraded by farming and erosion, often seem low in productivity, and stands suffer heavy damage from blister rust, white pine weevil, and occasionally snow and ice. The traveler will pass many overstocked plantations whose tending has been ignored. Overstocked, their trees do not develop deep full crowns and many will die from suppression. The forest floor in these plantations accumulates a fragrant brown mat of needles, littered with cones. Until the stand opens up by the death and blowdown of larger trees, little vegetation covers the forest floor.

Replanting after harvest cutting has been unusual in the Northeast until very recently. In the 1970s, several Maine paper companies began establishing nurseries and implementing extensive planting programs. They hope to overcome the frequent long lag in establishing new stands after clearcutting. Extensive research programs are working to develop improved tree varieties and management techniques for planted forests.

Plantations are costly to establish and yield their returns slowly. They will produce high-cost wood for their owners. On the other hand, they allow the use of selected strains of trees and their linear layout facilitates low-cost thinning and harvesting. Plantations are not likely to occupy a

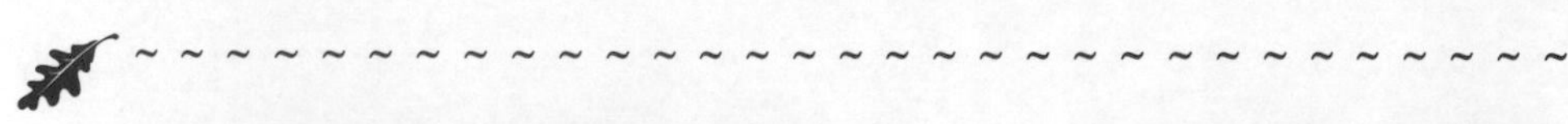

Black cherry stand, near Pigeon Run, Allegheny National Forest.
Courtesy of USDA-FS.

Skidder at work in oak stand in Pennsylvania's Ridge and Valley Province.
Courtesy of Penn State University, Agricultural Information Services.

noticeable part of the region's forest landscape in coming decades, though they will be increasingly prominent in a few localized areas.

Connections

The Northeast forest has emerged into the late twentieth century from the cumulative decisions of tens of thousands of landowners and dozens of private and public agencies. The presettlement forest was shaped by wind, naturally occurring insects, and occasional fires, supplemented in local areas by Indian burning and clearing. The forests of 1920 were depleted in area and in quality, and they had sustained a burden of repeated burnings

American chestnut, North Colebrook, Connecticut, 1912, before this species was virtually wiped out by the chestnut blight. This tree is two feet in diameter. *Courtesy of Thomas G. Siccama, G. E. Nichols Collection, Yale University.*

and sloppy cutting. Their species composition had been drastically modified, and average age much reduced. They have increased in area and volume since then, supplying improved opportunities for water supply, recreation, visual amenity, wood products, and energy production. Outside of the Wild Forest, the natural disturbance regime has been almost to-

Plantations, Tioga County, New York, 1948. *Courtesy of Leland J. Prater, USDA-FS.*

tally replaced by economic forces of cutting, by forest management practices, and by the effects of tens of thousands of land use decisions on individual acres each year.

Many of the nation's wetlands have been badly abused in the past. Some of the Northeast's forest types, such as cedar or red maple swamps,

TABLE 2.2. Wetlands Areas of Northeastern States, 1980s

State	Surface Area (1,000 A)	Estimated Wetlands (1,000 A)	% of Area
Maine	21,258	5,199	24.5%
New Hampshire	5,955	143	2.4%
Vermont	6,150	220	3.6%
N. New England	33,362	5,562	16.7%
Massachusetts	5,284	588	11.1%
Connecticut	3,206	173	5.4%
Rhode Island	777	65	8.4%
S. New England	9,267	826	8.9%
New York	31,729	1,025	3.2%
Pennsylvania	29,013	499	1.7%
New Jersey	5,015	916	18.3%
Mid-Atlantic	65,757	2,440	3.7%
Total Northeast	108,386	8,828	8.1%
Total U.S.	2,313,666	275,614	11.9%

Source: T. E. Dahl, *Wetlands losses in the U.S. 1780s to 1980s.* USDI, Fish and Wildlife Service, 1990.

Protecting the Pinelands' water resource is a key priority.
Courtesy of Pinelands Commission.

yellow cedar bogs, and elm-ash-maple bottomlands are in wetland situations.[13] Detailed data on forested wetlands are not available, but by some definitions, forested wetlands are extensive (Table 2.2 above). These areas are important for wildlife habitat and to the hydrologic cycle. They can be easily damaged if care is not taken in their management.

The Global Commons touches this forest, bringing concerns about the effects of atmospheric pollutants and longterm climate change. Much of the forest's expansion in area is owed to the region's *Economic Interdependency,* which enables it to import much of the food, raw materials, and energy that it consumes. Now that the region's forest has likely reached its maximum extent, concerns about its future habitat value are being expressed, based on the new metaphors of landscape ecology and conservation biology. The balance of this book offers an introduction to these issues.

3 Northeastern Forest Dynamics, Past and Future

Introduction

Understanding the geography of the Northeast's forests is essential to understanding their past, their present importance, and their importance for the future. In addition, it will be useful to understand something about how these forests function ecologically. Changing ways of measuring the forest show the interaction between society's perceptions about the forest, the techniques of measuring them, and the kinds of data assembled by the experts. Next, we review the complex of disturbances that shaped the presettlement forest and the effects of land use change, cutting, and fire. Finally, we describe forest management and other factors that will affect the forest in the coming half century. The wildlife resource is profoundly affected by all of these changes. In the last section, we suggest how forest management must change in light of current understanding of forest dynamics.

Better understanding of forest ecology brings with it a need to measure more aspects of the forest, and better forest measurements often highlight new problems and challenges for scientists. These concepts are building blocks of the Worm's Eye View of the forest, which enriches our understanding of the invisible flows of matter and energy within forests. They relate closely to the Eagle's Eye View, which seeks to understand how forest landscapes change over time. The countless interconnections between flows and stocks of water, organic matter, nutrients, and energy within ecosystems are only beginning to be understood. The complexity of the relationships, and their variations over time, challenge scientists attempting to measure them.

Broadening Concepts for Measuring the Forest, 1630–1993

The way the forest is measured is based on what observers feel is important, on the measuring tools available, and on what traits science or the

The size, timing, and arrangement of cutting units are important. In sensitively managed landscapes of the future, large clearcuts with square boundaries will not be seen. *Courtesy of Maine Department of Conservation.*

marketplace tell us are important. Over the years, the ways of measuring the forest have become ever more complex and have broadened to include more and more traits of forest ecosystems.

In the 1600s, William Wood and other writers on the region's forests noted the "commodities" of the land and the uses of its trees and bushes. Their description of the forest was limited to making lists of species. By Moses Greenleaf's time in 1829, this had hardly changed; his book's section on the forest is still little more than a species list.[1] Since, in his view, the forest would give way to farmland clearing, there would have been little point in making exhaustive measurements. In the 1880s through the turn of the century, experts began developing timber volume measures for forests of entire states. At the time, they concentrated on what the market wanted. So they estimated the volumes of pine or spruce.

As the forests were better known and became important items on their owners' balance sheets, and as statistical techniques came into wider use, actual "cruising" began, in which trained crews measure randomly selected plots as a basis for estimating the timber on large areas. A review of Maine's forest resource was prepared by the Forest Commissioner in 1902 which relied heavily on fieldwork by experienced scalers and woodsmen. In early reports on forest conditions, analysts had to rely on estimates such as these. Then in 1928, Congress authorized the USDA-FS to conduct periodic Forest Surveys to measure the forest areas, volumes, and growth in the various states and to prepare from those estimates a full reckoning of the nation's timber budget. From these reports, most of the resource information in this book has been drawn.

Measures of the forest under this Forest Service program have continued to respond to broader concepts of forest values. The basics are still the economic values. For example, in the State of New York, the commercial forest land in 1993 was estimated to contain:

	Billion Cubic Feet
All Timber	23.1
Sawtimber	13.8
Poletimber	8.1
Growing Stock	21.9
Rough	0.8
Rotten	0.2
Salvable Dead	0.2

In the 1970s, growing interest in wood energy use and in the science of forest growth led the timber cruisers to sharpen their pencils and develop ways to measure the total forest biomass. They found that in 1993 New York State's forests continued the impressive total of 1.1 billion dry tons of tree and shrub biomass, distributed as follows:[2]

	Million Dry Tons
Timber	972
Salvable Dead	9
Saplings	95
Seedlings	13
Shrubs	12
Total	1,100

Experts have been able to reckon up the weight of total forest biomass including foliage, shrubs and understory plants, forest floor organic material, and even roots. This approach enables scientists to understand how forest communities store, accumulate, and cycle organic matter and nutrients. The components of tree biomass in New York in 1993 were:

	Million Dry Tons
Growing Stock	603
Branches	78
Foliage	28
Stump and Roots	183
Cull Trees	78
Total	971

More recently, the changing Global Commons has led investigators to attempt to map the flows of carbon through the earth's carbon cycle,

Ecology and Management of Northern Hardwood Forests in New England

The northern hardwood type is identified by a predominance of American beech, yellow birch, and sugar maple, occurring either singly or in combination. There are a variety of associated species, with some of the more important being red maple, white ash, Eastern hemlock, paper birch, quaking and big tooth aspen, Eastern white pine, red spruce, and northern red oak. Within the United States the range extends south from Maine to the Southern Appalachians, and west to Ohio, upper Michigan, and Eastern Wisconsin.

Wildlife Habitat

There are 338 nonmarine (inland) wildlife species in New England: 26 amphibians (salamanders, toads, frogs), 30 reptiles (turtles, snakes), 220 birds, and 62 mammals. Out of this number, 266 use the forest at least part of the time and the others are restricted to water or nonforest (e.g. fields). Only 16 of the forest-users stay completely in coniferous forests, e.g. the three-toed and black-backed woodpeckers or the spruce grouse, leaving 250 species that sometimes use hardwood or hardwood-conifer stands. Most of these species are not very specific to a forest type, so it is not surprising that 215 species are listed for northern hardwood stands: 61 percent birds, 25 percent mammals, and 14 percent amphibians and reptiles. Only a few of these species are game animals or even well-known songbirds. But they serve several functions, as: food for larger animals; insect, seed, and vegetation eaters; and seed dispersers and buriers. Numbers of species do not vary greatly by stand-size classes, but a forest that maintains all sizes and ages of timber, including some temporary or — better yet — permanent nonforest openings, can support approximately 215 species as compared to an all-aged forest with 160 species.

Potential numbers of wildlife species in northern hardwood stands by stand size-class and type of wildlife. Many species are found in several stand size-classes

	Stand size-class					
	Regeneration	Pole stands	Saw timber	Over-mature	All aged	All
Amphibians	8	16	19	18	18	19
Reptiles	9	12	12	12	11	12
Birds	96	72	82	90	84	131
Mammals	45	40	47	47	47	53
All	158	140	160	167	160	215

Within a northern hardwood forest, wildlife species occupy several different types of habitat conditions for purposes of breeding, feeding, or cover. High, exposed perches are used by hawks and eagles and other birds, low perches by flycatchers and several sparrows. Many warblers are canopy feeders. Typical tree-bole users are owls and woodpeckers. Large, dead, and partially-decayed living trees are more use-

ful than small. Many hawks, owls, and smaller birds utilize small patches of softwood trees existing with a northern hardwood stand.

Potential numbers of wildlife species found using certain habitat factors in northern hardwoods. Many species use more than one factor

Habitat Factor	Number of Species
High, exposed perches	7
Low, exposed perches	13
Tree canopies	89
Tree boles	40
Midstories (10-30 feet tall)	18
Shrub layers (3-10 feet tall)	81
Overstory softwood inclusions	60
Upland herbaceous vegetation	47
Wetland herbaceous vegetation	15
Decaying logs	3
Humus layers	53
Below ground	38

Upland and wetland herbaceous vegetation supports a variety of birds, amphibians and reptiles, and small mammals; these layers are best developed under low-density canopies or in openings. Decaying or hollow logs are used by amphibians and reptiles, as well as small and large mammals. Ground surface and below-ground dwellers also include a variety of amphibians and reptiles, mammals, and birds. The general message is that there is a great variety of wildlife habitat in northern hardwood stands and a myriad of opportunities to influence habitat conditions through silvicultural practice.

Source: Hornbeck and Leak, 1991. *Ecology and management of northern hardwood forests in New England.* USDA-FS Gen. Tech. Rept. NE-159. June. p. 1, 22–23.

identifying its sources, pools, and sinks. The forest is a significant item in the global carbon budget. Deforestation has been identified as a significant source of carbon increasing the CO_2 content of the atmosphere. Conversely, because Northeastern forests are expanding both in area and in volume, they are sequestering carbon as their biomass builds up. Experts are arguing about how to respond to the atmosphere's increasing carbon content, and planting and growing trees have received a share of the attention. Also, there is a consensus among climate change experts that additional use of wood for energy, if it substitutes for hydrocarbons, can reduce carbon emissions.

Forest survey experts have calculated the carbon storage and accumulation patterns of U.S. forest ecosystems. They find that in almost all

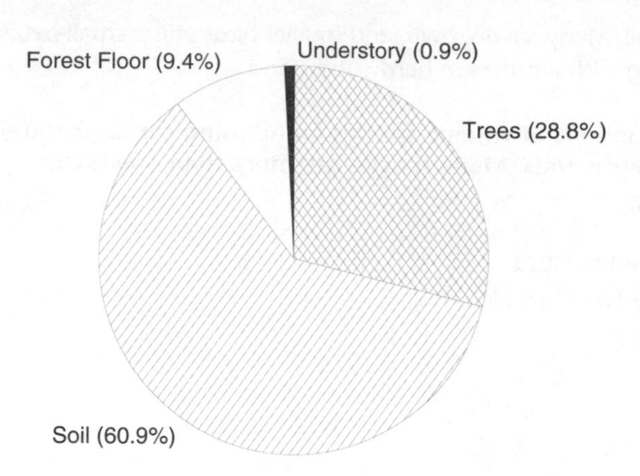

Figure 3.1 Distribution of Stored Carbon Per Acre, New York, 1987
Source: R. A. Birdsey, Carbon storage and accumulation in U.S. forest ecosystems, USDA-FS, Gen. Tech. Rept. W0- 59, 1992 , p. 20.

forest types, more than half of the ecosystem's organic matter is stored in the soil. It reaches the soil after being broken down by biological and chemical processes in the forest floor, and in some areas, with the aid of the feeding and digging of a myriad of worms, insects, and small animals, as well as the annual dying of roots and rootlets. The trees themselves are the second most important storage pool, followed by the forest floor (Figure 3.1) The experts calculated that the forests of New York State contained 1.4 billion metric tons of carbon in 1987, including soil and forest floor pools. Between 800 and 1400 pounds of carbon per acre per year were added to this carbon stockpile.

Not only has the more complete and detailed accounting for forest biomass and its carbon content become more important, so too has the measurement of forest traits important to wildlife.[3] Working with a Maine wildlife biologist, Forest Service analysts analyzed 1982 Maine forest survey data to characterize wildlife habitat conditions. They presented detailed tables by geographic divisions of the state for variables such as indexes of the amount of edge habitat, number of fruit and nut producing trees, trees with cavities for cavity-nesters, understory species by browse preferences and degree of browse utilization, and a detailed picture of the forest by stand size class and by areas of stands. As an example, they summarized forest traits of importance for deer habitat (Table 3.1). Using this information, biologists can understand the distribution of key habitat elements around the state, and make judgments about how they are likely to change in the future. As these surveys are carried forward into the future, they will provide a valuable basis for management decisions.

TABLE 3.1. Selected Habitat Components Describing Aspects of White-Tailed Deer Habitat, Maine 1982

	State Average	Range Across 9 Survey Units
Percent forest area	88.8	75.9–96.6
Edge indices (Total)	32.6	16.9–45.3
Percent timberland Area		
Sawtimber, conifer	33.0	21.5–41.3
Sapling/seedling	11.2	6.6–20.0
Stands less than 100 acres	53.5	30.2–86.3
Browse potential, 1000		
Sapling, seedling, shrub stems/A.	10.7	8.2–14.4
Browse potential, percent with observed use	17.4	9.5–23.1

Source: R. T. Brooks, T. S. Frieswyk, and A. Ritter. *Forest wildlife habitat statistics for Maine 1982.* USDA-FS NEFES Res. Bull. NE-RB-96, 1986, p. 3.

Disturbance Patterns and Forest Dynamics: Presettlement Forests

Ecologists and foresters have taken a deep interest in disturbance patterns and how they affect the way natural forests are shaped. It turns out that the species composition and age and size structures of unmanaged forests are powerfully shaped by the forces that kill individual trees, patches of trees, or whole large areas of forest. The wildlife inhabiting these forests, in turn, depend on the stand structures created by these natural forces.[4]

In all forests, individual trees succumb to mortality from old age, insects and disease, or windthrow. When this occurs one tree at a time, it is referred to as gap replacement. If the predominant disturbance over long periods occurs as the death of individual trees, then forest composition will come to be dominated by species like beech and hemlock that can survive for long periods in understories and then grow to full size when released by death of overtopping trees. Some of these gaps emerge slowly as an overstory tree gradually succumbs to decline or old age; others occur overnight as a tree is blown down or struck by lightning. In other instances, mortality occurs in larger patches consisting of a number of trees. When this occurs, trees demanding more light will sustain themselves in the canopy. Finally, massive "catastrophic" or stand-replacement events can occur. Examples would include the 1938 hurricane in New England, or the 1950 "Big Blow" that damaged forests extensively in the Adirondacks. In these instances, the overstory is removed over large areas, and understory trees may be killed by fire or smothered under the down tops. Scientists often speak of return periods of disturbances, which would be the average interval between, say, blowdowns on an average acre. Actually, there is a wide scale of sizes and time scales of disturbances, and they can vary from the low swampy areas to exposed ridgetops.

Examples of how these disturbance patterns functioned in the presettlement forest have been constructed by scientists carefully measuring longterm weather data and reconstructing early forest conditions from surveyors' notes, reports of early timber cruisers, and travellers' accounts. In one especially thorough example, Whitney reconstructed the forests that ultimately became the Allegheny National Forest in Pennsylvania. He found that prior to settlement, the forest of that region was dominated by hemlock and beech; pine was scarce. He concluded that in that area, the return time of major stand-replacing storms or fires was on the order of 1000–2000 years. These conditions suggest that presettlement forests had time to develop uneven-aged structures dominated by a few species.[5] (In all-aged stands, trees of numerous age classes from seedlings to mature emergents are represented, in contrast to even-aged stands, which are dominated by a single age group.)

In other work on remaining Pennsylvania virgin stands, a slightly more complex picture emerges. In many stands, pines displayed traits of essentially even-aged stands, suggesting that their presence was due to occasional catastrophes with uncertain return periods.[6] Given that pine rarely lives beyond 450 years, at least a portion of the landscape must have experienced stand-replacing disturbances with return periods shorter than that to sustain pine stands.

In the Adirondacks, intervals between major stand-replacing disturbances are long. With the region's cold climate, winter kill occasionally affects individual trees and small patches. Fire is not a major contributor to the natural disturbance regime. Windthrow can at times be severe, however.

In Northeastern Maine, Lorimer conducted a similar analysis to reconstruct presettlement forests there.[7] He found that the average recurrence interval of major fire was 800 years, and of catastrophic windstorm was 1,150 years. These are a multiple of the life cycle of major forest species. This suggests that uneven-aged stand structures would have been common in the presettlement forest. This result is supported by analyses of the few remaining stands that have been measured, and by the modest presence of short-lived or early-successional species in such stands. Yet short-lived species did occur in these forests, suggesting a diversity of dis-

Old growth spruce, Squaw Mountain Township, Maine, 1902.
Courtesy of USDA-FS.

turbance factors at work on differing time and spatial scales. In Lorimer's study area, species accounting for more than 10 percent of forest composition included the spruces, beech, balsam fir, hemlock, and yellow birch.

In Northern Vermont, Siccama studied the presettlement forest composition.[8] He found that the predominant tree was beech, a long-lived highly tolerant tree. Beech made up 40 percent of the composition. In the present second-growth forest, however, it was only 3 to 5 percent, showing the dramatic change in composition in the second growth.

Current Impacts of Disturbances

In contrast to studies of the historic impact of disturbances on forest structure, there has been little systematic analysis of the effects of such disturbances in the present forest.

In the Big Blow in 1950, more than 2 billion board feet of timber were blown down, including several remaining virgin stands. This event created new patches of even-aged forest. Again in July 1995, a major storm blew down 22,528 acres on private land alone, with additional losses in the Forest Preserve lands. Some stands lost were old-growth. An estimated 600 miles of trail were blocked. Other examples come to mind. In the 1980s, a storm blew down a long swath of timber on state lands in North Central Pennsylvania, totalling 10,000 acres. Most of this area was

Forest Fires in New Jersey: 1880s

In Atlantic County it was estimated that the area burned over in the early part of 1880 was some 50,000 acres. The fires were started by sparks from railroad trains, the burning of brush in the clearing of land, by tramps, and other causes not known. The estimates of loss vary, and might average three dollars to the acre for wood land, not including the damages done to vineyards and buildings. Eight dwellings and twenty-nine barns were destroyed. In the absence of water, the progress of the fire was checked by throwing fresh earth. It was found difficult in many cases to render the roads available for stopping the progress of the fire, on account of their narrowness, which, clear of combustible material to the width of three or four rods, they might have proved of much greater service in arresting the fires.

This county suffered from forest fires in 1820, 1829, 1832, 1833, 1857, 1858, 1859, 1865, and 1872.

Burlington, N.J. — In the spring, during an unusual drought, which continued some four weeks, a most disastrous fire occurred in the timberlands of this county, sweeping over about a third part of the forest area. They originated from locomotive sparks, coalpits and clearings, and, as some suppose, from incendiaries, who wish to buy the dead timber cheaply for coaling. The estimated area of the burned district was 40,000 acres, and the loss from $10 to $30 per acre, except in cedar timber, in which the value may range from $100 to $1,000. The loss may be estimated at $130,000. Besides standing timber, there was a loss in charcoal, wood and rails, that might amount to $15,000; some buildings, usually not valuable, fences, stocks, &c., were burned, worth perhaps $9,000 more. So much loss has been occasioned in this county by incendiary fires, that the owners of timber lands in this county have organized a timber-protection company. Depredators have been cutting and selling timber from lands of non-resident owners, and to prevent this, individual owners and this society are offering rewards for their detection, and punishing the guilty when convicted, and it is thought that some of these fires have been set in revenge.

salvaged. And in January 1998, a regionwide ice storm struck parts of Northern New York, Northern New England, and Southern Quebec. Damage to trees was widespread, though highly patchy and variable. In local areas patches of complete blowdown as large as 1,000 acres occurred.[9]

History of Human Disturbance

The early history of forest disturbance, clearance, and regeneration in the Northeast is intimately connected to the rivers, with which the region is liberally endowed. The major streams from the St. John on the Maine border to the Ohio in Southwestern Pennsylvania provided avenues for exploration, for rafting logs, and for developing trade routes. At tidewater

A newspaper account dated Atsion, May 14, says:

> The fierce forest fires now burning in this region alone are reported as destroying over 100 square miles of arable, berry bog and timber land and vineyards. Great Republic, a prosperous village, is now in a state of desolation. Monday night's fire destroyed no less than twenty-two buildings in and about the village. At Bennett's mills, two houses in the path of the flames were consumed. From Great Republic the fire took a southeasterly course towards the iron mills and Cedar Lake, and skirting about Hammonton and Ellwood, spread towards Mullicus River on the northeast and Great Egg Harbor River on the southeast.

A correspondent in Burlington County, New Jersey, writes:

> I have had under my care large tracts of growing timber, and have had very little loss from fires. Our plan is to burn over our tracts early in spring, before the sap starts — burning up all the underbrush, leaves, &c. We have found that a fire in February does not hurt the growing timber; and should summer fires come in from adjoining property that has been neglected, they have nothing to feed on but very green leaves, &c., and they are much more easily controlled. Even if they go through the woods they find little to burn; they get up but little heat, and do but little damage.

Source: F. B. Hough, 1882. *Report on forestry.* Washington: Government Printing Office. p. 158–159.

they offered millsites where sawmillers could trade with the world. Farther upstream, they offered waterpower sites that have shaped the location of the region's major cities and towns, and hence its highway and rail systems, to this day. For many years, the timber resources of a state were reckoned up according to the major river watersheds, instead of by counties and administrative or other subdivisions as they are today.

The region's present forests have been largely created — by intention or by chance — by human action. This is not to deny the basic influence of soils and climate, or the importance of wind, fire, and insects, but merely to assert that human efforts have broadly shaped today's forest. In the future, human action will be even more important.

Cutting and, in some places, fires had a staggering impact on the region's forest. According to Sargent's 1880 Census Report, fires damaged 964,000 acres of northeastern forest or cutover land each year, including an astonishing 685,000 acres in Pennsylvania alone. In the summer of 1903, 637,000 acres burned in the Adirondacks. In New Jersey in 1872, one fire covered more than 15 square miles. In the White Mountains, the fires of 1903 burned 80,000 acres. The cumulative effect of fires on this scale is difficult to imagine, considering that the annual average area burned today is only about 30,000 acres. The damage to soils, waterways, and wildlife would have been enormous. It is easy to see why early foresters placed such a priority on fire prevention and control. If the 1880 fire cycle were averaged over the 48 million or so acres of forest then existing, the entire forest would have burned in only 50 years! One result of the abusive cutting and extensive fire damage was that in its 1920 Capper Report, the Forest Service rated 14.3 million acres, or 30 percent of the forest area of the time, as barren, unproductive, or good only for fuel.[10]

Almost one-third (20 million acres) of the region's current forest area has seeded in on abandoned fields (Figs. 3.2 and 3.3). This secondary forest is in some places more than a century old but remains profoundly affected by its old-field origin. At the other extreme, the acreage of undis-

Hardwoods on an abandoned field, Derry, New Hampshire.
Courtesy of American Forest Institute.

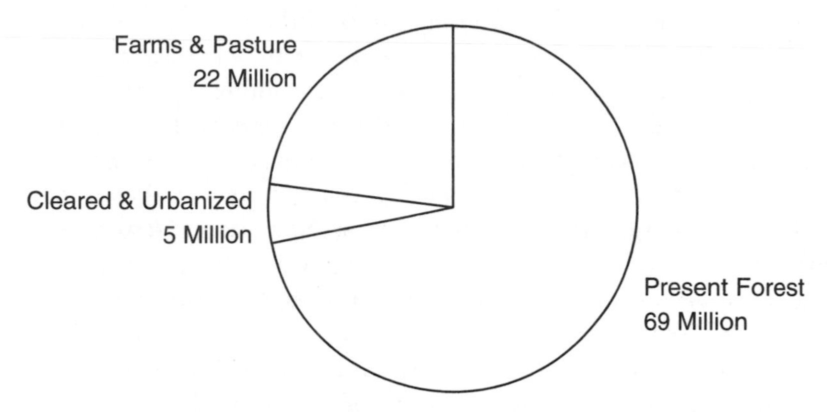

Figure 3.2 Fate of Original Northeastern Forest
Total = 96 million acres

turbed ("virgin") forest is nominal, probably less than the area planted by human hands to pines and spruces. Sometime in the 1950s or 1960s, the acreage of cleared land still in farms in New England and New Jersey fell below the cumulative area converted to subdivisions, roads, and shopping centers. This was a significant point in the area's land use history. In the three Mid-Atlantic States as a whole, farmland area remains far larger than developed area.

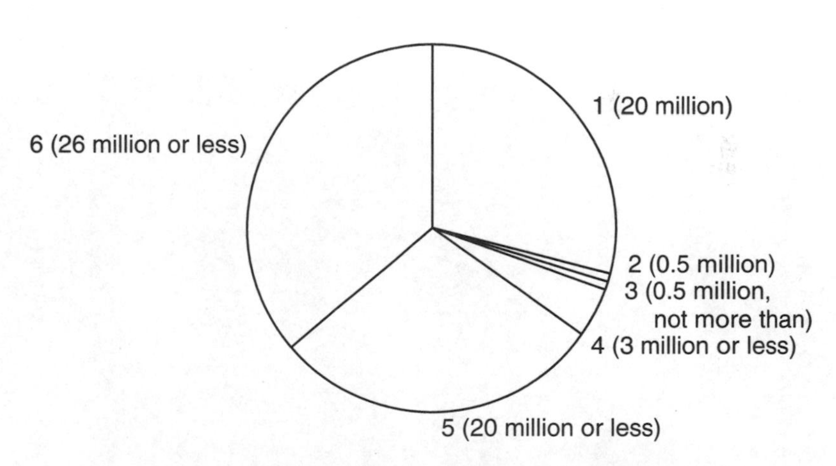

Figure 3.3 Disturbance History of Current Northeastern Forest
Total = 70 million acres

1. Regrown from Farmland (effect of cutting: variable)
2. Planted
3. Essentially Virgin Forest
4. Managed in Intensive Even-Aged Regimes
5. Heavily Influenced by Post-1940s Cutting
6. Minimally Influenced by Post-1940s Cutting

Providing energy has profoundly affected the region's forests. About 10,000 acres per year are stripmined in Pennsylvania. In local areas, forests were cleared or covered with gob piles in coal mining. In other areas, the oil industry's wells, roads, and pipelines are prominent. In the nineteenth century, the fuel needs of the iron furnaces prompted the ironmasters to own large forests that they cut periodically for wood, until coal replaced wood as fuel. Large areas in Pennsylvania, New York, New Jersey, and Southern New England were managed in this way. These areas are now in largely even-aged stands of sprout origin, less than a century old.

The lands that persisted in forest through these years fall into two categories — areas that were once heavily cut but then ignored until recently, and a much larger area of the industrial forest that has been subject to periodic harvests at a relatively low level for more than a century and a half. In Connecticut, for example, about ten times as much timber was harvested in the 1860s as in the 1960s — and that from a forest area of half the extent. Cronon notes an estimate that New Englanders alone likely burned 260 million cords of firewood between 1630 and 1800.[11]

Cutting continues to be an important force. Forest Survey data for New England have been analyzed in detail for the kinds of cutting that have occurred,[12] and the results may be broadly applicable to the whole Northeast. Between the surveys of the early 1970s and the early 1980s, cutting occurred on 30 percent of the region's acreage. The average cut-

Pine stand in an old field, Waterboro, Maine. *Courtesy of Bill Leak, USDA-FS.*

ting rate for basal area was 1.3 percent per year. (Basal area is the cross-sectional area of trees, in this case live ones, measured at breast height.) The cutting rates varied, however. On 4 percent of the region's timberland, 80 percent or more of the basal area was removed, while on 79 percent of the harvested land, removals were less than 20 percent of the basal area. The effects of these cuttings on quality of residual stands was not rated. If these relationships hold in the future, then, we might expect that on average in a 30–40 year period, most of the unreserved forest will be lightly cut at least once.

Not only have fire, cutting, regrowth of old fields, and windthrow affected the extent of the current forests, but their composition as well. In few places were the influences simple or uniform. In Western Pennsylvania, for example, cutting practices changed over time, with different influences on the forest. As Whitney put it: "Three successive attacks on the forest — the selective cutting of the white pine and the hemlock, the clearcutting for chemical wood and overbrowsing by deer — have left their mark."[13]

In central New England, Foster documented the complexity of influences shaping present forests:

> The land use pattern that emerges on examination of the Prospect Hill tract is exceedingly complex and variable in scale and intensity over time. Each of the individual parcels comprising the tract was owned by an average of 13 individuals in the 160-year period . . . Although the details vary, each forest stand was cut repeatedly, most were cleared, and all were subjected to varying agricultural use, including grazing.[14]

The forest landscape has been modified in ways more subtle than clearing and reseeding. According to J. J. Dowhan and R. J. Craig, 25 percent of Connecticut's flora, 30 percent of its fish, 7 percent of its mammals, and 4 percent of its birds were introduced from elsewhere. Truly, many of the creatures of Connecticut's landscape owe their presence there to human action. According to Williams, more than 25 percent of the plants in Pennsylvania and New York are introduced; it is estimated that the figure is 37 percent for Massachusetts.[15]

Regionally, the decline of farming has led to a noticeable decline in the birds of fields and openings, and an increase in forest dwellers. While the wolf and several other species have vanished from the scene, beaver and turkeys are returning to areas where they had been long forgotten. Due to better controls over pesticides, expanding forest area, and restoration efforts, many species of predatory birds are stable or increasing. Deer are so abundant in many areas that they damage crops and tree regeneration.

Mature sugar maple, North Colebrook, Connecticut, 1912. *Courtesy of G. E. Nichols Collection, Yale University.*

Black bear survive in the wooded hills of New Jersey, and with 8,000 roaming the woods of Pennsylvania, their population there is at historic highs. The fisher is once again abundant in the woods of Massachusetts.

Forest Management and Harvesting as Disturbance Factors, 1990–2040

The complex of natural forces shaping the region's forest was massively changed during the period of settlement and cutting between the 17th and early 20th centuries. The reductions in cutting pressure since 1910 and the reversion of farmlands since 1850–1880 have restored a more mature and more extensive forest than the region has known in 150 years. Yet today, as wood demands escalate and competing demands for land make themselves felt, the likely effects of cutting and active forest management practices are generating rising concern. So we should look quickly at the kinds of effects that may be expected in the future. While these estimates are all subject to many uncertainties, they are probably reasonable in terms of order of magnitude.

Virgin white pine stand, Colebrook, Connecticut, 1911. *Courtesy of G. E. Nichols Collection, Yale University*

Estimates of management activity and disturbances likely to affect the region's forests over the coming half century, can help place the various forces in perspective. The management activities should not be added up, because they overlap in many instances (Table 3.2). The management activities will be concentrated in the Industrial Forest and selected areas near it, so their impact on forest habitat and visual quality there will be more than this regionwide comparison suggests. The most important factors

~ ~

TABLE 3.2. Forces Affecting Northeastern Forests Regionwide, 1990–2040

Total forest area	Annual acreage	Areas affected in 50 years	Percent of total in 50 yrs
Urbanization		2–4 million	2.8–5.8%
Leisure lot development		2–4 million	2.8–5.8%
Conversions back to farmland		Uncertain	
Planting	30,000	1.5 million	2%
Stand improvement	63,000	3.2 million	5%
Herbicide treatment	50,000	2.5 million	4%
Harvest cutting*	1,000,000	50 million	72%
Fires	30,000	1.5 million	2%
Windthrow	60,000	3.0 million	4%
Natural processes only	5 million or more	5 million	7%
Global Commons	70 million	70 million	100%

*Defining harvest cutting to include only cuts removing more than about 40 percent of basal area at one time to avoid double counting with Stand Improvement.

Source: Author estimates; see also Irland and Maass, *Regional perspective: forest conditions and silvicultural trends in Northeastern USA,* For. Chron. 70(3): 273–278.

affecting the forest, then, are likely to be harvest cutting, land use change, and the Global Commons. The forest by the mid-21st century will be almost entirely shaped by these three forces.

The amount of forest and its visual character will be profoundly shaped by development. The estimates in Table 3.2 for urbanization and leisure lot development are speculative but, given recent history, could certainly occur over 50 years in the absence of more effective land use regulation. Not only will development remove forest directly, but "shadow conversion" will effectively diminish or remove many forest use opportunities on at least an equal if not larger area. Shadow conversion occurs when the sale of houselots on one acre of a farm or woodlot compromises management options on many more adjacent ones, as the new homeowners object to manuring or logging. Farmland expansion is not suggested, but given trends in world food supplies and yields, there is more than a possibility that farming will reclaim once again some of its lost pastures and fields. The regional effects of conversion and shadow conversion will outweigh those of intensive forest management practices other than har-

vest cutting. Also, effects of fire and windthrow will be of similar magnitude to those of intensive management.

The point of these estimates is the incredible complexity of the forces that will shape the forests of 2040 and beyond. Most important will be how these influences are managed and controlled. This in the long term will be as important as the total acreages affected by any one of them. The specific quality of their effects, and their arrangement over space, will be critical.

Wildlife in the Northeastern Forest

The history of wildlife conservation is interwoven with that of forest conservation. The retreat of forests before farming and lumbering meant a major shrinkage of habitat for most forest dependent wildlife. During the same periods, overhunting was rampant. The result was the complete extirpation of some species from the Northeastern landscape. The passenger pigeon is generally believed to have been extinguished by loss of habitat.[16] At the same time, species of open farming country flourished. As game laws and wildlife refuges were established, and as the forest cover was reestablished over 17.5 million acres, some wildlife species have rebounded dramatically.[17] In the Adirondack Park, 43 plants and 12 animals have disappeared since settlement. But the peregrine falcon, raven, bald eagle, moose, and lynx have been reestablished through reintroductions or on their own. Also, 23 species of animals and plants have been introduced.[18]

The region's wildlife resource is important to millions of anglers, hunters, birdwatchers, and other enthusiasts. It is also a part of the enjoyment of the landscape for many other rural residents and recreationists. Finally, it is widely believed that the health of wildlife populations can be an indicator of the health of the entire ecosystem. In the world of wildlife there has been a trend of broadening inclusiveness of measurement, of policy, and of public consciousness. At one time, only spectacular birds and mammals and game animals were of concern. Ways of thinking about wildlife populations have broadened through several stages:

- *deer and ducks:* lands are managed for featured species, and these are what is counted and studied;
- *multiple use with nongame wildlife:* in addition to the specific management objectives for the land, other species are identified and given consideration; and
- *biodiversity.*

Corresponding to each of these ways of thinking about the wildlife resource is a different approach and philosophy of management.

In managing for individual game species, agencies and landowners will flood lands to expand marshes for ducks, push over trees to create browse openings for deer, or carefully cut patches in the forest to regenerate particular tree and shrub species or create suitable stand conditions.[19]

In multiple use management with allowance for nongame wildlife, efforts should be made to identify occurrences of rare or unusual species of plants, animals, or natural features, and protect them by suitable means. This is done within a primary management emphasis for timber, recreation, or the featured game species. In such a case, snags might be deliberately left for cavity-nesting birds, or potential raptor nesting trees near waterbodies might be protected by seasonal no-cut policies. As another example, in managing for the American marten, research suggests maintaining a substantial "matrix" of interconnected areas of mature softwood forest in the landscape over a long period of time. Subject to strict rules on patch size and shape, cutting can occur within this landscape as long a substantial matrix of mature forest is maintained.[20]

Fisher. *Courtesy of Maine Department of Inland Fisheries & Wildlife.*

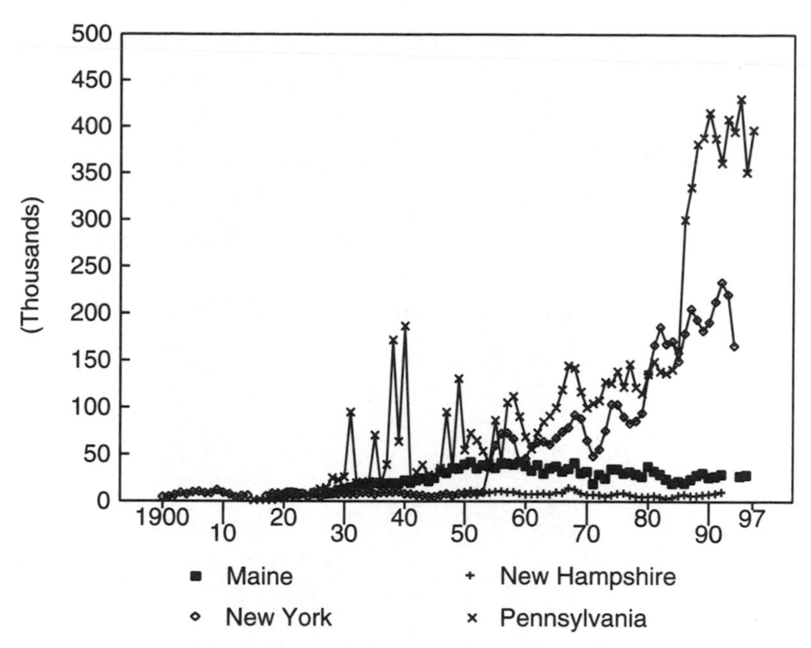

Figure 3.4 Deer Kill Trends, Selected States
Note: New York data 1900–1953 are for Adirondacks only; 1900–1911 is all species; 1912–1953 is adult male only; 1954–1994 is all species, ages, and regions.
Source: State fish and game agencies.

A Habitat Generalist: White-tailed Deer

The regional population explosion of white-tailed deer is one of the nation's conservation success stories (Figure 3.4). Outside of a few remaining forested areas, deer were nearly absent from the region's landscape a hundred years ago. The harvest fell to about 1,000 per year in Pennsylvania early in this century. With controls over hunting and the recovery of habitat, deer numbers multiplied. Deer are adaptable users of early successional habitats and landscapes of mixed farming and young forest. In the Adirondacks, the deer take was from 7,000–12,000 around World War I, but rose to 47,000 by 1970.[21] In Maine, the legal kill was less than 6,000 in 1919 and peaked in the 1950s to the 1970s.

Deer were abundant in the young forest that followed the wave of heavy cutting around the turn of the century. As the young stands mature, their value as deer habitat decreases. Across the northerly end of the region, habitat is marginal for deer. In the Adirondacks, northern Vermont and New Hampshire and most of Maine, wildlife managers attempt to retain winter cover as much as possible since this is more limiting than browse supplies. These different habitats must be nearby and not separated by large clearings or developments. In Pennsylvania, the central

Deer surveys Adirondack grandeur in A. F. Tait's *Autumn Morning, Raquette Lake* (1872).
Courtesy of Adirondack Museum.

Adirondacks, and other areas, deer numbers are so high that they prevent forest regeneration and they probably damage populations of birds and other species dependent on forest understory shrubs. The annual loss of deer to highway mortality in Pennsylvania is 43,000.

At present, the region's deer herds are at or even above the levels biologists consider their habitat can support. In New Jersey, management goals are to reduce populations over a significant part of the area; this is also true in large parts of Pennsylvania. Maine's herd is close to its range's carrying capacity.[22]

Land and Water

Especially important is the land-water connection represented by riparian, or streamside, habitats. Mismanaging these areas results in elevated water temperature, stream sedimentation, and loss of fish habitat. The alders, cottonwoods, or silver maples growing along the stream provide important wildlife habitat and travel corridors, and their leaffall into the stream supplies organic matter to the aquatic food chain. For these reasons, protecting riparian corridors is of high interest to wildlife and fish biologists and to forest managers.[23]

Runoff from sloppy logging operations damages water quality and can destroy fish habitat. *Photo by Roger Lee Merchant.*

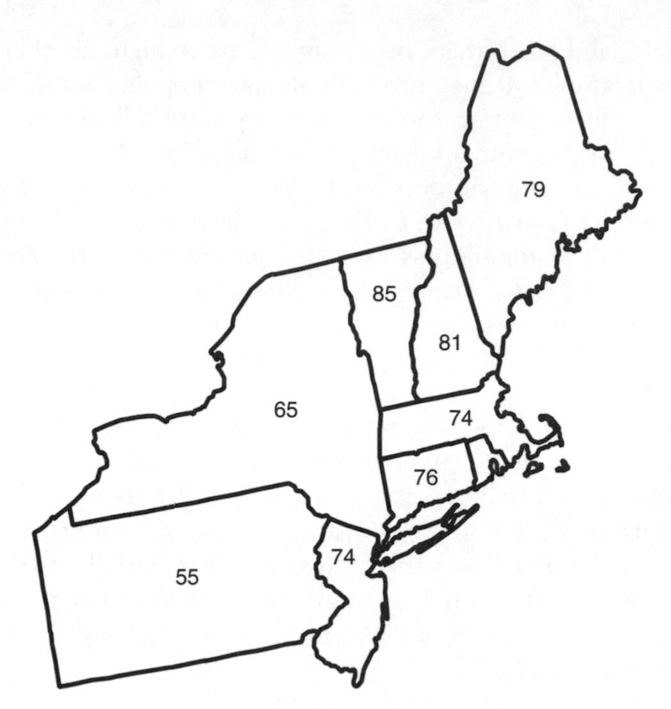

Figure 3.5 Recent Declines in Neotropical Migratory Birds
(number of species declining)
(Rhode Island was not included)
Source: National Fish and Wildlife Foundation, Partners in Flight, NTMB Conservation Program, n.d., p. 1.

Songbirds: Indicators of Regional Biodiversity

Ornithologists and birdwatchers gradually became aware during the 1970s that populations of certain songbirds were declining in the region (Figure 3.5). Of greatest interest has been a group of forest songbirds known as neotropical migratory birds (NTMBs). As this name suggests, these birds migrate to Caribbean or South American wintering grounds each year. In some instances, population declines have been severe. Regular censuses at specific sites showed this clearly. When researchers studied detailed datasets like the Breeding Bird Survey, they found a more complex picture. In particular, forest patches larger than 100 acres do not show population declines. Many of these birds increased from the 1960s to the 1970s, and of those, some declined again into the 1980s. Tropical deforestation has probably been important for some species. In local instances, overpopulations of deer have apparently affected habitat for intermediate canopy-nesting songbirds.[24]

Experts are coming to understand the common elements in the life his-

Forest Management and Early-Successional Forest Songbirds

Most woodlands in New England are privately owned. Would the aging, extensive forests of New England, particularly Northern New England, be fragmented by even-age management, specifically clear-cut harvesting? Two lines of evidence from managed public lands suggest that they would not. First, regeneration occurs rapidly and closed-canopy sapling stands form within 7–10 years after clearcutting. The interfaces between even-aged stands (internal edges) are ephemeral and do not support distinct bird communities as to field/forest edges. We have found no evidence for increased rates of predation on artificial nests along these internal edges. Predation rates on artificial nests, which are elevated in fragmented forests, were not elevated in seedling/sapling or poletimber stands compared to rates in mature northern hardwood stands in extensive forest.

Secondly, all species of birds found in old-growth or virgin northern hardwood stands are also found in mature managed stands. In New England northern hardwood forests, four distinct breeding avifaunas occur in seedling, sapling, poletimber, and mature stands. No species are unique to old-growth stands, nor are there differences in breeding bird composition among even-aged sawtimber, old growth, or uneven-aged stands. Furthermore, a distribution of size-classes ranging from regenerating to mature stands provides breeding habitat for approximately twice as many bird species as does an extensive, uneven-aged hardwood forest. Among small mammal communities, all species found in mature stands are also found in younger stands.

Management to provide early-successional habitats is necessary in view of recent declines of such habitats. In 1950, about 30 percent of the New England forest was in the seedling or sapling stage; by the 1970s these stages represented 14 percent, and by the 1980s 8 percent of the forest cover. Hay crop acreage has declined 46 percent in New England since 1966. The decline of early-successional habitats and the aging of forests in the Northeast have implications for all wildlife species.

Source: Excerpted from R. DeGraaf, "The Myth of Nature's Constancy — Preservation, Protection, and Ecosystem Management," presented 23 March, 1993, North American Wildlife and Natural Resources Conference, Washington, D.C. (references deleted).

tories and habitat trends affecting the species that have been shown to be declining. They are learning how land use, forest clearing in wintering areas, farming practices, habitat fragmentation, isolation, patch size, and forest composition affect the health of bird and mammal populations. On the basis of this knowledge, they are learning how to modify forest practices and landscape management prescriptions to conserve forest wildlife. It has often been assumed that all forest management will be hazardous to these bird populations. Yet recent research is showing that conservative forest management does not harm their abundance or diversity.[25]

Changing songbird populations have fostered interest in forest dynamics. *Source: U.S. Fish and Wildlife Service.*

A Recovering Top Carnivore: The Bald Eagle in Maine

Our national bird, the bald eagle, experienced dramatic population declines during this century due to eggshell thinning caused by pesticides, habitat destruction, illegal hunting, and water pollution. Because it is a widely appreciated top carnivore, uncommon in the East outside of local areas, there has been strong public and legislative interest in its restoration. In Maine, the population has recovered well since the early 1960s (Figure 3.6), though its reproductive rate is still low. Eagle populations and reproductive success are carefully monitored. Major landowners have signed protection agreements for nesting sites. The progress seen thus far, however, is not secure. In Maine, the Department of Inland Fisheries & Wildlife reports: "Changing land uses, mostly along coastal and other waterfront properties, threatened more than 30 percent of all occupied eagle nests in Maine during recent years."[26] Yet again, to secure the gains achieved by past conservation programs will likely require more restrictive and intrusive controls over private land use than have been used in the past. Already, a combination of regulations and voluntary agreements is in place in efforts to protect nests.

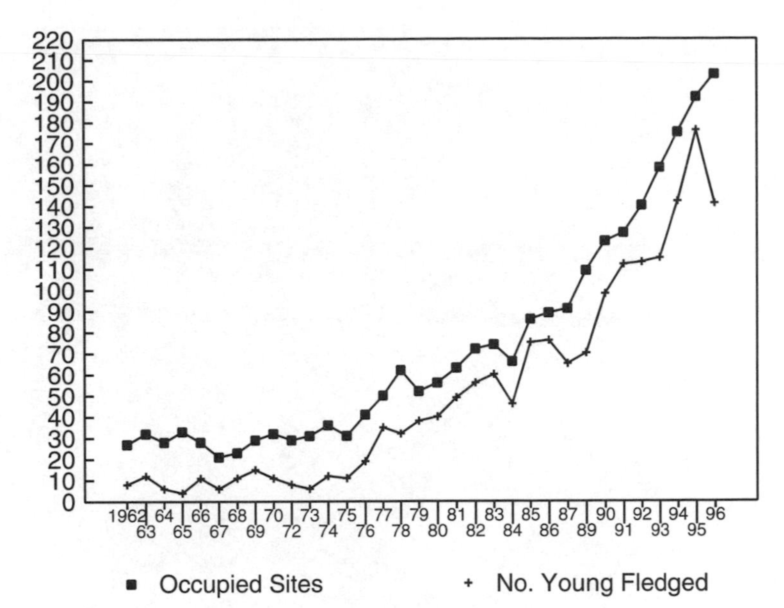

Figure 3.6 Maine Bald Eagle Occupied Sites and Number of Young Fledged, 1962–1996
Source: Maine Dept. of Inland Fisheries & Wildlife, annual reports

Landscapes for Wildness

In managing for biodiversity, some experts and advocacy groups argue that modern disturbed forest landscapes contain far too much habitat that is suited to abundant generalists like white-tailed deer, and far too little that is suited to rare species like pine marten, loons, or other animals that require large areas of undisturbed habitat. They advocate reintroducing animals like the timber wolf that inhabited the presettlement forests. As yet, endangered species have not played a major role in forest management in this region.[27]

In its purest form, forest management for biodiversity would aim at restoring old growth conditions to at least a portion of the landscape. The old growth areas would be large enough — perhaps 50,000 acres or so — to provide extensive "interior" habitat (in contrast to "edge" — the boundaries between distinctive habitats), and to provide ample elbow room for rare species with very large home ranges, like the timber wolf. Intrusions by roads into such areas would be discouraged, or if already present, eliminated. Within these areas, to the extent that active management is permitted at all, it would be in efforts to mimic the disturbance patterns that occurred in the presettlement virgin forests. Generally, these are considered to be the individual tree gap formations and small wind-

The bald eagle, top predator and symbol of wildness, is recovering in Maine, but its position is far from secure. *Courtesy of Maine Department of Inland Fisheries & Wildlife.*

throw patches that lead to stands composed of multi-aged mixtures of tolerant species such as sugar maple, beech, and hemlock. A new discipline, called Conservation Biology, has arisen to analyze these regional-scale and longterm habitat management and restoration issues. Some fairly ambitious proposals have emerged from the application of conservation biology concepts.[28]

Some 5 million acres of northeastern landscape — the Wild Forest — are in land units managed for wildness as a principal if not sole objective (Chapter 8). In these areas, principal management concerns will be with recreation, resource protection, and research. Yet even in Forever Wild areas like the Adirondack Forest Preserve lands, National Forest Wilderness, and Maine's Baxter Park, management questions arise when fires, exotic pests, or major blowdowns threaten the retention of natural values. For example, following the Big Blow of 1950 in the Adirondack Park, salvage operations were authorized by the legislature even on Forever Wild lands. This was done to avoid a replay of the slash-fueled fires that had swept the region 50 years earlier. Similarly, in a fire in Baxter Park in 1977, there was much controversy surrounding the Bureau of Forestry's decision to employ heavy equipment to build firelines to contain the fire.

So wilderness areas will continue to raise management questions for agencies and the public.

Ecosystem Management

On state and National Forests, wildlife lands, other parts of the public estate, and on some private properties, conscious strategies of ecosystem management are in early stages of development. In these strategies, efforts are made to identify a Desired Future Condition at a landscape level and to tailor all management practices to move the landscape slowly toward that condition.[29] Desired conditions include maintenance of riparian vegetation, avoidance of adjacent cutting units for prescribed periods of time, and specified levels of dispersal of cutting units and roads. Careful attention is given to the location, size and shape of cutting units and the amount of canopy removed. Efforts may be made to retain dead trees or even create them if they are a limiting habitat element, and to increase the amount of dead woody debris on the forest floor. Plans for such management can become quite complex even for small properties. The number of special area designations, rules, and constraints involved becomes so high that larger organizations conduct management planning with the aid of computer-based mapping systems known as Geographic Information Systems (GIS). Using such systems, planners can project specific landscape conditions a hundred years into the future, under alternative management policies.[30] As computer capabilities increase and costs decline, this capability is within reach for more and more forest and wildlife managers.

Given conditions left behind by 75 years of forest expansion and maturation, managers of these lands are likely to focus on the two extremes of the age/condition distribution. The first extreme is the situation in which extensive, unbroken areas of mature forest need the introduction of some diversity through small patch cuts and partial cuts to regenerate edge and understory vegetation. Practices such as these are in use on state wildlife lands and private reserves around the region.

The opposite extreme is the situation in which it is desired to create more mature habitat conditions consisting of multi-aged stands containing many large trees. Foresters cannot claim to be able to duplicate old growth forests. But they can devise cutting practices that will enable us to move stands toward a large-tree condition faster and more profitably than if left alone. In addition, management practices aimed at growing large trees of selected species for their clear wood may well be profitable for private industrial owners as well.

In the Industrial Forest, the contrast between industrial woodgrowing, modified by constraints for maintaining key deer habitat and protecting riparian strips, and management for biodiversity is the most dramatic. Many wildlife experts, environmental groups, and citizens are concerned that wide use of intensive management practices will reduce wildlife habitat values. They are also concerned that practices of short-rotation management will prevent the restoration of large blocks of mature, multi-aged forest stands. The science for understanding these issues is all too sparse at present. We know that intensive treatments like planting and herbicide treatment will be applied to limited areas in a patchy manner — there will be no vast landscapes of pure spruce plantations, anywhere. Also, it is not certain how to motivate small landowners to manage actively for wildlife.[31] Wildlife biologist M. L. Hunter, Jr., believes that there is a place in forest landscape management for creating new stands on a variety of spatial scales, and not just for stands managed by gap or patch dynamics theories.[32]

Overview

Northeastern forests have been shaped by different forces at different periods. In presettlement time, winds, lightning fires, and natural tree death and re-establishment were the most widespread forces. In local areas, native American farming, woods burning, and plant utilization practices affected small acreages. During the colonial period and up to about 1880, farm clearing converted ever larger areas to open spaces, and rural industries plus fuelwood cutting maintained secondary forests at young ages. In the unsettled regions, incursions for logging were few until what Cox et al. called the "Floodtide of an industry"[33] that washed over the region after the Civil War. This period of logging virtually eliminated remaining Wild Forests except in very limited areas. Across larger portions of the region, logging not only removed mature forest, but fires in its wake destroyed soil organic matter, injured waterways, and set back natural ecological cycles. At about the same time, the peaks of both farming and industrial wood utilization were reached and then receded. Farming and logging dramatically reshaped the ownership and composition of the forest and powerfully affected the wildlife resource.

Since World War I, the region's forest has increased by almost 18 million acres, and because of lower cutting pressure, its average age has increased significantly.

As we look to a new century, we see how the forest's condition is affected by forces of the Global Commons, and how its migratory birds are affected by forest management and land uses on other continents. In addition, new scientific understandings and new technologies enable us to visualize management strategies that can implement management goals at any point on a spectrum from intensive fiber farming to absolute preser-

Blowdown, Tionesta State Forest, Pennsylvania, illustrates natural forces shaping the forest. *Courtesy of the USDA-FS, NEFES.*

vation. Our growing awareness of the Eagle's Eye View points to a need to consider forest, water, and habitat concerns on a landscape scale, over different ownerships. This creates new technical and institutional challenges. Our Bequest to the future is being shaped. How these forces vary around the region is discussed in the succeeding chapters that examine the region's Five Forests.

A Damp and Intricate Wilderness: The Industrial Forest

THE INDUSTRIAL FOREST includes the forests that are managed primarily to produce industrial crops of wood. These forests of spruce, fir, pine, and hardwoods stretch from the Allegheny Plateau of Pennsylvania through the Adirondacks and then to the Green Mountains, across Northern New Hampshire, and over much of Maine. Industry ownership is locally important in New York and Pennsylvania. In Pennsylvania and Maine, industry ownership has expanded considerably since the 1950s, while it has shrunk in Vermont and Massachusetts. Most of this region remains "a damp and intricate wilderness," as it was when Thoreau visited the Maine Woods in the 1840s and 1850s. About 15 million acres of the region's forest is operated primarily for industrial wood production, with about 10 million owned by industrial firms (Table 4.1). But the industrial forest includes many smaller owners, since their management options and economic situation depend so heavily on the forest industry.

Not only does this forest provide wood for industry, but wood covers the cost of ownership, protection, and management. At the same time, these remote lands represent a heritage of wildness that is significant to the region's Cultural Heritage and its Bequest to the future. Thus, these largely private lands are heavily affected by important public interests.

Five characteristics distinguish the industrial forest. *First,* it is the region where forest structure and composition have been least affected by human actions like farming that removed forest cover for any length of time. *Second,* it is a forest primarily owned in large tracts by sizable corporations, wealthy individuals, and public agencies. Many of the owners are not living or headquartered in the counties where the land lies. Thus, the industrial forest is the region's lead example of absentee landownership, though by no means the only one. *Third,* most of the industrial forest is consciously managed for renewable wood crops to support wood

TABLE 4.1. Forest Industry Ownership of Commercial Timberland, Northeastern States, 1997

	Thousand Acres
Maine	7,298
New Hampshire	513
Vermont	227
Northern New England	8,038
Massachusetts	71
Connecticut	0
Rhode Island	0
Southern New England	71
New York	1,220
Pennsylvania	613
New Jersey	0
Mid-Atlantic	1,833
Region	9,942

Source: USDA-FS Website.

processing plants. These plants provide some local communities with 40 percent or more of their employment. Much of the region's wood industry employment lies outside the industrial forest. In some instances, as in the paper industry in Massachusetts, New York, and Pennsylvania — those jobs may not depend on locally harvested wood, but on wood or pulp brought from elsewhere. *Fourth,* the industrial forest consists of vast blocks of land with few local settlements, and limited local government. People often enter the industrial forest from outside to work there. Tens of thousands of people earn their livings in and near this forest, cutting and hauling the trees, servicing the logging industry, and running the mills. Because of the vast distances involved, logging camps persisted in the remote Maine woods until the late 1980s, though they had virtually vanished elsewhere in the United States. *Fifth,* millions of seasonal visitors enter the industrial forest to hunt, fish, or otherwise enjoy its wildness. Regionwide, the following percentages of Northeastern residents 12 years and older participated in activities that are important in the industrial forest:[1]

Motorboating and other boating/watercraft	12%
Outdoor swimming, not in pool	36%

Fishing	25%
Hunting	9%
Camping in developed campgrounds	13%

These five features are responsible for many of the important trends, controversies, and public concerns over the industrial forest.[2] These mostly privately owned forests contain such significant public values as headwaters of major streams, extensive coldwater fish habitat, habitat for wildlife, and important biodiversity values.

History of the Industrial Forest

Until the 1850s, the region was largely unexplored and was harvested heavily only along its fringes and several major rivers. Up till then, its mar-

Industrial forest: private lands near Raquette Lake, New York. *Courtesy of U.S. Geological Survey.*

gins were slowly receding before expanding settlement and landclearing. Between the 1850s and the turn of the century, the Maine Woods, the Adirondacks, and the Allegheny Highlands were heavily cut for merchantable spruce, pine, hemlock, and hardwood. Much of that timber moved by water to distant mills — down the Penobscot, Connecticut, Hudson, Delaware, Susquehanna, and Ohio. The region's timber capital was simply liquidated, as fast as the market would permit. Large, ancient trees, of sizes that will never be grown again, were felled. Despite heavy cutting — often followed by fire — reseeding, sprout regeneration, or residual stands of small saplings reestablished a forest quickly. Occasional depredations of spruce budworm and other insects and diseases caused locally heavy mortality.

Maine, an early national leader among lumber-producing states, was eclipsed before the Civil War by New York and Pennsylvania. New York led the nation in lumber output in 1850, sawing 20 percent of the nation's total. By 1920, it cut only 1 percent. Lumber production peaked in Pennsylvania in 1889, and in Maine in 1909.[3]

Much of the early lumbering in Northern New England was done by New England capitalists. In Northern Maine, New Brunswick operators were active even into the mid-nineteenth century. By the 1870s, absentee ownership of Bangor sawmills was common. Similar patterns emerged as New York and Albany investors developed mills in upstate New York, and Philadelphia and Pittsburgh investors controlled sawmilling and tanning in the Pennsylvania woods.

Between the 1890s and World War I, ownership of the industrial forest was radically reshuffled, as the large softwood sawlogs were gone by then. The emerging paper industry arrived from its birthplaces in Massachusetts and Eastern Pennsylvania. By 1900 the old lumber names were being nudged aside by new corporate giants familiar today from the Fortune 500: International Paper, Great Northern, Champion, and others. These firms, based in New England and New York, were augmented by smaller ones like St. Croix Pulpwood, Hollingsworth and Whitney, Finch Pruyn, and Glatfelter. Several grew from old, established paper firms, like S. D. Warren of Westbrook, Maine, that predated wood pulp paper. Others were founded by aggressive businessmen, as were Great Northern and St. Regis (now Champion International). At times, an entire small town was built in the wilderness to house the mill workforce.[4] In parts of Pennsylvania, the oil and coal companies became major landowners, creating holdings which persist today and from which timber has always been an important by-product.

The newly arrived paper corporations bought land from the lumbering families, the tanners, or the old iron families. The paper mills could use the smaller trees and the species that were of little value for lumber. Even during the Great Depression, they bought land. During the 1960s and 1970s, new corporations arrived on the scene, from the South and West. They bought established companies — land, mills, and all. These newcomers, which included Diamond International, Georgia-Pacific, and Scott, completed the transformation of the industrial forest into a region held largely by national and multinational corporations. They often brought new ideas and invested capital in new mills. They rebuilt the old mills, which had grown obsolete because of high pollution, low machine speeds, aged and inefficient boilers, narrow papermachine widths, and obsolete products. They often built new sawmills. For example, Georgia-Pacific's local employment in Eastern Maine nearly doubled from 1963, when it acquired its land and mills, to 1980.

Throughout this period, a small group of timber-owning families resolved to continue in the timbergrowing business and not sell out or industrialize. In Maine, names like Coe, Coburn, Pingree, Webber, Dunn, Carlisle, and others have long been prominent in the Maine woods and in local and state civic life.

In many areas, the 1920s were years of hard scratching for wood to keep mills turning. Then came the Depression, and lumber and pulpwood requirements fell. Nationally, the paper industry grew rapidly from the late 1940s to the 1970s, but most of the growth passed the Northeast by. In the 1970s, growth returned to the region's paper industry, and large sawmills sprouted anew in the industrial forest (Chapter 11).

Currently, none of the major paper and lumber companies can supply all of their wood needs from their own land. They buy half or more of their wood requirements from the large nonindustrial or small owners. Most of the large industrial owners sell large volumes of their wood, because of the location of the land, or because their mills do not use certain species.

Imbedded in the industrial forest are many scattered units of public land (Figures. 4.1 and 4.2). Except for the National Forest wilderness, the Adirondack Forest Preserve, Baxter Park, and other state parks, these units are generally managed for multiple uses, including industrial wood production. These public lands depend on the region's wood industry for their markets. Regionally, they contribute a small flow of timber, but they are important to nearby mills. The region has not yet seen the kind of battles over public-land timber policy that have resounded throughout the West, but an accelerating flow of litigation, appeals, and advocacy suggests that this may be changing.

In the softwood areas, economic and emerging environmental changes undermined the remaining river drives in the 1950s and 1960s. The drives tied up too much capital for too long, and the loss in breakage,

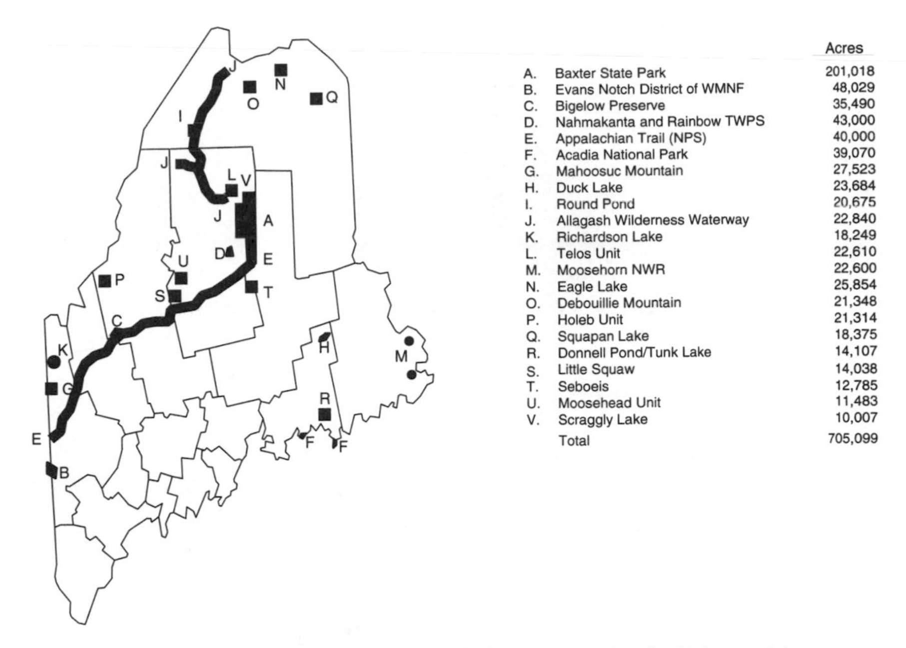

Figure 4.1 Major Public Ownerships, Maine Woods (Schematic)
Source: Maine Bureau of Parks and Lands.

scraped wood, and "sinkers" grew more worrisome to pulpmill accountants. The rivers would not float hardwoods, which grew on millions of acres and which the pulpmills were beginning to use. In these areas, hardwood logs for lumber and veneer mills were the principal products.

The region's rivers were once choked with wood for several months of the year. The demands of log driving led the lumbermen to modify the natural drainage of the entire region to improve water flows. They blasted out boulders to speed log movement. They created or raised lakes with dams, some lakes more than once. As the region's lands passed into control of paper companies, log driving retained its importance but the rivers were prized for producing hydroelectric power at larger downstream dams. All the region's major papermills rely heavily on water power, so the upstream waters of the region, as well as its forests, were firmly controlled for the log driving and power needs of the paper industry.

The environmental impacts of the log drives were severe. Not only did the drives simply occupy the riverway for part of the year, the logs damaged riparian vegetation. The waterway modifications disrupted fish habitat and migrations. The bark and fiber, plus the "sinkers," covered the bottoms of streams and piled up in deadwaters and behind dams. Some of this material is still there, stifling habitat for fish, aquatic insects, and plants.

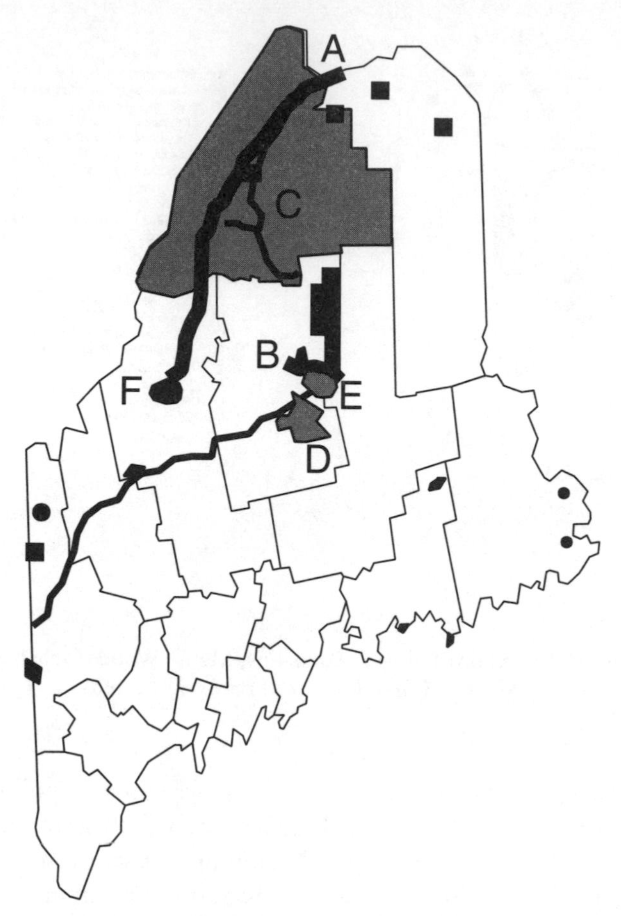

Figure 4.2 Major Private/Cooperative Recreation Management Efforts in the Maine Woods (Schematic)

A. St. John River P-RP, a negotiated regulatory scheme to protect the river corridor. Recreation management is by North Maine Woods, a private organization.
B. West Branch Penobscot, with P-RP and state-held scenic easement.
C. North Maine Woods, a private landowner cooperative managing campsites and access points.
D. Katahdin Iron Works — Jo-Mary area, managed by North Maine Woods.
E. Debsconeag Lakes area, managed by Great Northern Paper Company.
F. St. John Ponds, also managed by Great Northern Paper Company.
(Tribal lands, managed under separate fish and game laws, not shown.)
Source: Maine Bureau of Parks and Recreation.

By the 1890s, the log drives were waning everywhere. The lumber statistics at Bangor end in 1905. Great Northern's pulpwood drives continued on the Penobscot to Millinocket until the late 1960s. St. Regis drove sawlogs on the East Machias until the 1960s. Brown Company drove the

Androscoggin into the 1960s, and the last drive on the Kennebec was in 1976. Elsewhere, the Upper Connecticut saw its last major sawlog drive in 1915, the Hudson in 1924, and the Susquehanna in 1901.

By the early seventies, bulldozers appeared and began pushing main-line haul roads into the northern backwoods. The silent fir flats of "black growth" and the peaceful lakeshores were invaded by massive diesel log-hauling trucks, carrying huge loads of tree-length wood. In some areas, the roads are built entirely by the landowners, who share their cost and control their use.

As the economic pressures against river driving rose, environmentalists worked to end the drives. Maine passed a law outlawing river drives by the end of 1976. In the fall of 1976, the Kennebec Log Driving Company completed the last drive in New England, bringing pulpwood to Scott Paper Company's mill in Winslow. The log drives then passed out of existence, ending an era in the region's history.

Pennsylvania's Industrial Forest: A Sketch

The industrial forest of Pennsylvania displays a very different history of ownership and use, but the region remains largely forested all the same. For present purposes, we will treat this forest as equivalent to the USDA-FS's Allegheny and North Central survey units (Figure 4.3). It contains 37 percent of the state's forest land. This region was heavily stocked with hardwoods, pine, and hemlock prior to logging and settlement. The period of logging and subsequent fire dramatically altered its composition. Today, out of 5.8 million acres of commercial forest in the two survey units, only 400,000 acres (7 percent) consists of softwood types.

Only 9.5 percent of the sawtimber volume is in softwoods, of which hemlock comprises 64 percent. The northern hardwoods, including 563,000 acres of the valuable black cherry type, account for 49 percent of the present acreage.[5]

The Allegheny National Forest lies at the westerly end of this region, in the headwaters of the Ohio Valley. Logging and land use change dramatically changed forest composition there, according to a detailed analysis by Gordon Whitney:[6]

Type	*1800*	*1986*
Allegheny hardwood (black cherry, maple)	6.8%	53.0%
Hemlock-northern hardwood (hemlock, beech, birch, maple)	83.4%	15.8%
Oak	4.9%	18.2%

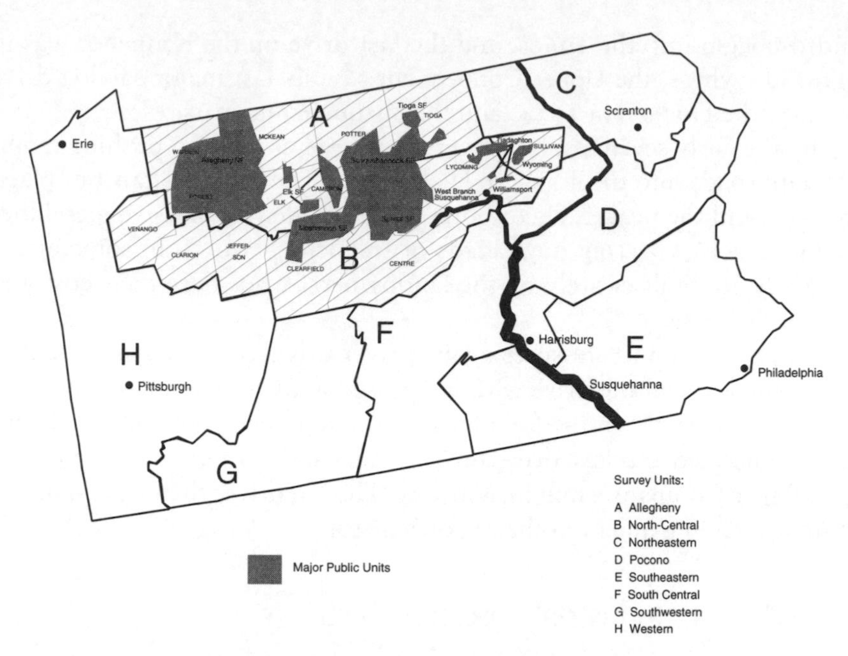

Figure 4.3 Pennsylvania's Industrial Forest

The softwood logs from most of this region went down the west branch of the Susquehanna. Williamsport was at one time the world's leading lumber port. The log drive at Williamsport peaked in 1873, at 318 million feet; the drives were largely over by 1901. As Sargent's correspondent described conditions in this region in the early 1880s:

> Active lumbering operations on the West Branch of the Susquehanna were begun in 1850, when the boom of the Susquehanna Boom Company was constructed at Williamsport. At this place the greatest part of the lumber on the West Branch is sawed. . . .
>
> Williamsport is situated on the north or left bank of the West Branch of the Susquehanna, and for 2 or 3 miles along the river side are ranged the mills and lumber yards of the thirty-four lumber companies operating here. . . .
>
> The greatest prosperity or fullest development of the business was attained, as will be seen, in 1873. After that year, with the steady decrease of the supply of pine and the consequent increase of expense in securing logs, the annual stock steadily diminished until 1877. During the past three years the increasing demand for lumber has stimulated the operators to greater activity, but more than to this cause the recent gain in the yearly stocks is due to the substitution of

hemlock for pine, the ratio of hemlock to pine being at present as 1 to 4, although the average for the last seven years is but as 1 to 10. As the supply of pine lumber is exhausted, hemlock will be more and more handled until it will become the most important timber of this region. . . .

I found it more difficult to obtain information of the extent and limits of the hemlock woods of Pennsylvania, and of the amount of the standing timber and the annual crop of hemlock, than I did to get the same facts respecting the pine. Lumbermen agree that there was originally far more hemlock in this state than pine, and they speak of it now as inexhaustible, which is not strictly true, for it is doubtful if it holds out to supply the increasing drain made upon it by tanneries and sawmills for more than twenty-five years to come. Large quantities of hemlock have been wasted. Much that grew intermingled with the pine has died after the pine has been removed, partly from exposure to fuller sunlight and summer drought, and partly to forest fires induced by and following lumber operations. In the early days of the tanning industry of this region, when hemlock lumber was esteemed of little value, and whenever of late years the lumber trade has been so dull as to offer no inducement to send to market the trunks of the trees felled for their bark, large quantities of these have been left in the woods to decay. Now, however, with a good market for hemlock lumber, tanning companies owning hemlock lands, or the contractors who furnish the tanneries with bark, buying for this purpose stumpage from the proprietors of the timberlands, often own sawmills in the timber region, and cut and ship this lumber to market by railroad. . . .

On the summit of this divide (between the Susquehanna and the Allegheny) the forest had a truly northern aspect, except that we missed the spruce, not seen in Pennsylvania. The dark foliage of the hemlock mingled with sugar maples, beeches, and birches. For many miles above Lock Haven it was a second growth which occupied the hillsides, a thin growth of white oak, chestnut, locust, etc., which had followed the lumberman and forest fires. Considerable second-growth white pine was seen in a few places, but on this none of the present generation seem to set much value, and I have yet to meet any one in the state who gives a thought to encouraging and preserving such growth. To consume the forests as speedily as possible, satisfied with what can be realized from them in the operation, appears to be the spirit which rules this region. . . .

Through Elk, the southwestern corner of McKean, and the southeastern corner of Warren runs the Philadelphia and Erie railroad.

> Along the line of this road, as it passes through this portion of the timber belt, are located the largest tanneries of the United States. These are consuming the hemlock of this region at an enormous rate, and, in addition to the vast amount of bark which they consume, large quantities are shipped out of the region by railroad. The first important tanneries of Warren county were established 12 or 15 years ago, and at the present rate of consumption the hemlock of this county can hardly hold out 20 years longer. The land, after the forest has been removed, is excellent for agricultural purposes throughout this region, and on all sides pioneers are making themselves farms. These men prefer to begin in the undisturbed forest rather than locate on the slashes, because they can pay for their land with the hemlock bark which it yields; and from a radius of 15 miles bark is drawn and sold at from $4.50 to $5 a cord to the tanneries. On an average, four trees yield a cord or ton of bark, the equivalent of 1,000 feet of lumber, board measure. In Warren county from 5,000 to 6,000 acres of hemlock were cut down in 1880, and there is no possibility of this growth being renewed, for every foot of slashed land is eventually burned over, and sometimes the burnings are repeated until the soil is nearly ruined for agricultural purposes. . . . [7]

During the 1920s, Pennsylvania state foresters rated the condition of the state's forests. Their results probably describe this portion of the state. Of 11.5 million acres of forest other than small woodlots, they found that 300,000 acres were "burned until practically no forest growing stock remains." Brush lands occupied more than 10 percent of the forest area. Timber described as "merchantable" (using a liberal definition), stood on only 32 percent of the acreage.[8] Yet these dismal conditions were a material improvement over previous decades.

Since the 1920s, the forest has rebounded dramatically under low cutting pressure and with the control of fire. By 1989, the area's forests were 59 percent in sawtimber by acreage, compared to less than one-third in the 1920s, and 54 percent in 1978. From 1978 to 1989, sawtimber volumes increased by 42 percent for all species, 58 percent for softwoods, and 57 percent for black cherry. This was on a forest land base that declined slightly. Pennsylvania was the nation's leading hardwood lumber producing state, turning out a billion feet annually.

Whitney documented the dramatic increase in age of the forests of the Allegheny National Forest, which is roughly representative of the region.[9]

Age	*1923*	*1986*
0–30	46%	5%
31–50	39%	6%

Railroad logging in Pennsylvania, nineteenth century. *Courtesy of Delbert Sives Collection, USDA-FS, NEFES.*

51–70	8%	45%
71–90	3%	38%
91–100	0%	6%
<101	4%	1%

Six of Pennsylvania's 15 north central counties were 81 percent or more forested in 1978; the area is now 77 percent forested. This rebounding forest has supported increased production of hardwood lumber, and exports of cherry and other logs to Europe and Japan. The forest also supports a population of deer, which helps make Pennsylvania one of the East's major hunting states. The deer are so numerous that they consume new tree seedlings so rapidly as to make adequate regeneration almost impossible to achieve. Fencing is a costly remedy, and intensive hunting is unable to do the job except on a local scale. The bear kill in Pennsylvania has been at all-time highs.

As the forests of this region matured, a number of waves of pests have passed through the forest, inflicting mortality and value loss. At times, a

Impact of deer browsing is evident in exclosure experiment, Pennsylvania.
Courtesy of USDA-FS.

good deal of the harvesting was driven by salvage needs. The gypsy moth outbreak of the 1970s and 1980s caused extensive damage to oaks. Currently, the beech scale/nectria complex is moving into the state and beginning to kill beech. A poorly understood decline is affecting sugar maple, beech, and red maple.[10] In addition, advance regeneration of trees is sparse to nonexistent, grasses or other unnatural understory plants interfere with regeneration, and many desirable understory plants are also growing scarce. Scientists are attempting to untangle the various factors underlying these conditions.

In this forest, now restored to beauty and productivity, Pennsylvania landowners, governments, and citizens now face decisions as to how this forest is to be managed and sustained for the future. In the late 1980s, Pennsylvania developed a strong program to foster more local processing of its hardwoods. The program was just getting underway when it encountered funding cuts, and the wood industry saw its markets shrink in the global recession of the early 1990s.

Penn State University forestry economist Charles Strauss and his students have shown that these forests present society with a win-win opportunity. Their detailed studies show that future timber output can be boosted significantly, but only if a significant increase in cutting occurs

Deer and Pennsylvania Forests

By 1890, market hunting had made deer so scarce that they were rarely seen in their native habitat, and such sightings as did occur rated front page coverage in the local newspaper.

However, public concern for deer eventually led to an increase in deer populations. Pennsylvania appointed a game commission in 1896. Salt licks and dogs were forbidden for deer hunting in 1897, and market hunting was prohibited after about 1900. Game refuges were established beginning in 1905, and in 1906 deer were imported from other states to restock depleted areas. Perhaps most important of all, beginning in 1907, hunting was restricted to bucks. . . .

As early as 1917, a few biologists began to warn that the deer population would soon be too large unless measures were taken to reduce further increases. The warnings were ignored. By 1922, deer were causing serious damage to agricultural crops and forest reproduction. In 1923, new laws permitted farmers to kill deer of either sex at any time of the year if they were damaging crops, and the state began providing free materials to farmers who wanted to fence their crop lands. . . .

Large deer kills (as high as 186,575 in 1940) during the seasons of 1931, 1938, and 1940 checked the deer irruption and eventually brought the deer populations down from their peak levels, but not before severe overbrowsing had occurred. Vegetation less than 6 feet tall was completely eliminated in many areas, and available food supplies were badly depleted. . . .

During the peak years, local populations as dense as one deer on 5 to 6 acres were believed to be common. On study plots established by the Pennsylvania Department of Forests and Waters, every wood plant less than 6 feet tall had been completely destroyed or hopelessly injured at these densities. Tree planting had to be suspended in many localities, and continued clearcutting under such conditions would almost certainly have resulted in the elimination of tree growth.

By the early 1950s, deer populations in the state leveled off, then began increasing again in the 1960s. In the heavily forested Northern Allegheny Plateau, damage to tree reproduction continues, and small deer size persists.

Source: D. A. Marquis, *The Allegheny Hardwoods Forests of Pennsylvania,* USDA-FS Gen. Tech. Rept. NE-15, 1975.

now. The increased cutting would have to be carefully applied, especially to remove pulpwood and smaller size trees. Under such management, average age of the forest and the total inventory would still continue to increase. Poor pulpwood markets, however, make such a program infeasible for the moment.

> The current trend in pulpwood consumption suggests a potential consumption of 230–260 million cubic feet during the 1980s, about

Forest Management and Wildlife Conservation in Western Pennsylvania

On June 21, 1993, the Western Pennsylvania Conservancy announced a major land acquisition — the 11,300-acre President Oil property situated between Tionesta and Oil City.

Located adjacent to the Allegheny River at the village of President, the tract is highly prized for its wildlands and high-quality trout streams. Encompassing over 17 square miles, the property protects more than four miles of frontage along the Allegheny River, now part of the Federal Wild & Scenic Rivers System.

The property has many natural amenities, but perhaps the greatest are its numerous high-quality trout streams. In addition to several unnamed streams, the tract contains a five-mile stretch of famed Hemlock Creek, which supports wild rainbow, brown, and brook trout, and 3½ miles of Porcupine Run, which is home to native brook and rainbow trout. The land also includes portions of Reese Run and Norway Run, two streams that the Pennsylvania Fish and Boat Commission has recommended as Wilderness Trout Streams — the highest designation in the state . . .

Because of its valuable timber, the property carried a very high price tag of $8 million. It was apparent from the outset that, if the Conservancy was going to acquire the property, it would have to sell part of the timber to raise the needed funds. After a thorough analysis, the Conservancy negotiated a timber contract with Industrial Timber & Land Corporation (ITL) of Beachwood, Ohio. Under this contract, a division of ITL Corporation would manage the timbering operations for 10 years.

The Conservancy worked closely with the Pennsylvania Game Commission and Pennsylvania Fish & Boat Commission to establish strict timbering guidelines to protect key ecological features, such as the exceptional trout waters, and provide for longterm regeneration of the forest.

Under terms of the agreement, no trees will be harvested on 1,700 acres of the property which have been set aside as buffers for streams and a protective river corridor along Rt. 62. The Conservancy has provided 200-foot buffers along named streams and 100-foot buffers along unnamed streams. This far exceeds the 50-foot guidelines recommended in the Pennsylvania Department of Environmental Resources' Forestry Manual.

To protect against soil erosion, the timbering agreement also minimizes the number of haul roads, skid roads, and log landings, emphasizing instead the use of existing roads.

A consulting engineer and forester are being retained by the Conservancy to regularly monitor the timbering operations. They are working closely with ITL to develop timber masterplans for each forestry plot in advance of the actual cutting to ensure that all of the environmental objectives are met.

Another factor is extensive gypsy moth damage that has occurred in recent years. Tree mortality has been very heavy in over 30% of the property, and it is anticipated that the trees will remain susceptible to continued gypsy moth infestation. A large tract adjacent to the President Oil property was clearcut by its owner, a major timber company, rather than contend with gypsy moths, and regeneration of new trees on these thousands of acres has been very good.

The property is immediately available for hunting, fishing, hiking, and other outdoor recreation.

In 1998, the Conservancy sold the property outright to ITL, retaining a stringent conservation easement that maintains these restrictions for the future. The restrictions would apply to any future owner. Conversion of the land to developed uses is prohibited with only minor exceptions.

Source: Conserve XXXV11(1), Jan. 1994; and WPC press release.

> half of the region's available harvest. Timber not harvested will simply accumulate as added inventory. Surplus stocking will only serve to reduce the aggregate growth and potential quality of the existing forest and lengthen eventual rotations. In turn, the extension of rotations will delay gaining a better balance in age classes.[11]

Only 9 percent of Pennsylvania's industrial forest is owned by forest industries, but this area contains 85 percent of it. It also contains 55 percent of the state-owned commercial forest land, and 100 percent of its National Forest acreage. Its ownership pattern differs markedly from the Maine Woods since 34 percent of the forest is in public ownership (25 percent by the state). Yet dealing with land use, forest practices, and other forestry issues will be difficult, since 54 percent of the forest land is owned by many small private owners. The Commonwealth has in progress a number of programs aimed at ensuring future health and sustainability of its forests.[12]

Harvesting the Timber: Technology and the Forest

In the industrial forest, methods of harvesting the crop remained little changed from colonial times until the age of the motor truck. Some areas of the mountains were entered with logging railroads by the 1880s, but the other features of logging operations changed little.[13] Generation after generation of woodsmen followed the same seasonal cycle. Enter the woods in the autumn, often by canoe or boat. Build camps and prepare "tote" roads, skid trails, and log decks. Build or repair splash dams, bridges, booms, and other facilities. On the earliest snow, tote supplies into the camps by oxen and begin cutting. Axes were standard until the late nineteenth century, when they were replaced by one or two-man saws. For large pines and spruce, long two-man "misery whips" were used. Oxen or horses skidded logs on sleds or directly over ice roads and snow to landings by streams or lakes. At ice-out, the drives began, ending many miles downriver. Occasionally due to low water, the drives would "hang" upstream, leaving the mills without wood and in danger of bankruptcy.

Oxen hauling logs over snow. *Courtesy of John Springer,* Forest Life 1858, *Forest History Society.*

Loggers struggling to free a logjam, 1850s. *Courtesy of John Springer,* Forest Life 1858, *Forest History Society.*

Timber cruisers mapped wildland towns, seeking large spruce and pine. The markets required large logs, so small trees were left. Since logs were hauled away from the stump over snow, little soil was disturbed. Young trees growing in the shade were often preserved. Large areas of fir or hardwood growth were generally bypassed.

Thus, winter logging, as practiced in northerly spruce-fir areas, was an ecologically benign system of exploiting the forest, apart from its effects on waterways. Experts believe the heavy cut of spruce reduced the representation of that species in the forest. It was an age of careless land exploitation and little thought for the future. Much of this cutting was wasteful, as noted by Austin Cary: "Trees are killed that are not utilized. Stumps are cut high and valuable lumber left in the shape of long tops to rot. . . . Much is due to mere force of habit in our lumbermen. In the old days they learnt wasteful habits."[14] But the forest's floor and its longterm productivity were scarcely disturbed during this period of exploitation for lumber. Despite the sloppy cutting, residual trees left behind earlier continued to grow, sometimes rapidly. Many towns were re-entered 10–20 years after being "cut out," and loggers took an abundant harvest once again, though the trees were smaller than in the first cut.

Changing utilization standards and better logging techniques enabled loggers to go beyond what had been taken in the first cut. A gentleman named Mr. Pringle supplied Charles Sprague Sargent with a report on New York's forests for his 1880 Census volume. Pringle noted:

> Thirty or forty years ago it was thought that all the accessible spruce in the Upper Hudson above Glens Falls had been harvested, but there is today nearly as much sawed at Glens Falls as there was at that time. At that time nearly all the timber standing near this river . . . had been cut. Such as stood 5 or 10 miles back from these streams . . . or higher up the mountain slopes, would not pay the cost of hauling to the larger streams; but it is this timber which now furnishes the present supply. Logs are now driven on streams which were then thought incapable of being driven.[15]

By the late 1890s, railroad logging became common. It was the only way to reach timber too far from drivable streams and to get out the hardwoods, which would not float. The high costs of laying track demanded that operators remove every last stick. From the White Mountains to the Adirondacks to Pennsylvania, the railroad loggers stripped the hills, drainage after drainage. The huge piles of slash were easily set afire by the sparks from locomotives and "hotboxes." Thus, in the rail-logged areas, residual stands were sparse, soil organic matter and understory seedlings were often consumed by fire, and the likelihood of natural reseeding of desirable softwoods was small. In an ironic vicious cycle, the high risk of fire restrained anyone from practicing any form of sustained yield management or conservative partial cutting.

In these ways, differences in landownership, markets, and technology shaped the forest that grew up after the waves of cutting up to World War I. By World War I, the machine was entering the industrial forest. Handsaws and horses still brought the logs to roadside, but steamdriven loghaulers and trucks moved wood to the landing. In the 1920s, woodsman "King Ed" Lacroix built a railroad between two Northern Maine lakes to move wood to the Great Northern Paper Company mill on the Penobscot. Lacroix's locomotives still lie rusting in the woods, attracting the curious Allagash canoeist or bush pilot flying by. By the early fifties, the chainsaw made its appearance.

In parts of the North Woods, however, the horse reigned supreme for skidding logs well into the twentieth century. Horse operations on major industrial holdings were common until the 1960s in Northern Maine. The trails used in horse operations are still visible from the air in millions of acres of the North Woods. Horse logging required a lot of land to grow hay, and a lot of labor to cut it and handle the horses. The increasing labor shortage in the woods during and after World War II, the move away from river driving, and a desire for less seasonality in wood production led the companies to turn to machines for skidding.

The skidder came into its own only in the 1960s. Earliest models lacked the size, power, and toughness for the region's rocky, steep, and often boggy woods. The skidder made possible a tremendous boost in each woodsman's weekly output. A rising volume of wood could be cut with ever-fewer workers. Skidders can operate in wet, boggy ground or in the snow, enabling cutting to proceed over much of the year. The cutting and hauling are suspended during "mud season," since roads become quagmires, incapable of supporting huge trucks, and skidders bog down in the mire. In many areas, a late fall shutdown before "freeze-up" is also necessary.

Two men with chainsaw and skidder can cut and pile roadside 100 cords of wood (a pile the size of a modest house) per week in good timber and good weather, though the average for a year's work is much less. The skidder enabled loggers to increase their output and helped some earn a better living. But the high costs of financing and operating skidders left them more vulnerable to the frequent downswings in the pulpwood business. Nonetheless, rising wages, safety issues, and a labor shortage have fostered still heavier mechanization in the woods.

Changing mill technologies have had a major impact on the markets for timber and on the ability to manage the industrial forest. "Chip-n-

Total tree utilization raises many ecological concerns.
Courtesy of Maine Department of Conservation

saw" and other systems capable of making framing lumber from small trees have significantly increased the stumpage value of spruce and fir and of smaller trees. New processes using hardwoods for pulp have allowed the increased use of hardwoods. During the postwar decades, mills began using smaller trees and species of trees that had formerly been left behind. Machines for chipping trees in the woods have made it possible to use low-grade hardwoods and small softwood trees. Significant increases in use of wholetree chips for fuel have also promoted increased use of chippers. These technologies have transformed the opportunities for forestry. It is a great tragedy, however, that these opportunities to improve forest management have so rarely been taken.

After the railroad era but before the skidder, the woods were rarely clearcut, for reasons unrelated to silviculture but based on product requirements and logging costs. Most of the region's forest stands simply did not contain many trees worth cutting until the 1960s. So the "logger's choice" commercial clearcut was the rule. The skidder, unless carefully handled, will rut the soil, sweep aside tiny young trees, and scrape bark off the trunks of trees to be left behind. As the need for wood grew in the 1970s, industry discovered that skidders make wonderful clearcutting machines. Clearcuts of up to one square mile began appearing. Environmentalists protested and some loggers argued that industry was liquidating the forest.

No sooner had skidders become well-established in the industrial forest than the largest companies began testing giant harvesting machines.

These machines are capable of still greater wood production per man-day. Though they are costly and demand skilled maintenance personnel to keep them running, they help companies keep a large volume of wood flowing. Some are capable of cutting a tree, clipping off its limbs, and laying it in a bunch with its neighbors. Some of these giant machines actually damage the soil less than skidders and cause fewer ruts, because of their huge tires. They are used primarily for clearcutting, but are used also for strip clearcutting. Some operators are experimenting with them for partial cutting. Since these machines have such high hourly operating costs, they must harvest a lot of wood each day to pay their way.

As the wave of budworm-caused mortality subsided, land managers began deciding, for a variety of reasons, to shift emphasis from clearcutting to partial cutting. In some instances, this shift was urged on by public opposition to clearcuts and by actual or threatened programs of regulation. To harvest wood economically in partial cuts, different machines were required. They had to enable an operator to quickly handle smaller stems, and to remove them from the stand with minimal damage to residual trees, soil, and roots. Machines capable of doing this are appearing in greater numbers in the woods, and are allowing foresters and landowners a much wider range of choices in cutting practices. At the same time, these machines are reducing the use of chainsaws for felling, limbing, and bucking, thereby eliminating many arduous and dangerous tasks in the woods.

Massive Koehring feller-forwarder carries whole trees to the landing on Great Northern Paper Company land. *Courtesy of Maine Department of Conservation.*

Strip cuts, Maine woods 1980s. *Courtesy of Maine Department of Conservation.*

Cutting and Growing Trees: Debate Over Forest Practices

Clearcutting has attracted media attention and critical comment from ecologists, environmentalists, members of the public, and politicians. New science, bringing the perspectives of the Worm's Eye View and the Eagle's Eye View, is being brought to bear on traditional questions. Legislation and zoning to ban clearcutting on public and private lands have been advocated. Yet, many foresters point to conditions common in New England where clearcutting is a sound means of regenerating the forest. Many hardwood stands are cluttered with overmature and deformed trees and lack advance regeneration, and clearcutting may be the most practical way to start over.

In parts of the region, patch cuts or clearcuts are used in efforts to regenerate valuable hardwoods like black cherry and white or yellow birch. Since clearcuts also successfully regenerate red maple sprouts and other trees considered undesirable for timber, they must be used with care. Many spruce-fir stands consist of overmature fir trees that have stopped growing. They may be starting to blow down and are often severely damaged by spruce budworm or balsam wooly aphid. Partial cuttings often do not work in such stands because there are few vigorous trees capable of surviving, or because the stand cannot be kept dense enough to resist windthrow. Also, desirable advance regeneration may be absent. Clearcutting can remove these low-quality stands and start a new cycle of regeneration. Many northeastern species are adapted to regenerate in understories and can be released by careful removal of overstories. These cuttings look like clearcuttings, though technically they are often termed "overstory removal" cuts. Since foresters realize that they are releasing regeneration, they often do not realize that the public perceives these as clearcuts. Design of logging practices to minimize damage to seedlings and saplings is a rapidly emerging area.

The discussion of clearcutting has become so polarized that the issue has become a common test of political correctness. If you oppose it, you're an environmentalist. If not, you're an exploiter. This is why it is hard to find balanced information on the subject. Foresters and environmentalists opposed to clearcutting cite the soil erosion that can be caused by skidders, soil compaction, the destruction of advance regeneration, the loss through volatilization and erosion of nitrogen and mineral nutrients, the long period of brush, grass, or fern competition that can suppress desirable regeneration, and the impact on aesthetics and wildlife habitat.[16] Where a single management practice becomes, by itself, the criterion for distinguishing good from bad forestry, the debate becomes too narrow. In fact, in recent decades sloppy or thoughtless partial cutting and high-grading have done a good deal more to degrade the timber productivity of Northeastern forests than has clearcutting.

Large clearcuts have stirred controversy across the Northeast. *Courtesy of Maine Department of Conservation.*

The selection of a cutting method is complex (Table 4.2). Many conditions of soil, current stand composition, markets, management objectives, and environmental constraints must be considered. Reducing the choice of cutting methods by ignoring these complexities will not lead to better forests in the future.

Beyond mitigative measures, one way to avoid the impacts of clearcutting is to employ alternatives. Some have advocated that these effects justifying banning clearcutting and turning exclusively to alternatives.[17] A comprehensive view of alternative cutting methods is not possible here, but some comment is necessary.

Abuses of clearcutting are highly visible, while abuses of other cutting methods are not. Logging a selection cut on wet soil with poor skidtrail layout can result in as much erosion as clearcutting. Mishandled selection cutting over several rotations can undermine stand productivity and eliminate desirable species.[18] Poorly managed selection cuttings often lead to root and stem damage that can sap future productivity. Managing without clearcutting and cutting the same total volume means that more acres must be logged to obtain a given harvest volume, and more roads constructed.

Insect and disease considerations are also involved. Maintaining extensive areas in uniform cover of mature trees can provide favorable habitat for insect outbreaks. The spruce budworm is the best local example.

Before considering what role clearcutting should have in forestry, it is necessary to gain a balanced view of the costs and technical limits of the alternatives, and the extent to which the alternatives depend on specific favorable circumstances for their success. In any specific situation, a list of factors such as those in Table 4.2 are considered in choosing a cutting prescription. The many complexities, uncertainties, and judgments involved need to be considered in designing public policy responses to the forest practices issue.

It is only occasionally that large clearcuts or overstory removal cuts are required by the biological facts of regeneration. More frequently, clearcutting is chosen because of its lower logging costs, its administrative practicality, and by the absence of sufficient desirable windfirm growing stock to support alternative cutting methods. Clearcutting is more often dictated by practical constraints in local situations than by its intrinsic silvicultural desirability. These constraints can be difficult or impossible to overcome. Certainly, clearcutting is frequently used by owners simply as a means of turning forest capital into cash quickly. Also, during the region's periodic land booms, many of the "clearcuts" seen at roadsides are actually real estate development projects.

TABLE 4.2. Considerations in Setting Silvicultural Prescriptions

1. Existing Stand Conditions
    - Species composition
    - Size and age distribution
    - Age
    - Tree health and size of crown
    - Quality and species of overstory
    - Windfirmness
2. Access, Stumpage Values, Markets and Logging Costs
3. Costs of Required Treatments After the Harvest is Complete
4. Season of Operations
5. Type of Equipment To Be Used
6. Climate and Soils
7. Existing Advance Regeneration
    - Presence and stocking
    - Species
    - Condition
8. Damaging Agents
    - Provision for wind hazard at stand edges
    - Pests of regeneration
    - Pathological rotations
    - Avoiding regeneration of high risk species (e.g., fir)
9. Whether the Desired Species Require Abundant Sunlight and Exposed Soil for Germination and Seedling Success
10. Management Objectives
    - Immediate treatment goals for stand harvest
        - regeneration
        - stand improvement
        - insect or disease
    - Longterm objectives for the stand
        - rotation age
        - product objectives
        - likely regime of intermediate treatments
        - likely follow-through immediately following cutting
11. Regulatory Constraints
12. Multiple Use Considerations and Constraints
    - Wildlife
    - Aesthetics
    - Water quality or quantity; erosion hazard
13. Whether Soil Scarification is Desirable or Undesirable
14. Practicality of Prescription, Supervision, Recordkeeping, and Management

Source: The Irland Group, *Clearcutting as a management practice in Maine forests,* Augusta: Maine Dept. of Conservation, 1988.

Professional foresters, scientists, and environmentalists debate the merits of clearcutting at great length and have not formed a consensus on its role in the industrial forest. Some of its strongest proponents — and strongest opponents — are the woodlands managers of major landowners. The debate over clearcutting will not be soon resolved.

Some landowners practice site preparation and planting, often using genetically improved seedlings. When these plantations can be carefully managed, they may produce twice as much wood per year as the natural forest.[19] On some larger industrial holdings, herbicides are used to temporarily suppress competing vegetation to speed growth of conifers. This practice, while involving a small acreage, is highly controversial. Periodic efforts are made by environment of groups to ban the use of herbicides in forest management in a number of northeastern states. A moratorium on forestry use of herbicides has been adopted in Vermont.

Even more difficult than the debate about clearcutting is the debate about the quality of cutting practice in the partial cutting that accounts for most of the annual acreage harvested. Defining what is meant by good practice and bad practice is easy at the extremes — it takes no trained expert to see the difference. In a wide range of middle ground, it can be difficult to judge. Further, there are examples in which partial cutting seems to be working fine for several decades, after which serious problems suddenly become evident. Many foresters believe, whatever the complexities, that on average, cutting in the Northeast is depleting stand quality and

Cut-to-length harvesting machine increases management flexibility and, properly used, can reduce soil damage. *Courtesy of Hahn Machinery Company.*

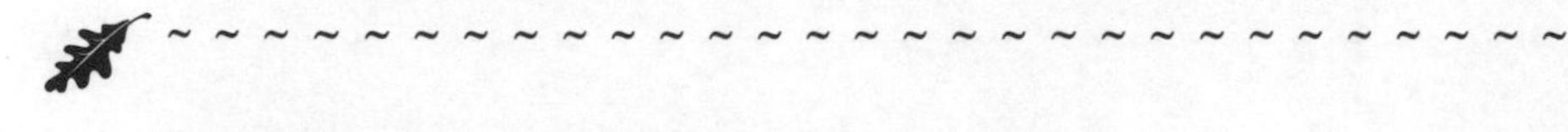

The Approach to "New Forestry" in Baxter State Park's Scientific Forest Management Area

The SFMA Landscape

The Scientific Forest Management Area (SFMA) is located in the northwestern corner of Baxter State Park and, at 29,537 acres, constitutes about 14% of the Park's 202,064-acre total. The area includes five ponds, ranging in size from 13-acre Lost Pond to 331 acres of Webster Lake. The SFMA is divided approximately in half by Webster Lake and 6 miles of Webster Stream. In general, the terrain is flat to gently rolling. In the 1980 inventory, prior to management activity, the area was 80% softwood by volume, principally spruce and fir . . .

Forest Operations

As mentioned above, a forest system approach suggests that forest operations be dispersed over the landscape. Traditionally, forest management activities have progressed over landscapes in segments determined by the pace of road construction. As an area is accessed, it is usually harvested, requiring more road construction and then more harvesting and so on. Dispersing activities over the landscape requires two things, more miles of road construction per year and less harvesting per mile of road. Both of these are difficult economic propositions and, when combined with the lower per-acre removals usually associated with New Forestry thinking, day-to-day forest management requires discipline and the willingness or ability to forego currently available revenue in lieu of greater future revenue. SFMA operations attempt to construct roads at the greatest rate that finances will allow. Accessed areas are divided into management blocks from 5 to 60 acres. Somewhere between 5% and 10% of the area accessed by roads will not be harvested in the first 15 to 20 years that access is available. These areas serve a number of roles. They act as controls against which other areas with a variety of silvicultural treatments can be compared and, since these areas are not harvested, they generally provide areas of high stocking and mature canopy to mix in with areas in which harvesting has somewhat lowered stocking and perhaps canopy heights. From an economic standpoint, these blocks can also serve as reserves of merchantable timber with a number of flexible options for access at appropriate points in the future. . . .

The retention of large stems, particularly red spruce stems, is a constant question for us. On many areas within the SFMA, there exists a regular but thin stocking of large spruce and white pine that, for whatever reason, survived both the spruce budworm outbreak of the nineteen teens and the most recent harvest of the late 1940s. These spruce hold great current value as wood products and also have higher than average risk, as generally taller heights and older age leave them more vulnerable to wind pressure, insects and disease. At the same time, these tall, large trees provide a unique and valuable component of forest structure that, once removed, will take quite some time to replace. These considerations lead to careful evaluation of these stems on a tree-by-tree, stand-by-stand basis. In general, increment cores

have shown us that these larger trees are growing at real values of 1% to 2% per year and the risk of leaving such a tree from an economic standpoint may not be worth the return offered by this modest growth. In the end, a subjective judgment must be made in measuring this cost of deferred revenue against the value of a healthy forest system.

Source: J. Bissell, 1994. *The approach to "new forestry" in Baxter State Park's scientific forest management area.* In, D. B. Field, Proceedings Second Annual Munsungan Conference, *The Triad Concept for Maine's Future Forest: A Model For Harmony or Discord?* January 5–6, 1994, Univ. of Maine at Orono. ME Agr. & For. Exp. Sta. Misc. Rept. 388. pp. 45, 47, 48.

value far more often than it is improving it. Several surveys are underway to get more specific facts.

If we could overcome these problems, it is hard to design regulatory programs to control the worst of the abuses. Writing a rulebook for forest practices can quickly yield a volume exceeding 100 pages in length that still leaves questions unanswered. The administrative task of overseeing individual cutting jobs — as many as 8,000 a year regionwide — is daunting and costly. In the current political climate, the prospects for improvement by this route seem remote.

The problem is undoubtedly serious. What can be done? First, field surveys specifically documenting the quality of cutting practices are needed so that the problem's existence can no longer be denied. Then, the significance of the costs in future productivity need to demonstrated. Finally, forestry professionals need to be willing to criticize exploitive management, and to refuse to associate themselves with it. It has also been suggested that the industry's willingness to close ranks with the worst offenders has been counterproductive. Whatever is done, however, oversimplified, punitive "solutions" will in fact only make the problem worse.

The debates about overuse of clearcutting, high-grading, herbicides, biomass chipping, and short-rotation management have entered legislative halls as state forest practices acts are debated. They have been aired in appeals of National Forest Management plans and individual timber sales. With the host of scientific, practical, and political issues involved, the question of forest practices in the Industrial Forest may never be fully settled. In Maine, a two-year referendum battle over clearcutting finally resulted in a 1998 legislative compromise that may not settle the matter politically.[20]

Conflict on the Fringes

The margins of the industrial forest meet the suburban, rural, recreation, and even urban forests. A number of major papermills are in locations

New Forestry in Eastern Spruce-Fir Forests: Principles and Applications to Maine

What's "Wrong" with the Present Forest?

1. There are few old-growth stands in the subboreal forest and accessible parts of the true boreal forest. . . .
2. A few commercially valuable species (e.g., white pine, red spruce, and yellow birch) have been greatly reduced in certain stand types through preferential high-grading and disease. . . .
3. Some desirable aspects of current harvesting practices (incomplete clearcuts, etc.) happen largely by default, not by design, thus creating an inherently unstable forest management situation. . . .
4. Extensive clearcutting and the associated road systems have created a fragmented landscape in some regions. . . .
5. The region has only a few formal reserves, and they do not adequately represent the region's ecological diversity. . . .
6. There is limited land formally dedicated to multiple-use management, in which non-timber values are weighed equally with forest products. . . .
7. Not enough land is producing high timber yields. . . .

Conclusion

We conclude by proposing an Agenda for Action. Specifically, we recommend:

1. Widespread professional acceptance and political support of the Triad concept of land allocation and management. . . .
2. Development and implementation of New Forestry-based silvicultural systems on lands not specifically dedicated to high-yield timber management or preservation. . . .
3. Greatly accelerated research on both stand-level and landscape-level effects. . . .
4. A revamped view by land management professionals of what constitutes "good" forestry. . . .

Source: R. S. Seymour and M. L. Hunter, Jr., 1992. *New forestry in Eastern spruce-fir forests: principles and applications in Maine.* University of Maine College of Forest Resources, Maine Agricultural Experiment Station Misc. Publ. 716. p. 6–9, and 29.

now surrounded by suburbia. Small private owners sell wood to the major industrial landowners, and the industrial owners hold scattered parcels in the midst of rural landscapes. Three conflicts that have been especially bitter in the industrial forest fringe have been recreational and industrial development, taxes, and cutting practices.

In the land market booms of the early 1970s, and later in the mid-1980s, several landowners were briefly tempted into the resort development-land sales business. In Vermont, Weyerhaeuser opened the Jay Peak ski resort. On the shores of Maine's Moosehead Lake, Scott Paper opened the Squaw Mountain ski area with a sizeable lot-sales program. This venture lost money so badly that Scott donated it to the state, which has struggled with it ever since. In 1969, the International Paper Company proposed a massive ski resort-land sales development in southern Vermont. The local public reaction was overwhelmingly negative. The governor personally talked the company out of pursuing it, and to prevent future developments of that kind the Vermont legislature passed a model land use law, Act 250.[21] As the 1970s wore on, more and more major forest products companies took financial losses in recreational land developments. Several New England resort-land sales operations were in bankruptcy or serious financial straits, including Haystack and Mount Snow in Vermont, Evergreen Valley in Maine, and Bretton Woods in New Hampshire. The lure of easy money faded; no new major resort schemes appeared. The large landowners were spared further conflict with their neighbors over recreational land use. Later in the late 1980s, a new wave of development, emphasizing remote subdivisions, emerged; this is discussed below in Chapter 9.

Within and on the fringes of the industrial forest are the milltowns. Some are cramped, grim, sad places. Three-story, smoke-stained, shingle-sided tenements line narrow streets in tight valleys. Drab main streets lie in the shadows of smokestacks and pulpwood piles. Such milltowns have been decaying all across the Northeast, as paper, textile, leather, and steel plants have closed. In these communities, industrial development means economic opportunity and is eagerly supported.[22]

A second source of friction on the fringes of the industrial forest has been taxes. The forest tax problems experienced in Maine provide the classic example, for there the stakes are highest and the conflict is at its most intense.

In 1973, Maine adopted a new form of tax for its forest lands.[23] Called the Tree Growth Tax, it was based on the view that forest property taxes should be based on land productivity and not on current market value. The annual ad valorem property tax, proponents argued, discriminates against longterm investments like timbergrowing, which take a long time before harvest. Moreover, rising land prices due to speculation and residential development produce land values that cannot be supported by timbergrowing revenues. Further, periodic reassessments would escalate tax burdens to unpredictably high levels, rendering investment in timbergrowing unnecessarily risky.

The tax assessment for each property is based on the U.S. Forest Service estimated annual growth rates for softwood, mixedwood, and hardwood for each county. Those growth rates are discounted for access and marketability, and then multiplied by current stumpage prices. The result is converted to a capital value representing the present worth of the land for growing wood. The rules have been changed several times, but today the assessments are multiplied by a millage rate that is based on the cost of state services to the unorganized territory (where local government does not exist).

In most of Maine's vast unorganized territory, no major tax problem arises. But where local governments exist in small towns and plantations on the fringes, the local taxes must be based on Tree Growth Tax assessments, which are determined by the state tax assessor. In the early 1970s, Tree Growth Tax assessments were at times higher than ad valorem assessments, since many rural tax assessors appraised woodland at nominal market values. But the rising land values of the 1970s changed all that, and re-evaluations shook the property tax structure of town after town. Many small towns, owned 50 to 90 percent by paper companies, found that residents paid taxes on $100 to $150 per acre assessments on their forty-acre woodlots, while paper companies paid taxes on $40 per acre assessments. For small towns of forty to a hundred inhabitants, with low incomes and high unemployment, struggling to maintain municipal services, such a situation was intolerable. Periodically, municipalities and rural legislators have campaigned hard to change or abolish the Tree Growth Tax, but they have been unable to overcome the resistance of the large landowners. More to the point, the opponents of the Tree Growth Tax have failed to convince the legislature and many state officials that they have an alternative that will not harm forestry in the long run.

In the other states of the Northeast, the same problems exist. For example, a recent survey in New York, Vermont, New Hampshire, and Maine found that for 11.6 million acres enrolled in preferential tax systems, a total property tax shift of $9.7 million occurred; just under half was reimbursed by the states. Various preferential tax systems exist in efforts to mitigate the tax problems of rural landowners, but enduring solutions to the inherent basic conflicts of use value taxation have not been found.[24]

The conflict, then, still festers. It is not easy to see how it will be solved. But it must be solved, for a stable, predictable tax regime is a crucial requirement if the owners of the industrial forest are to invest willingly in its maintenance and improvement. They must be convinced that their investments are secure from future confiscatory tax increases. At the

same time, an equitable solution to local grievances about use value taxation must be found.

Finally, towns at the edges of the Industrial Forest have clashed with landowners over cutting practices and management practices. In various parts of the region, they have attempted to regulate logging through local ordinances, to control it by the use of bridge permits or other devices, or to ban clearcutting or herbicide treatments within their borders.

These conflicts reflect a complex and unfortunate love-hate relationship between forest landowners and their neighbors. Whether the owners are public or private is of no consequence. But for the industrial landowners, local feelings can run especially high. Most local communities depend heavily on the forest and the mills for their livelihood. Many families have worked in the woods and mills for generations. There are few other jobs. The dependence on the industry is clear. The industry's property tax breaks anger their neighbors. Many families include workers whose pulpwood contracts were abruptly cut off by companies whose woodyards were full. Such cutoffs often meant financial hardship. In some places, decades ago landowners drove off squatters; memories can be long.

So a list of grievances has accumulated in many towns. In part, this reflects the normal human tendency to resent any large, absentee landowner. In part it flows from a history of paper company arrogance and abuse of

Intensively managed stands are growing wood for the future on industrial lands.
Photo by author.

Mechanical delimbing improves both productivity and safety.
Courtesy of Georgia-Pacific Corporation.

power, and the desperate, powerless feeling of their neighbors. We rarely love those on whom we depend for our livelihoods, especially when we have no choice of employers.

This local resentment undermines forestry. In its most visible form, it results in the condoning of widespread timber thievery. Local residents do not think it a crime to steal paper company wood or butcher stands on firewood permits: "They've cheated us; they've got more than they can manage, haven't they?" Stories are told of loggers earning a living for years without ever paying for stolen timber. At its worst, the resentment flares in "Allagash lightning strikes" that keep landowners nervous and Maine Forest Service firefighters working overtime. These concerns are often acted out in the town meeting, the press, and in legislatures. In local town meetings, wider interests and longer-term concerns may not always receive attention. In the state legislatures, many rural towns feel that their distinctive problems and interests are not fairly considered.

Working in the Woods

Woods work has never been very comfortable, pleasant, financially secure, or financially rewarding. The industrial forest in its earlier years was harvested by a temporary workforce consisting of farmers and nearby residents who regularly worked in the woods, supplemented by drifters. They would disappear into the woods in the fall and return in the spring on the log drive, after a winter of eating beans and hardtack, working ten-hour days in all weather, sleeping in shabby overcrowded shanties. The

A forester inspects an overstocked stand of spruce and fir in Eastern Maine. *Courtesy of Georgia-Pacific Corporation.*

shiftless among them gave loggers their bad reputation by drinking, gambling, and whoring away their winter's wages in the gin mills of a dozen lumber towns.

A few farseeing organizations have put loggers on their own payrolls and provide fringe benefits. A company logger on a mechanized operation earns fringe benefits and sits in a cab on a huge logging machine, listening to the radio while working. This life is as far from the nineteenth century logger's as an astronaut's is from Orville Wright's. But most loggers work in an illusory independent status denounced by industry critics as "pulpwood peonage." Since the loggers are legally contractors and not employees of the companies, they have no collective bargaining rights.

Logging is inherently dangerous, and never more so than in today's age of chainsaws, skidders, and loaders. While the trees are not large by western standards, they are quite capable of killing the unlucky or unwary. Logging has one of the highest accident rates of any industry. Although the Occupational Safety and Health Administration has made logging a target industry, singled out for special attention because of its poor safety record, it has in fact received little attention from the safety authorities.[25] Mechanization, however, has steadily eliminated many of the more dangerous jobs.

Moving the wood. Scenes from river drive on Vermont lands of International Paper Company, 1930s. *Courtesy of International Paper Company.*

The logger's high accident rate imposes severe burdens of injury, loss of life, and family hardship. It also exacts a financial toll through astronomical workmen's compensation insurance rates. A basic source of high accident rates is the piece-rate system of payment, which encourages loggers to push work too fast and take too many shortcuts. Officials of insurance companies have stated that the most positive safety step the industry could take would be to eliminate piecework.

Because it is arduous low-wage work, logging in the industrial forest has always been plagued by labor shortages. In the 1840s, Maine contractors were already bringing in workers from the Maritimes. In World War II, German prisoners-of-war were even put to work cutting pulpwood in the Maine woods.[26] Because of the dense settlements in Quebec on the edge of the isolated woods of Northern New Hampshire and Maine, and because of the log trade to Quebec, more French-Canadian workers have worked in the industrial forest since World War II.

The Spruce Budworm

Industrial forest management can be profoundly affected by natural pests. The most dramatic problem facing the managers of the industrial forest from 1975 to 1985 was the spruce budworm. This insect is a native of the spruce-fir forest of Northeastern North America. Periodically, it erupts in regionwide outbreaks capable of killing up to 40 percent of the spruce-fir volume over large areas. In stands of mature fir of low vigor, mortality can reach 100 percent. Typically, these outbreaks are followed by regeneration and regrowth of spruce-fir forests. But there follows a long period of tight supplies of large timber. In the past, this was not a problem, since the industrial capacity using the wood was at a low level. Today, the wood industry supported by this resource is capable of using all of the forest's growth. Failure to control the outbreak, then, could mean costly dislocations, readjustments of mill production processes, and possible reductions in capacity in some areas. The entire wood-using industry from Quebec to Northern New Hampshire, Maine, and the Maritimes was deeply concerned with this outbreak.[27] In New York, the outbreak was far less severe.

The outbreak of the 1970s and 1980s was serious, because the forest was not yet fully developed with logging roads, and it was seriously overbalanced in mature age classes as a result of light harvesting pressure from 1920 to 1970. The result was a huge area of vulnerable forest, too large to spray, and so vast that it would take decades to harvest all the high-risk trees.

Another budworm outbreak is likely to erupt in twenty to forty years. By then, however, the forest will be fully accessible. The industrial capacity now in place will assure markets for the wood. Much less of the forest area will be in overmature age classes. This alone will make the next outbreak easier to handle. With the early warning methods being developed by scientists, extensive spraying may not be necessary in that future outbreak.[28]

A Paper Plantation?

In 1974, Maine's industrial forest was characterized in a highly critical book as a "paper plantation."[29] Similar resentment is found throughout the Industrial Forest. The image is meant to suggest the oppressive, unaccountable power of a plantation overseer over his tenants, together with the owner's absolute control over the state legislature.

The industrial forest has been a paper plantation in the sense that its principal product is paper, but the growth of sawmilling and wood energy has changed that significantly. In the past, the major landowners have received considerable deference in state legislatures, as do large landowners in any state, be they farmers, house builders, railroads, or paper companies. But they no longer have everything their way. The companies are occasionally at odds with the state forestry and regulatory agencies.

In the backwoods of Maine, local government does not exist: the territory is "unorganized." In 1969, Maine established the Land Use Regulation Commission to act as a zoning agency for the state's privately owned wildlands.[30] This triggered a ten-year battle over the issues of land use control and the minutiae of day-to-day zoning problems. The agency's planning and zoning functions are slowly winning more acceptance, but forest landowners and the few families living in its jurisdiction never like to hear "no" to their development proposals. The commission is often forced to make planning, zoning, and regulatory decisions based on inadequate information, which does not enhance its popularity. It has never been given the staff or resources to do the necessary work. In early years it was plagued by a host of political and managerial problems. Similar conflicts are being dealt with in the Adirondacks, the Pine Barrens, and elsewhere in the region.

Overview and Themes

Following the liquidation era, change in the industrial forest occurred at a slow pace. Logging, transportation, and marketing systems evolved gradually, involving trucks, saws, chainsaws, and skidders. Today, both horses and giant logging machines work in the woods. For a long period, forest landowners faced limited markets and no real pressure for change. Since the 1960s, however, change has accelerated. The budworm outbreak, new technologies, the burst of mill shutdowns and expansions, the growing log trade to Quebec, and public concern over management practices and taxation all contributed. Assertive loggers demanded attention for their grievances. And the legislature and bureaucracy were no longer uniformly sympathetic to the landowner's concerns. Political, economic, and technological change has been disorientingly rapid in the past two decades. There is no sign that the pace of change will slacken.[31]

As later chapters argue, although local conditions vary, it will be impossible to meet projected needs for wood from the region's industrial forest on a sustainable basis without a major increase in intensive management, upgraded utilization practices, and application of landscape management principles. The region's forest owners, and managers — and governments — have never before faced such a challenge. How they will respond will test our region's science, its landowners, its government institutions, and its political culture. This issue will interweave the metaphors of Economic Interdependence, the Town Meeting form of politics, and the Global Commons and Worms' and Eagle's Eye Views of the forest. We must recognize that as a society we are only at a very early stage of accomplishing this.

5 Rockwood's Pasture: New England's Suburban Forest

On the Monday after Thanksgiving, 1979, carpenter Ralph Anderson and his crew began clearing a lot near the 1828 church at Rockwood Corner, Belgrade, Maine, for my new saltbox house. Anderson was participating in the latest cycle of land use changes in the forests of central Maine. The area was wilderness in the 1790s, a farm in the 1800s, went back to a rural woodlot, and was cut several times up to the 1960s. This one and a quarter acres became an island of the suburban forest in the winter of 1979–1980. Ralph Anderson repeated the labor of clearing and housebuilding first done, then abandoned, by farmer Rockwood and his descendants generations ago. Chunks of the birch, fir, and maple cut by Anderson's crew were burned in my living room woodstove. Along this road, three other houses appeared in two years.

Rockwood never lived to see his farm grow trees and later be sold by his descendants, who had moved away. He never imagined that one day his pasture would become part of the suburban forest, that it would grow tomatoes in a tiny plot by a house occupied by two state employees who worked by day in Augusta, twelve miles away.

The suburban forest displays the slow loss of economic, amenity, and wildlife values of the rural landscape. Since the 1940s, these losses have escalated as land use patterns became more wasteful of land. Still, a large forest of some 10 million acres remains in suburbia. Until the 1970s, the woods of suburbia were valued primarily as a green backdrop for housing. The 1980s resurgence of fuelwood for space heating, however, shifted this situation somewhat. Today, wood cutting is more and more widely accepted, setting off new tensions between local citizens. The increased acceptance of wood harvesting, while far from universally shared, represents a major social change in suburban areas. Town and state governments will continue to mediate between shifting coalitions of landowners, conservationists, real estate groups, loggers and wildlife groups. The region's Town Meeting culture must address new and baffling challenges. Following a general review, this chapter will examine how the suburban forest has developed in New Jersey and Connecticut, to illustrate the forces at work in this region.[1]

Suburbanization

The forest most familiar to the region's residents is the suburban forest. Even on the fringe of large cities, trees set the tone of the landscape. The suburban forest is the result of a new cycle of land use change — from farm to forest to tract development. A rigorous definition of the suburban forest would be based on studies of commuting patterns, settlement densities, and land use patterns. The suburban forest includes the undeveloped brush and forest overwhelmed and bypassed by suburban-exurban sprawl. It consists of the fraction of an acre behind the house, the overlooked pasture and farm woodlot, the water company forest, the Audubon Society preserve, and perhaps the state park or forest. Its boundary fades gradually into the rural landscape of the farm forest.[2] The nation's largest metropolitan area is here — the Philadelphia-New Jersey-New York-Connecticut region of 25 million residents — 10 percent of the nation's population.

The outer boundaries of suburbia are diffuse and spreading rapidly. We can estimate that some ten million wooded acres, or one acre of out of every seven, lie within the suburban forest. For this discussion, we need not fix the limits with precision. The suburban forest is now the home of a major — and increasing — portion of the region's population. As the central cities lose population, settlement spreads. Ring highways sprout electronics factories, and rural land prices catch up with the suburbs, as interstate highways increase practical commuting distances. Lakefronts and hillside "viewlots" become part of the suburban fringe.

The spread of suburbs, mostly into abandoned farms and regrown woods, has been a feature of the region's urban geography for generations (Figure 5.1). Spawned by the desires of middle-class families for pleasant surroundings and promoted by trolley lines and freeways, suburban settlement spread along the coasts and rivers, radiating outward from the cities. In 1810, only the immediate Boston area within New England had a settlement density greater than ninety persons per square mile. By 1850, this level of density had been reached in Northern Rhode Island; Cumberland County, Maine; Middlesex, Bristol, and Norfolk counties, Massachusetts; and New Haven and Hartford counties, Connecticut. By 1930, this level of density had been reached in all but four counties in

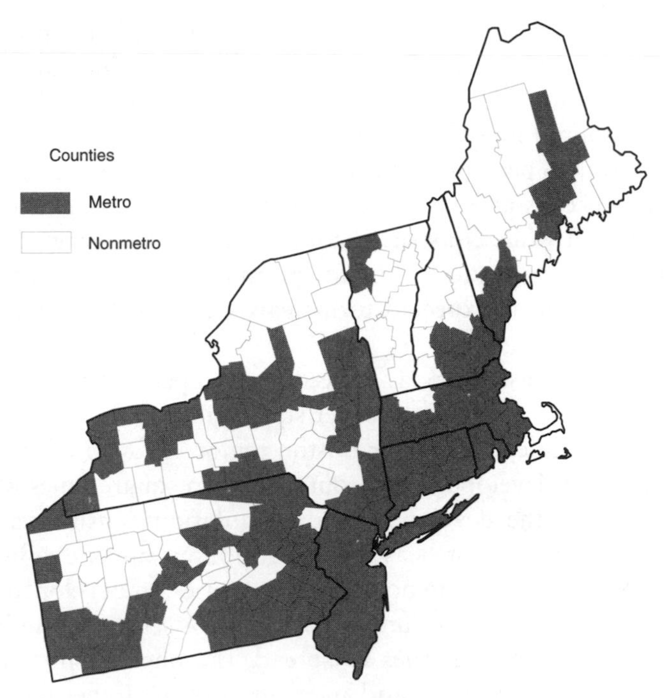

Figure 5.1 Metro and Nonmetro Counties, 1990
Source: U.S. Office of Management and Budget.

Southern New England, plus Hillsboro and Strafford counties in New Hampshire and Androscoggin County in Maine.

Commuting patterns continue to shift. With increasing emphasis on home offices, and the spread of office complexes to suburbia, the outer edge of suburbia merges into a thinly settled, diffuse "exurbia." These exurban areas sprout small subdivisions of "executive homes," often surrounding golf courses. These serve high-end homeowners seeking prestige locations away from the congestion and overdevelopment that has overtaken the first-generation suburbs of the 1950s.[3]

In ninety-six Northeastern counties from 1950 to 1960, 0.22 acres was removed from rural uses for each person added to the area's population. In the Standard Metropolitan Statistical Area (SMSA) counties in the region, the ratio was 0.20 acres per person. The ratio was much higher in non-SMSA counties: 0.40 acres per person. In a later sample of fast-growth Eastern counties land consumption was 0.22 acres per additional person, up from 0.16 acres in the 1960s. In eastern Massachusetts, population growth and suburbanization consumed one-half acre per person from 1960 to 1970, compared with one-eighth acre in the pre-World War I years. Development becomes increasingly hungry for land. During the 1964–1990 period, in the three-state urban region surrounding New York City, the amount of developed land increased by 60 percent, while population rose only 6 percent. In the Chesapeake Bay region, recent land consumption was 0.65 acres per additional resident, and one county actually lost population while constructing hundreds of expensive leisure homes. According to Horton and Eichbaum, population growth of 20 percent regionwide by 2020 is expected to increase developed acreage by 60 percent.[4]

Population densities today exceed 10,000 persons per square mile in the cities, and are roughly 1000 per square mile in suburbs. In the early 1960s, for example, the greater Boston region of 2500 square miles was only 54 percent developed, and contained 110 square miles of dedicated open space. On the developed land, population density was 5400 per square mile.[5] Other Northeastern metropolitan areas exhibit the same characteristics — moderate population densities on developed land while the bulk of the land remains undeveloped, bypassed by expanding sprawl.

As development continues to spread, the phenomenon of "shadow conversion" becomes more and more important. Experts estimate that when an acre of farmland is set off as a houselot, an additional 2 to 3 acres are compromised for future farming. For forestry and recreation, the ratio may be even higher. Thus, scattered sprawl directly affecting a modest acreage of land may diminish potential rural land uses disproportionately on the surrounding lands whose vegetative cover and ownership have not yet changed.

The sprawl of American suburbia, so wasteful of land, time, and energy, depends critically on the interdependent international economy. This interdependence enables suburbia to draw its energy and material needs from distant continents.

Farm, Forest, and Suburbia in New Jersey

New Jersey, "The Garden State", has the most valuable farmland in the Northeast on a per-acre basis. This state's suburban homes have the highest average value in the nation, so the amenity value of the forest is recognized. Major Green Acres bond issues attempt to secure available bits of land for the future. This is the only Northeastern state with no nonmetropolitan counties. Land use pressure would be expected to have a dramatic effect on the forests here. Due to distance from major mills, its forests produce only nominal amounts of wood products.

The spread of suburbs from Philadelphia and New York caused New Jersey's population to rise from 2 million in 1950 to 7.4 million in 1980; growth thereafter was much slower, reaching 8 million by 1997.

There are two detailed studies of land use change in New Jersey. First, is a detailed study performed at the University of Pennsylvania for the nine-county Philadelphia-Trenton metropolitan area for 1970–1980. While it is dated, it gives us useful insights.[6] The region includes about 2.4 million acres of land. This metropolitan area saw major land use changes even though population growth was modest. Using a sample of aerial photographs, the authors detailed land use changes in this region. Their findings included:

- In New Jersey, 11 percent of the projects exceeded 10 acres, and accounted for 77 percent of the land converted from rural uses (p. 11).
- A total of 121,485 acres were converted in use (p. 14).
- Seven percent of the land developed in this period came from woodland.
- 9,200 acres were converted from woodland to urban uses, or 920 acres per year.
- About 2,000 acres were in New Jersey (App. C), for 200 acres per year.
- 80,000 acres of the total had formerly been in farming or old fields.
- Fully 6.8 percent of the undeveloped land at the beginning of the period was developed during the decade, despite the fact that the region's population declined (p. 28).

This region contains abundant additional farmland that can be converted to residential and other uses. There is a premium on wooded lots, especially for high-end development that accounts for an increasing share of activity in the region.

A second study was conducted by the USDA-FS for the New York-New Jersey Highlands, the broad hilly belt along the highlands in the north central portion of the state (Michaels, n.d.). Counties included were Bergen, Hunterdon, Morris, Passaic, Somerset, Sussex, and Warren. The region was 48 percent forest in the mid-1980s. Land use changes were measured for 1968–1984 using aerial photography. They were then projected to the year 2010. In the New Jersey portion of the area, 417.7 acres of forest per year were converted to other uses from 1968 to 1984. From 1984–1986 to 2010, the analysts projected that 689 acres per year of forest would be converted to other uses. In this area, forest was predicted to be 54% of the total land converted.[7] These studies cannot be extrapolated to the entire state, but they provide useful background information. These

Suburbia engulfs the forest, New Jersey, 1992. *Courtesy of U.S. Geological Survey.*

local examples show that landclearing can occur as land use patterns change, even when population growth is slow or negative.

In New Jersey, there remain 900,000 acres of land in farms, and 809,000 acres of crop and pasture land (1992), of which only 394,000 was harvested in 1996. From 1952 to 1992, farmers sold 330,000 acres of forest, 75 percent of their 1952 holdings. Despite a population increase of almost 4 million from 1952 to 1992, New Jersey's commercial forest area shrank only slightly, from 2 million to 1.9 million acres. Publicly owned forests increased dramatically, however, from 181,000 acres to 464,000 acres by 1992. That so much forest was retained here is due to the concentration of development on farmland as well as the return of marginal pastures to woods.

Changing Landscape

The bypassed lands will continue to be developed, often in multi-unit apartments or office complexes. The remaining suburban forest will be under increasing pressure for development as land prices rise and the energy costs of commuting translate into higher price premiums for close-in land.

Strip-zoned and unplanned sprawl continue to erode the suburban forest's significant aesthetic and amenity values.[8] The blight of signs, plastic pizza parlors, and chain stores spreading along main streets, and the splattering residential development along rural highways and lanes have stranded the forest, cutting it off visually for passersby. Even if only psychologically, the forest is cut off for use by walkers and birdwatchers as well. The ecological impact of habitat fragmentation, predation by household pets, and reductions in water quality is significant. From the Eagle's Eye View, the suburban forest is becoming a less and less hospitable habitat.

The suburban yard and street tree resource provides enormous economic values. Real estate brokers, landowners, and local tax appraisers all agree that trees add value to suburban real estate. Recognizing these values, many cities employ foresters who specialize in the management, improvement, and protection of street and park trees. A lucrative business is carried on in trimming, removing, planting, fertilizing, and spraying sub-

Farmland, returned to woods, then cleared for house lots, shows land use change. *Courtesy of Rhode Island Division of Forest Environment.*

urban yard trees. Pests such as the gypsy moth and the Dutch elm disease require costly and often controversial management programs and have given rise to the systematic study of urban pest management.

Suburban hinterlands of coastal cities and major river towns were cleared of their forests early for farming and to supply building materials, fuel, dunnage, and shipbuilding timber. Remaining woodlots were cropped frequently because of high demand. Then for many years, the forests of suburbia lay fallow. Many of the trees were small and of low quality. Builders found low-cost western lumber more suitable than native pine. The region's small sawmills found the suburban forest a high-cost and inhospitable business environment, and melted away. Western lumber companies could deliver carefully graded lumber to tract housing builders by the shipload and trainload, which the tiny local mills could not do. The small size of the lots and the reluctance of owners to allow cutting made it harder for loggers to find timber and for mills to buy it. Since the wood consumption of the region is largely imported anyway, manufacturers needing lumber bought from outside, except for small local specialty producers like lobster-pot makers and pallet plants.

These land use changes have converted the suburban forest into a patchwork of ever-smaller pieces. The progressive fragmentation of woods reduces habitat for songbirds and other wildlife species that require extensive areas of forest for breeding territories or feeding range.[9] In the Eagle's Eye View, suburban habitat values are disappearing fast.

The amenities in suburbia, however, are all strictly for private or local consumption. Virtually every vacant lot has a "No Trespassing" sign; town beaches are restricted to residents. High amenity areas, like beaches and shorelines, use abundant "No Parking" signs to shoo curious outsiders on their way. Suburbia is inherently oriented toward private amenity, residential, and investment values, and not toward commodity production or public uses. Until recently, many suburbanites have opposed the use of forests for producing materials, or for many forms of recreation, seeing them only as a green backdrop. For a time, the fuelwood revolution seemed to be changing that, though low fuel oil prices have momentarily reversed the trend. But in the future, the suburban forest will be increasingly recognized as it once was for producing fuelwood and other essential raw materials.

Suburbs and Forests in Connecticut

The history of forest, land, suburbanization, and forest industry in Connecticut illustrates the dynamics of land use change in the suburban forest (Table 5.1). Despite its location in Boston-Washington megalopolis, and

TABLE 5.1. Forest and Farmland Area of Connecticut, 1600–1997

Year	Forest Acres	Percent	Farm Acres	Percent
1600	3,010	96	—	
1700	2,130	68	—	
1800	1,644	52	—	
1860	923	29	2,504	81
1900	1,276	41	2,312	75
1920	1,489	48	1,899	61
1945	1,907	61	—	
1969	—		540	17
1970	1,823*	60	—	
1977	1,806*	60	470	15
1987	1,776*	57	410	13
1997	1,815*	59	380	12

* Commercial forest land only.

Note: May not total 100 percent because of urban and other land uses and the inclusion of woodland in farm acreage.

Sources: E. V. Zumwalt, *Taxation and other factors affecting private forestry in Connecticut.* New Haven: Yale School of For., Bull. 58, 1953; N. Kingsley, *The Forest landowners of Southern New England.* USDA-FS NEFES Res. Bull. NE-41, 1976; USDA-FS, Analysis, Appendix 3; R. Sherman, et al., *Open-land policy in Connecticut.* New Haven: Yale School of For. and Env. Studies, Bull. 87, 1974; USDA, Agricultural Statistics, Washington, D.C.: GPO, Annual. 1970-1992 from Powell, et al., Res. Bull. RM-234. 1992 farmland from Agricultural Statistics.

an average (1992) population density of 677 persons per square mile, Connecticut's forest land area increased from 1,483,000 acres in 1913 to 1,815,000 acres in 1997. Most of the increase was due to the forest invading abandoned farm land, which was also absorbing postwar development pressures. During this period, the state's population grew dramatically, by 2.1 million people. The Connecticut State Planning Office has estimated that 172,200 acres were developed for residential, commercial, industrial, and transportation uses between 1960 and 1970. Development drew heavily on farms and forest plantations — for 105,000 acres. The remaining acreage came from brushland and forest.[10] The area of commercially available forest land fell from 1,973,000 acres in 1952 to 1,815,000 acres in 1997. This was largely due to shifts of land to other forest uses and not primarily due to conversion to urban uses.

Population growth has been only one of several influences on the remaining acreage of farm and forest land. Changes in landownership have been dominated by the marginal character of agriculture in most upland towns. Land use change has been dominated, statewide, by the reappearance of forest on abandoned pasture and crop lands not needed for intensive development. Most counties contain more forest land now than they did in 1913. Even at the extremes, the past coexistence of suburbanization and the forest is striking. Most of the twentieth century population growth in Connecticut occurred in New Haven, Fairfield, and Hartford counties, yet each experienced an increase in forest acreage to 1970.

The return of forest to Connecticut's landscape has been a major factor promoting recovery of many important wildlife species. Beaver, which vanished from the state in the 1840s, are now so common that their activities generate many complaints. The fisher has been reintroduced, and a small black bear population survives in the northwestern forests of the state. Following re-introduction in 1975, the wild turkey population has grown to 8,500. Not all species have thrived, however. Spreading development has reduced woodcock habitat, for example.[11]

Connecticut, a state long settled and farmed, has only 249,000 acres (14 percent) of its commercial forest land in public ownership. The bulk

Black bear. *Courtesy of Maine Department of Inland Fisheries & Wildlife.*

of this is in state forests and parks, mostly acquired between 1900 and 1940. Farmers own about 13 percent of the state's forests, a great change from their dominance in the 1880s. In 1952, farmers owned 670,000 acres of Connecticut forests; in 1992, they owned only 225,000. Forest Service surveys show that 66 percent of the 91,900 individual private owners in the state were white collar, executive, professional or retired.[12]

The water utilities, including both public and private organizations, own a total of roughly 150,000 acres. These tracts include the largest forest properties in the state. The Hartford Metropolitan District Commission owns about 32,000 acres, while the Southern Connecticut Regional Water Authority owns some 26,000. These lands produce regular timber harvests.

In 1993, there were an estimated 77,200 holdings less than 10 acres in Connecticut (Table 5.2). The amount of land in this size class, however, was still less than 10 percent of the state's privately held acreage. The acreage in tracts between 10 and 100 acres in size shows remarkable stability. The 500 acre and larger class actually increased its total acreage significantly. This means that much of Connecticut's forest still remains in parcels that can be effectively managed for timber, wildlife, and recreation. The skewed size distribution of parcels means that averages make a poor way of summarizing conditions and trends.

The subdividing process was accompanied by accelerated ownership turnover. In Fairfield County, 82 percent of a sample of thirty-eight parcels were sold at least once between 1940 and 1959. In the three southern New

TABLE 5.2. Connecticut Private Forest Ownerships by Size Class

	1946				1993			
Size (acres)	Owners	(%)	1000 Acres	(%)	Owners	(%)	1000 Acres	(%)
1–9	3,925	(11)	22.3	(1)	77,200	(76)	139	(9)
10–49	19,729	(56)	440.8	(26)	18,400	(18)	444	(29)
50–99	7,429	(21)	529.9	(31)	3,200	(3)	222	(14)
100–499	3,889	(11)	646.5	(38)	3,000	(3)	416	(29)
500+	70	(—)	75.5	(4)	200	(0)	333	(21)
Total	35,042	(100)	1,715.0	(100)	102,000	(100)	1,553	(100)

Sources: E. V. Zumwalt, *Taxation and other factors affecting private forestry in Connecticut,* New Haven: Yale School of For., Bull. No. 58, 1957; T. W. Birch, *The private forest landowners of the northern U.S.,* 1994, USDA-FS, NEFES, Res. Bull., NE-136 (1996).

England states, 43 percent of the owners, with 35 percent of the acreage, had owned their tracts less than ten years. One-fourth of the owners, with 29 percent of the land, had owned their properties longer than twenty-five years.[13] A 1975 survey of Connecticut forest landowners found that the average tenure was only seven years.[14] Ownership as brief as seven years is not long enough to develop an interest in forest management, much less to perceive the benefit of longterm forest husbandry. Since suburban land markets are blind to timber values, the market provides no incentives, except to the extent that rutted, slash-piled woodlots may be harder to sell.

Because of small parcels and past owner attitudes unfavorable to logging, a significant part of Connecticut's forest has been unavailable for industrial wood production, although small parcels are now producing an increasing amount of firewood. Even in this suburbanized state, however, almost 60 percent of the woodland remains in parcels larger than fifty acres and therefore still economically manageable. These lands belong to less than 8,000 people. Thus, only 8,000 people need to be contacted and encouraged to promote private forestry, wildlife management, and recreation management in this state.

Growth and Cut of Connecticut Forests

By the mid-nineteenth century, Connecticut's original forests had been greatly reduced in area. The remaining stands were being heavily cut. The cut for fuelwood alone was far greater than today's harvest for all uses. The woodlands were a sprout forest, where trees rarely grew beyond thirty years before being cut. In 1909, 95 percent of the hardwood acreage in Litchfield County was less than forty-one years old.[15] In the first decades of the twentieth century, Connecticut's forests lost the valuable, fast-growing chestnut to the chestnut blight. Chestnut then accounted for one-half of the state's hardwood timber inventory, and three-fourths of the state's 1909 hardwood lumber cut.

After the boxboard boom peaked around 1909, and as coal and oil replaced fuelwood, the drain on Connecticut's forests declined. In the same period, forest area was increasing dramatically. In contrast to northern New England and New York, no pulpwood market emerged. For southern New England as a whole, in the early 1970s, landclearing and other nontimber product removals accounted for 85 percent of the drain upon forest growing stock. Fully 56 percent of all timber removed from inventory was unused. Also for that region, total industrial wood production increased by 70 percent from 1971–1984, and rose 99 percent in Con-

necticut. In keeping with the maturing forest and higher wood values, sawlog production in Connecticut roughly tripled from 1952 to 1984.[16]

Low removals in relation to growth permitted Connecticut's hardwood inventory to more than double between 1952 and 1992. The softwood inventory increased over the same interval by 160 percent. Over this period, growth/removal ratios continued to be favorable for both hardwoods and softwoods. The resulting buildups in growing stock have made many areas overstocked and loaded with mature and disease-prone trees.

Forest Industries Since World War II

After World War II, Connecticut's primary forest products industry settled into a condition of stability based on harvesting a largely suburban and exurban forest. Wood-using firms in Connecticut are small. Only thirty lumber and wood products establishments (out of 152) employed more than twenty persons in 1972. The primary forest products industry of Connecticut today consists of about eighty-five sawmills, one mill buying pulpwood, several hundred persons employed in logging, and a collection of specialty operations such as turnery plants, post and piling plants, and pallet mills. Today, a number of Connecticut sawmills are exporting lumber and wood products around the world. The converting branch of Connecticut's forest industry is substantial, accounting for the bulk of the 9,400 or so individuals working in forest industries. The largest single converting activity is paperboard containers and boxes. The converting plants depend on raw materials from other states: paper and board for envelopes and boxes, lumber and cut stock for millwork and furniture.

Overall, employment in the forest products industry has risen slightly since 1947 (Table 5.3). In 1987, these groups accounted for about 2.8 per-

TABLE 5.3. Connecticut Forest Industries, 1947–1992

	LUMBER AND WOOD					PAPER AND ALLIED				
	1947	1958	1972	1987	1992	1947	1958	1972	1987	1992
Employment (thousands)	1.4	1.5	1.9	3.4	2.2	7.5	8.3	6.9	7.3	7.2
Establishments	113	153	152	242	205	91	107	99	97	95

Sources: Federal Reserve Bank of Boston, Economics Almanac, 1971; U.S. Dept. of Commerce, Bureau of the Census, 1972 Census of Manufacturers, 1974, and 1987 Census of Manufacturers MC87-A-7(p), Nov. 1990; 1992 Census of Manufacturers, MC92-A-7, Apr. 1996.

cent of the state's manufacturing employment, though the percentage is higher in the rural counties. The stability of wood products and paper employment in a suburbanizing state is striking, retaining a steady, if small, contribution to the state's rural economy.

Policy Conflicts

The suburban forest is the scene of many policy conflicts. These range from how to tax tiny ten-acre woodlots that are worth $220,000 per acre for development, to how best to suppress the gypsy moth, to how to preserve greenspace by controlling development. In some towns, active programs of flood plain zoning, land use control, farm land preservation, and open space acquisition are under way. Private groups continue to do their part. State agencies are at work on parks, river corridors, and the like, but tend to concentrate their own efforts outside the suburban forest, where the land acquisition dollar goes further. The story of these policy conflicts could fill a book.[18]

Suburbanites have been legally mandating, through minimum lot sizes and restrictions on multi-unit housing, the excessive consumption of open land for development. But they have also become increasingly conscious of the value of open space. State and federal governments encourage localities to acquire open space with bond issues and liberal subsidies. Originally the province of small Audubon groups, land trusts, and academics and bureaucrats, the open space movement seems to have made its point. In a few communities, such as Wilton, Connecticut, and Lincoln, Massachusetts, they have been so successful as to be accused of using open space land acquisition to defend elite prestige communities, to raise real estate values, and to prevent outsiders from moving in. Whatever the justice of such criticisms, it is clear that many northeastern communities, both through government and nonprofit group action, have worked vigorously for open space preservation.

The most difficult obstacles to open space preservation have been lost tax revenue and the perception that there is really no general public benefit from conserving open space. But the unbalanced tax bases of most suburban towns produce a striking result. Often, new development increases municipal costs more than it increases revenues. Residents in some towns have been persuaded that new development should be resisted and that open land is a good buy to prevent development. No one ever mentions that most existing residents also contribute less to tax revenues than to costs of services.

With the revival of local sawlog markets and the increase in fuelwood cutting, a major social change in attitudes toward the forest occurred. The

increased cutting, however, has raised fears of overcutting, aesthetic and wildlife habitat losses, and erosion. These fears have been joined to the continuing desire that a green backdrop be the prevailing forest use in suburbia. An increasing number of towns are enacting or considering zoning restrictions or actual bans on commercial timber harvesting. This included as many as 135 in Pennsylvania (1995) and 123 in New York (1995).[19] This spreading regulatory mood poses a severe challenge to the region's landowners and wood-using industries. By the mid-1990s, a nationwide reaction to regulation was in progress, with intense political debate on "takings" and what to do about them.

As forests spread and mature, and as wildlife laws and management programs show success, populations of many species have been rebounding in northeastern suburbia. Yet, as parcel fragmentation and posting increase, hunters and birdwatchers are less and less able to enjoy the benefits.

How can forest owners address the concerns that motivate these restrictions? They will have to convince suburbanites that harvesting wood can be consistent with other forest values, and that wood production plays an important, albeit limited, role in the local economy. In addition, loggers, foresters, and the industry will have to police themselves much better then they have in the past. In short, all concerned will have to do a much better job.

Historic Perspective

The first wave of land use change in the region was the farm clearing, which persisted until 1840–1880. The peak years of cleared land were estimated to be 1850 in Rhode Island, 1860 in Connecticut, and 1870 in Massachusetts, and 1880 in New York. Before these peak years, however, birch, red cedar, and pines were already appearing in old fields in older settlements where rural populations dwindled in response to urban opportunity and the opening of the West.

In tens of thousands of pastures like Rockwood's, the decline of farming returned millions of acres to woods from 1850 to the late 1960s. Even after the suburban boom began, farmland supplied most of the land required for housing. Many growing suburbs retained their wooded character. Even the major metropolitan regions are only 25 to 50 percent developed.

Suburban communities have forgotten the seventeenth century. Then, town and colony governments built public policies on the basic importance of fuelwood, sawtimber, shingles, barrel staves, and bark. As the forest regrew on abandoned farms, it continued to be heavily cut for fuel and boxboards. After World War I, however, the woods lay fallow, growing in size, stocking, age, and value for multiple forest uses. Suburbanites saw the forest as a green backdrop, a place to stroll, and little more. Many were hostile to wood-producing uses of the forest. For a period, public policy treated the suburban forest as open space or greenbelt — simply the absence of anything else. To violate such greenspace with the axe seemed unthinkable. The ability of the urbanized Northeast to import its corn from Iowa, its lumber from Canada, its oil from Iraq, and its steel from Japan led its citizens to believe that they had been liberated from dependence on natural resources. Yet their ability to enjoy this illusion was supplied by the widespread interdependence of the world economy, which supplied raw materials that in colonial times had been largely obtained within town boundaries.

Today, the fuelwood boom, though it slackened after 1985, has returned the commodity use of the forest to respectability.[20] A new willingness to accept timber harvesting and, hopefully an awareness of the need to conserve sawlogs for higher uses will emerge. At the same time, local and export markets for lumber and sawtimber have improved, and prices have risen dramatically. There are many exceptions, and all too many examples of woodlot butchery. An overreaction back to the greenspace

The wild turkey, once eradicated from the Northeastern woods, has been restored through the diligent efforts of state wildlife agencies. *Courtesy of Massachusetts Division of Fish & Wildlife.*

mode of thinking is still possible, and may be occurring in towns that are considering zoning restrictions on logging.

The Future of the Northeast's Suburban Forest

The northeastern population will grow in coming decades. While each town has its own reasons for setting minimum lot sizes, no government seems able to take account of the unplanned cumulative impact of small decisions that affect total land consumption.

To accommodate an additional population of 6 million by the year 2025 — a midrange of recent Census Bureau projections — the Northeast's suburban forest and rural landscape would have to give up the following amounts of land:

	Thousand Acres
Urban density (28 persons/acre)	210
Density of 1950s–1960s suburbanization (4 persons/acre)	1,500
Density of large lot suburbanization (1 persons/acre)	6,000

Probably little or no future development will be at urban densities. The main question is what level of suburban densities will prevail. The low-density scenario would see two-thirds of the current suburban forest developed and fragmented by 2025 except for that owned by governments, nonprofit groups, and wealthy country clubs. The outward boundary would spread far into the rural forest. The high density would allow a substantial resource of open land to persist inside the current suburban frontier and in the spreading exurban zone beyond.

Whether there will even be a suburban forest forty years from now depends on real estate markets and on public policy. Those market and policy decisions will not consider their regional consequences. They will accumulate from literally millions of subdividing, public acquisition, zoning variance, and tax policy decisions.[17]

So, there is a distinct possibility that despite its high value for open space, amenity, watershed protection, wildlife, and wood production, the suburban forest will largely vanish in another generation. Some might argue that little would be lost, since many of that forest's values are already severely compromised. Further, to the extent that those values survive, they have been held tightly in private hands by the suburban "No Trespassing" mentality.

The values of the suburban forest are substantial enough, however, to motivate a drastically expanded effort for their conservation. The aesthetic, wildlife, watershed protection, and wood-producing values of

Recommendation: Establish a Hudson River Trail

A Hudson River Trail should be created to bring a sense of the River to its residents and visitors. The Trail would extend from the Mohawk River south to the Battery in New York City, be as close to the River as possible, and be completed by the year 2000, on both banks of the Hudson. The Trail should be restricted to non-motorized use where not part of a heritage trail or scenic roadway. Easements for public access, and not purchase of land, must be the preferred method of acquisition. Eminent domain would not be used by the Conservancy to acquire lands for the Trail.

The Trail would utilize established trails through Urban Cultural Parks, thereby linking urban and rural areas, and should be the spine of a system of Greenway trails linking the River to the Catskill Mountains, Shawangunk Mountains, Mohawk River and the Adirondack Mountains through the use of rights-of-way along abandoned rail lines and canal paths.

The makeup of the Trail would be more inclusive than such traditional trails as the Appalachian Trail or the Long Path in New York and New Jersey. The Hudson River Trail would be a true pathway, connecting urban parks, scenic highways, riverfront parks, bikeways and, where possible, railroad rights-of-way. Its elements would be as diverse as Flirtation Walk at West Point, the Corning Preserve five-mile bike trail in Albany and a revitalized Esplanade/Waterfront Park in New York City. Its trails could include the newly formed Hyde Park Trail, wooded walks in Bear Mountain State Park, the city streets of Kingston, town parks of Athens and Coxsackie and pedestrian walkways over the Newburgh-Beacon Bridge. *(continued)*

these woods will increase. The last chance to conserve these values, however, is in the hands of the current generation of landowners, conservationists, and public officials.

To illustrate one issue, consider the wood production value of 100,000 acres of forest — less than a decade's recent land consumption in Connecticut. These acres, even under extensive management, could produce enough wood to heat 6,000 homes for the winter, or to employ about a hundred people year round in logging and sawmilling. One hundred jobs seems like nothing until you consider how starved most suburban areas are for manufacturing jobs and tax base. For another value, Connecticut's system of forests, parks, and trails is a major quality of life asset. Building on this system to increase and diversify outdoor recreation opportunities near where people live has the potential to produce enormous social and economic benefits. Similar exercises could be performed for all of the other values of the suburban forest.

The Hudson River Trail must have two overriding features:

- It must be created with patience and negotiation, not eminent domain. Public testimony convinced the Council that the Trail will become an accepted and jealously guarded part of the Hudson River shoreline, but not if its development is mandated.
- Property owners who allow their land to become part of the Trail must be protected not only from liability, but also from the burden of defending lawsuits. The surest way to do this would be to completely indemnify, through the State, all owners who permit their land to become a part of the Trail.

Establishment of Trail would begin with a Hudson River Trail Symposium, organized through the Greenway Conservancy, bringing together representatives of local governments, local and regional trail associations, not-for-profit organizations and the federal government.

The Council also recommends that the Conservancy utilize the resources of local school districts along the River and of existing Youth Conservation Corps to build the trail and maintain it.

The Trail must be available to the physically disadvantaged where topographically feasible.

Source: Hudson River Valley Greenway Council, 1991. A Hudson River Valley Greenway. February, p. 20–21.

Challenge for the Future

The suburban forest is a fortuitous clutter of bypassed real estate that happens to grow trees. As the examples of New Jersey and Connecticut show, the suburban forest was able to absorb massive population growth from the 1950s to the early 1990s. But in the face of future development pressures, and their magnification through shadow conversion, its very survival as a component of the region's land use pattern is now in doubt. By firm public and private action, however, much of it could be saved.

Economists say that one of the Northeast's last economic advantages is its high quality of life. The suburban forest is without doubt one of the key amenities contributing to this quality of life. If the suburban forest is lost, it will harm not only our present quality of life but our region's economic future. This is recognized in many places. A major example is the prolonged effort to preserve the Sterling Forest, a 17,500 acre parcel from whose hilltops the hiker can see the skyscrapers of New York. A complex deal brokered by the Trust for Public Land depended on state and federal funding.[21] This approach of case-by-case federal fundings may set a trend.

New Jersey ties for the national lead in proportion of its land in parks and wilderness. Still, its citizens are committed to open space, and have approved a $1.7 billion ten-year acquisition program. Late in 1998, the Clinton Administration began announcing policy initiatives to deal with sprawl, often under the rubric of "smart growth." Where this new interest will lead is impossible to foresee.

One thing is clear. The future survival and condition of the suburban forest is a land use issue. Its survival cannot be ensured by the traditional tools of forest policy — use-value taxation, fire control, landowner assistance. The suburban forest can be saved — to become an amenity and quality-of-life asset unique in the nation. It should be saved not just to grow fuelwood, as in the colonial era, or as a leafy backdrop for bungalows, as in the 1950s, or to keep people out, as in the 1970s. It should be saved for its broad conservation, amenity, production, and cultural values. There is a major question, however, if a region committed to a Town Meeting pattern of localized decisionmaking can meet this challenge while equitably dealing with a multitude of other social problems.

If Walden Pond and other public parks are all that remains of the northeast's suburban forest by the 2020s, it will be because of a failure of public and private vision and will. Higher taxpayer costs for a variety of public services will be an inevitable result. The losses to the region's future quality of life will be beyond calculation.

6 Stone Walls in the Woods: The Rural Forest

VISITORS TO UPSTATE New York and New England often remark on the oddity of stone walls rambling through the roadside woodlands. Many are surprised to hear that large areas have been out of farm production so long that they have grown several successive crops of trees.

Farmers have by themselves wrought the most extensive and visible changes in the northeastern landscape. They cleared up to half of the forest, of which much has since naturally returned to woodland and brush. A sizable fraction of today's rural forest, then, was once cropland or pasture. The vanished farmers left reminders of their passing in the stone walls, the forlorn and forgotten cemeteries by remote lanes, the leaf-filled cellarholes stumbled on by hunters, and surviving apple trees, now surrounded by sprout maple stands. The acreage of New England's rocky hills cleared by grueling labor and later abandoned far exceeds the area's current acreage of crop and pasture. In New England, farmland now occupies less land than houses and factories (Tables 6.1 and 6.2), but the balance is squarely in favor of farmland in the Mid-Atlantic states.

In most rural towns in the northeastern hills, longtime residents can recall the distant open view once seen from every hilltop. Today the views are blocked by dense woods. As one writer put it, the region's hills were engulfed by "a rising sea of woodlands."[1]

These rural towns are beyond the megalopolitan fringe now swamped by suburbia, and most are distant from the mountainsides and lakes of the recreational forest (Figure 6.1). On the fringes of the industrial forest from Maine to Pennsylvania, tiny villages linger after most of the land has been absorbed by paper companies and timbergrowing landowners. Farmers and rural residents are the principal landowners here, but much is owned by summer people, nonresidents, and investors. Still, this is the region where farms set the characteristic tone of the landscape. Of the 34 million

Widely reprinted etching shows early settler's cabin, Holland Purchase, New York.
Courtesy of Forest History Society.

acres of forest in the rural forest, farmers now own roughly one-quarter. The rural forest gains its character from past, as well as from present, farm ownership.

The history of the Northeast's rural forest is one of overextension and contraction of agriculture, and neglect of the woods because of poor wood markets. That story is told in histories of the region's agriculture, and reports of the state foresters and the USDA-FS.[2] The classic rural forestry problems are public access to the land for hunting, fishing, and walking, and the problems of promoting better forest and wildlife management. Of more recent vintage is the question of fragmentation of habitat by roads and scattered development. These stories are the work of this chapter.

As we see below, the acreage devoted to farming has declined. Yet as yields have risen and processing industries have grown, agriculture remains important to the region's economy, with $9.5 billion in farm marketing volume in 1995. Farmers themselves account for only 14 percent of total farm-related jobs. Counting all of the suppliers, processors, and distributors, the sector accounts for 23 million jobs, 17 percent of all jobs nationwide. Ranked by total farm-related jobs (counting processing and distribution), New York places second after California, Pennsylvania is fifth, and Massachusetts is 16th. As a proportion of total jobs, the farm-related sector is highest in the more rural states, ranging from 20.8 percent in Maine to 14.1 percent in Massachusetts, which has a large food

TABLE 6.1. Number of Farms and Land Uses, Northeast, 1992 (Million Acres)

State	Number of Farms 1992	Rural Cropland	Pasture	Forest	Other Rural	Rural Total	Developed 1987
Maine	7,100	447.5	111.2	17,556.5	705.1	18,820.3	507.6
New Hampshire	2,900	141.5	98.3	3,932.1	217.0	4,388.9	372.1
Vermont	6,900	634.6	349.1	4,138.2	74.8	5,196.7	207.7
N. New Eng.	16,900	1,223.6	558.6	25,626.8	996.9	28,405.9	1,087.4
Massachusetts	700	272.3	169.7	2,778.1	310.2	3,530.3	1,063.2
Connecticut	3,900	228.5	109.9	1,760.2	139.7	2,238.3	693.2
Rhode Island	6,900	24.9	24.2	392.8	29.7	471.6	161.4
S. New Eng.	11,500	525.7	303.8	4,931.1	479.6	6,240.2	1,917.8
New York	38,000	5,616.1	3,001.2	17,178.3	987.3	26,782.9	2,485.3
Pennsylvania	8,500	5,774.2	2,507.0	15,398.2	1,347.7	25,027.1	2,795.9
New Jersey	52,000	649.7	159.1	1,766.0	386.0	2,960.8	1,324.9
Mid–Atlantic	98,500	12,040.0	5,667.3	34,342.5	2,721.0	54,770.8	6,606.1
Total	126,900	13,789.3	6,529.7	64,900.4	4,197.5	89,416.9	9,611.3
Percent		15.4%	7.3%	72.6%	4.7%	100.0%	
U.S.	2,095,740						

Source: USDA SCS, 1994, Summary Report, Natural Resources Inventory, table 2; developed land from SCS/Iowa State SB 790 tables 1&2; and USDA Agricultural Statistics 1992 for number of farms.

processing sector.[3] Yet the region's economy is so diversified that the USDA's Agricultural Research Service finds not one single "farm dependent" county in the entire region. Although the region as a whole imports many farm inputs and products, the Economic Interdependence of the urban and rural community is still real, however little it is recognized.

Farm Expansion and Retreat

Agricultural decline, especially in New England, occurred from the earliest years, as a result of competition from other regions, soil exhaustion, overpopulation, and greener grass elsewhere. At maximum extent, farmers owned probably half of the region's total land area. Most of the farms

TABLE 6.2. Cropland, Pasture, Forest Uses, Northeast, 1992, Percentage Distribution of Rural Land

State	Rural Cropland	Pasture	Forest	Minor	Rural Total
Maine	2%	1%	93%	4%	100%
New Hampshire	3%	2%	90%	5%	100%
Vermont	12%	7%	80%	1%	100%
N. New Eng.	4%	2%	90%	4%	100%
Massachusetts	8%	5%	79%	9%	100%
Connecticut	10%	5%	79%	6%	100%
Rhode Island	5%	5%	83%	6%	100%
S. New Eng.	8%	5%	79%	8%	100%
New York	21%	11%	64%	4%	100%
Pennsylvania	23%	10%	62%	5%	100%
New Jersey	22%	5%	60%	13%	100%
Mid-Atlantic	22%	10%	63%	5%	100%
Total	15.4%	7.3%	72.6%	4.7%	100%

Source: Table 6.1 above.

of 1880 no longer exist, having been absorbed by other operators, left to regrow to woods, or converted to tract houses. The Northeast's farm population fell from 3.2 million in 1890 to 230,000 by 1990. The story of agricultural retreat is best told by examples from different regions and time periods.

Concord, Massachusetts, was settled in 1635. By the Revolution, the soil was declining in fertility under poor management. R. A. Gross chronicled the impact of overpopulation on family life, politics, and economic opportunity in Concord: "on the eve of the Revolution, Concord was a declining town facing a grim future of increasing poverty, economic stagnation, and even depopulation."[4] By the 1790s, the effects of this overpopulation were clearly evident. More than half of the young Concord men who had fought in the Revolution had already left town forever.[5] Many went to farmsteads in Northern New England or New York's Mohawk Valley.

Thoreau noted the evidence of declining settlement near Walden Pond:

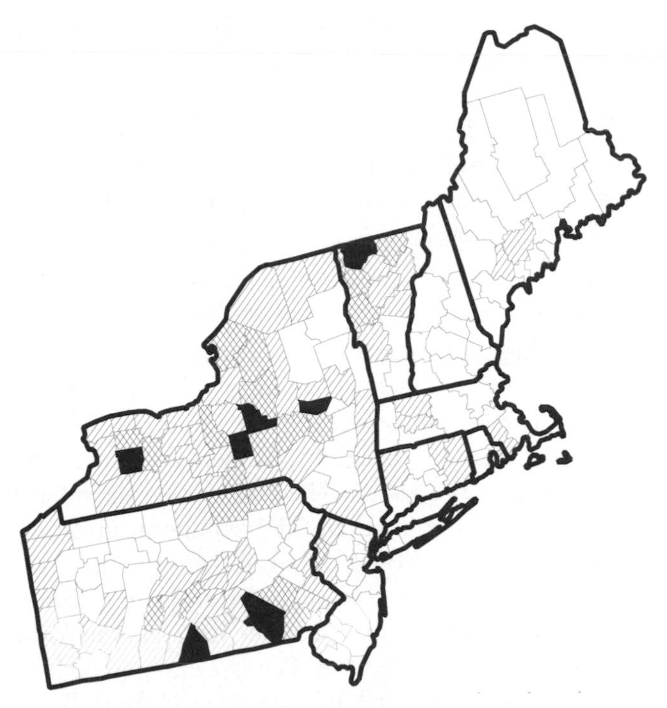

Figure 6.1 Dairy Cows Per 80 Acres in 1982
Source: J. F. Hart, The land that feeds us *(New York:* W. W. *Norton & Co.).*

Huge maple grows beside rock wall of abandoned field in New Hampshire.
Courtesy of American Forest Institute.

Now only a dent in the earth marks the site of these dwellings, with buried cellar stones . . . some pitch pine or oak occupies what was the chimney nook.

. . . These cellar dents, like deserted fox burrows, old holes, are all that is left where once were the stir and bustle of human life . . .

. . . Still grows the vivacious lilac a generation after the door and lintel and sill are gone.[6]

In a portion of North Central Massachusetts, the peak of farm clearing occurred about 1830; since then, investigators at the Harvard Forest found, the region has almost entirely returned to forest. Not only that, but 37 percent of its area has been placed into one form or another of conservation landownership, most notably the Quabbin Reservation.[7]

Despite the decline in old regions, farmers continued to open land in parts of the region into the 1880s. The Erie Canal opened in 1825, changing Northeastern farming forever. The competition from midwestern grain took away the attraction from grain farming after the 1840s. But shifts to sheep and dairying and to specialty crops like apples, potatoes in Maine's Aroostook County, tobacco in the Connecticut Valley, and truck gardening in New Jersey and Eastern Pennsylvania helped maintain farming. In many areas, much of the land was never cleared. By the 1890s, visible decline in farming was underway across New England and upstate New York.[8] As farming waned, forests began invading old pastures.[9]

In 1890, the State of Maine became concerned about farm abandonment and conducted a statewide survey to determine its extent and causes.

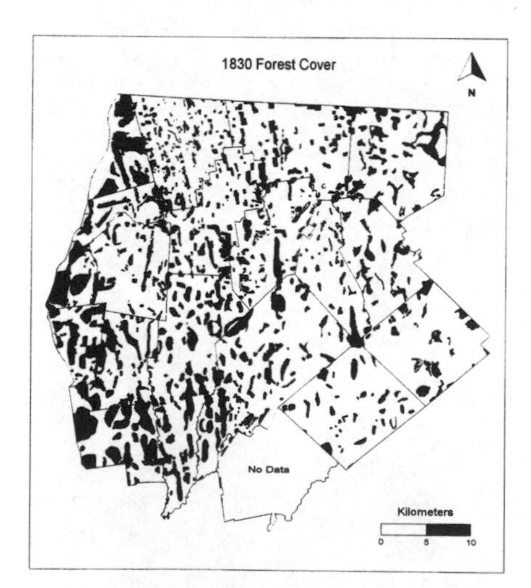

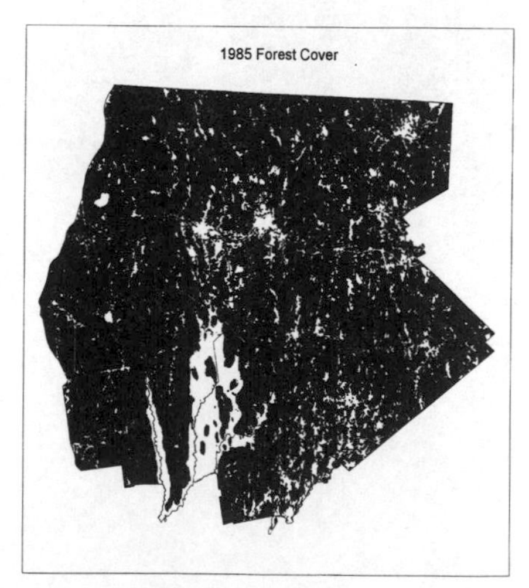

The distribution of forest cover at the approximate height of agriculture activity (1830) and in the modern landscape (1985) in the North Quabbin Region, Massachusetts. *Source:* Conservation Biology, *11(1) Feb. 1997, p. 231, by permission of David Foster.*

Western Maine farmscape near Newfield 1930's. *George French photo, courtesy of Maine State Archives.*

Nearly 3,300 abandoned farms, totaling 254,000 acres, were noted. The survey found that many abandoned farms were incorporated into neighboring farms.

In another town:

> Say 40 years ago, the inhabitants numbered about 1,000, and now think our census enumerator cannot give us 400. . . . Very many of the abandoned farms are well-fenced with good stone walls dividing fields, where trees are thickly growing, fast making timber of all kinds for another generation.[10]

Land cleared for farming reached its greatest extent by about 1880. Acreage in Northeastern farms slid slowly to about World War I, then plunged rapidly (Figure 6.2). From 1880 to 1930, land in farms fell by 20 million acres — to 54 million. Woodland owned by farmers fell by 6 million acres, while harvested cropland fell to only 21 million.[11] Cropland harvested shrank relatively little over this period; the bulk of the loss was in pasture and forest. Later in the century, however, cleared land shrank as well. According to Hart, from 1910–1959, every county in the Northeast lost cleared farmland, for a total decline of 20 million acres.[12]

Fast as was the decline of regional farm acreage up to 1930, the events since World War II have been dazzling. Cost pressures and technological changes led to the disappearance of most of the small part-time farms. The spread of suburbia, consuming land at new, higher rates per family, en-

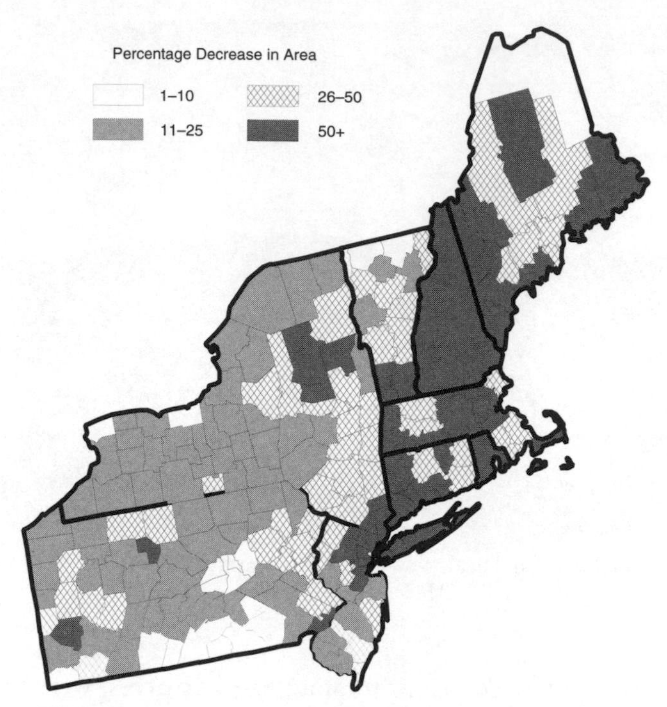

Figure 6.2 The Percentage Decrease in Agricultural Land from the Peak to 1930
Source: USDA Forest Service, 1933.

gulfed entire counties that had been sleepy farming communities in 1940. In that year, one quarter of the nation's population lived on farms. Wartime service showed rural young men and women a new world; many never returned to their farm homes.

In Connecticut, change was slow until about World War I. In 1910, more than two-thirds of Connecticut was in the cropland, pasture, wetland, and woodland of about 27,000 farms. By 1950, over 11,000 farms, containing 1.3 million acres, had disappeared. Suburban growth consumed only a small fraction of these farms, few of which were providing full-time livelihoods for their owners. In 1950, fully 45 percent of the state's farms were part-time, residential, retirement, or low-income operations. Farmland disappearance after 1950 was still more dramatic: nearly 800,000 acres in only twenty years. The number of farms fell to 3,900 by 1992. Losses were especially rapid in the suburbanizing counties of Fairfield and Hartford.

Examples of postwar decline elsewhere are common. From 1940 to 1974, in Maine coastal counties, farms fell from 19,000 to 3,000, from 33 percent to 9 percent of the area of those counties. In a detailed study of nineteen Maine towns, planners Joyce Benson and Paul Frederic found

Wild Turkey Recovers in Maine

Historical records document the existence of wild turkeys in coastal areas of Maine as far east as the Penobscot Bay areas. Unfortunately, the last of Maine's native wild turkeys disappeared in the early 1800s because of unrestricted shooting and extensive forest-clearing. The reversion of thousands of acres of farmland back to wooded habitat has greatly enhanced prospects for reestablishing wild turkeys into former ranges.

As early as the 1960s, Maine sportsmen began "thinking turkey." Fish and game clubs in the Bangor and Windham areas made attempts to reestablish turkeys into their areas using birds raised from part wild and part game-farm stocks. The Bangor stocking was unsuccessful, but the Windham population persisted in low numbers into the 1980s.

In the 1960s and 1970s, considerable work was done in other states to establish wild turkeys into former and new ranges of suitable habitat. Researchers noted the key to each success was to remove a small number of wild birds from one site and release them into suitable, unoccupied habitat. Maine too became involved in a similar program in 1977, when department biologists acquired 41 wild turkeys from Vermont and released them in York County. By the early 1980s, the York County population had become large enough to serve as a source of birds for new release sites. In the spring of 1982, 33 birds were captured in York County and released in Waldo County. In the winter of 1984, 19 additional birds were captured in York County and released in Hancock County.

The Waldo County release was successful and resulted in a stable population that appears to be increasing. Unfortunately, the Hancock County wild turkeys failed to produce a self-sustaining population. Illegal shooting of these birds is believed to be the major cause for this initial failure. Today, reports of adult wild turkeys in Western Hancock County are not uncommon; the birds appear to have crossed the Penobscot River on their own. . . .

Source: R. B. Owen, Jr. Wildlife division research & management report 1994. Maine Dept. of Inland Fisheries & Wildlife. p. 32–33, 38.

that farmland loss occurred rapidly from 1960 to 1980. Most of the loss was simply to abandonment of tillage — 72 percent of the land lost returned to woods. Conversion to other uses was most important near suburban areas. Most of the conversion occurred in small parcels, so that the cumulative effect is not readily seen by the public or by the farmers themselves.[13]

The patroonship system, absentee landownerships and Royal policies slowed settlement and development in New York and Pennsylvania until after the Revolution, so the settlement history of those states differs from the New England states. Settlement and farming in these states was pro-

Rural Pennsylvania's Nonagricultural Economy: Its Status and Direction

Pennsylvania is the most rural state in the United States, measured by number of people. More than 3 million people, over a quarter of the state's population, live outside urbanized areas. Any discussion of the rural economy in the 1990s, however, must start from a far different perspective than historically has been the case. Rural does not mean agriculture or farm, and nonmetropolitan or nonurban does not mean farm. One cannot understand the problems of rural economics and people today if rural is equated with agriculture, or with other traditional rural nonagricultural industries such as mining and forestry. We must start by recognizing the steadily smaller role farm work plays in the life and economy of rural and small town America. "The rural landscape is still farm and forests, but the people and their pursuits are overwhelmingly part of the nonfarm economy." This is particularly true for Pennsylvania.

The change in farm population provides an initial illustration of the meaning of the above statements. In 1950, the farm population was 6.7% of the state total, and about 23% of the rural total. By 1988, farm population had decreased by over 80% (from 705,000 to 132,000), and represented about 1% of the state total, and only 3.5% of the rural population. In comparison, for the nation as a whole, the farm population averaged 7.2% of the rural total in 1988. . . .

. . . Even 40 years ago only about a dozen Pennsylvania counties were classified as farm dependent. By the mid-1980s (actually, at least 10 years earlier), the number was zero. On the other hand, over half of our nonmetropolitan counties were classified as manufacturing dependent in the mid-1980s (30% or more of income from manufacturing), which is actually a decrease from the late 1970s. Much of this manufacturing is, however, resource-based; that is, food or wood processing . . .

. . . The farm financial crisis and declining farm numbers in the mid-1980s led to great concern for the health of rural economies. Even nationally, however, this was but a small part of a widespread economic decline in the rural economy. Rural job declines were not primarily in farm or agriculturally related industries. This was particularly true for Pennsylvania. Statewide, job losses in mining alone far exceeded those in farm and agriculturally related industries. . . .

What do these changes imply for rural Pennsylvania and its people and communities? A key implication involves the question I asked at the beginning — what does this mean for farm and agriculturally oriented people? Pennsylvania has maintained a relatively stable structure of small and medium size family farms, with more than half in metropolitan counties. Why? Farming is no more profitable in Pennsylvania than in other states. A main reason is the availability of off-farm employment.It has become increasingly clear that the maintenance of family farms and farm family income is related to the structure and vitality of the rural nonfarm economy. In almost 70 percent of Pennsylvania farm families, either the operator or the spouse works off the farm, and both have off-farm jobs in 30 percent of the farm families. Off-farm income accounts for two-thirds of taxable farm family income in Pennsylvania, with more than 80 percent of farm families earning more than half their taxable income from off-farm sources. The continued survival of farm enterprises throughout the state will be based on the availability of job opportunities in, or within a reasonable commuting distance of rural areas. As we have seen, these opportunities are more and more likely to be in nontraditional rural industries, particularly services.

Source: S. M. Smith, 1991. *Rural Pennsylvania's nonagricultural economy: its status and direction.* In John C. Becker, Proceedings of a Conference, Pennsylvania's Agricultural Economy: Trends, Issues, and Prospects. University Park, PA, March 20–21, 1991. Dept. of Agr. Econ. and Rural Sociology. p. 19–23.

foundly influenced by the development of transportation. The opening of the Erie Canal and other canals, the development of steam navigation on the Ohio and the Great Lakes, and the railroads all spurred farm-making in these states.

In New York, the number of farms peaked at 227,000 in 1900, when almost 23 million acres of land were in farms (not all tilled). By 1965, land in farms had fallen to 13.2 million. By 1992, land in farms had fallen still further to 7.4 million acres, about a third of the level at the turn of the

In much of Northern New England and New York, as in this New Hampshire site, old fields are colonized by red or white spruce. *Courtesy of Bill Leak, USDA-FS.*

Farm Landscape, Southeast Pennsylvania.
Courtesy of U.S. Geological Survey.

century. And still, on this smaller landbase, New York's farm output on 36,000 farms makes it a powerful farm state.[14]

In Pennsylvania, total land in farms reached 19.8 million acres in 1880, or 69 percent of the state's land area. At that time, 13.4 million acres were cropped each year on 213,000 farms. The number of farms peaked later, in 1900, at 224,000. Average farm size in Pennsylvania fell from 117 acres in 1850 to 81 acres in 1925, before slowly rising again. This reflects the more specialized and intensive operations fostered by good soils and nearby urban markets. By 1992, there remained 8 million acres in farms in the state (150,000 of them), with 4 million acres of cropland harvested. Thus, in just over a century, 9 million acres of cropland we released to forest and to urban uses.[15] From 1920 to 1992, Pennsylvania woodland owned by farmers shrank from 4 million acres to about 1.4

million acres.[16] Still, as the photo shows, forest and woodlot are virtually absent from intensively farmed areas on the best soils.

Three major forestry issues in today's rural forest are the fragmentation of habitat, public access for recreation, and the improved practice of forestry on these lands.

Habitat Fragmentation — The Eagle's Eye View

In the rural Northeast, forests have spread over the landscape for a hundred years, regaining perhaps 20 million acres from past farming alone, not to mention recovery from exploitive cutting and fire. As these forests have spread and matured, habitat suitability has shifted away from birds and mammals of open lands and forest edges and toward animals of the woodlands.[17] Yet now that the forest has reclaimed so much of the landscape, a new concern is emerging. This is the concern that the habitat value of this forest is measured not only by its aggregate area but by its unity and diversity and interconnections — that is, how it lays out on the land from the Eagle's Eye View.

Studies have shown that small woodlots, surrounded by open space harbor fewer species than do large ones. This is not the same situation as creating an opening in an extensive forest. But as leisure lots or homesites

Horse loggers still operate on rural woodlots throughout the Northeast.
Courtesy of Marshall Wiebe, Maine Department of Conservation.

are set off, as powerline rights of way are cleared, and as patches of woodlot are cut, the habitat value of the forest may change. The standard recommendation on this topic is to retain large blocks of undisturbed habitat, but this approach has limited relevance where landownership and use is already so fragmented. How to manage for landscape values in regions where the land is owned in 10–200 acre parcels presents a major challenge.

Public Access to the Rural Forest

The posting of private land against public use has disturbed those concerned with outdoor recreation, especially hunting and fishing, for decades. Considerable private land remains open to public use. The U.S. Department of the Interior estimated in 1979 that 32 percent of the noncorporate land acreage is open to the public (217 million acres) and 54 percent of the corporate land is open (40 million acres). Most of the land owned by forest products companies is open to the public.[18]

Land posting results from social change and social conflict. In the past, rural landowners in many regions accepted the right of others to cross their property for hunting, fishing, and travel. In the past, the hunters were neighbors, and the owner expected to hunt on neighbors' lands, so a supportive social consensus favoring public access to private land could exist. Today, the users are increasingly strangers from a distance. They at times damage roads with four-wheel drive vehicles, they hot-rod snowmobiles; or they hunt birds in the backyards of persons who disapprove of hunting. The consensus supporting public use wrinkles. When fences are broken and buildings vandalized, it collapses. The increasing population of rural regions means that residents are closer together, but even neighbors are often strangers.

Around the region, there is considerable variation in landowner reactions to different public uses of their land. The importance of these landowner preferences cannot be overemphasized. A change of 10 percentage points in the proportion of owners allowing a given activity can nullify the effect of millions of dollars spend on land acquisition, leaving the general public no better off than before. In Pennsylvania, for example, between 1978 and 1992, the acreage posted increased by 14 percent to reach nearly half of the state's private forest land.

Posting is likely to continue to increase, based on the social trends now visible in rural America. To preserve access to land and water in the face of changing landowner attitudes will require that the major conflicts be addressed. User groups and public agencies will have to assure landowners that vandalism, littering, and noise will be controlled. Own-

ers may have to be paid for allowing recreational uses. Users will have to accept "corridoring" along designated paths and corridors that avoid gardens, tree plantations, and homes. Snowmobile groups have successfully used this approach in many states.

In states where citizens have enjoyed common law rights of access to seashore, to great ponds, and to undeveloped wildland, the basis in custom exists to rebuild support of public use. A possible model is the program of public footpaths in England, in which the public is secured the right of transit over private lands for recreation. The footpaths are well-marked and provided with occasional parking places off small rural lanes. Thus, a medieval system of public rights necessitated by the three-field system of farming and the lack of public roads has been turned into a basis for recreational walking by visitors from afar. The rights are supported by a public consensus. Another example is the right of access to open land guaranteed to citizens by the German constitution.

In efforts to prevent the loss of public access through posting, most states have enacted laws limiting landowner liability and otherwise promoting public access.[19] These laws are probably of greater importance in reducing the anxiety level of large owners than they are in promoting public use on small ownerships.

Forestry in the Rural Forest

One might assume that farmers would have the time and inclination to manage their woodlots carefully. Forestry surveys, however, show that the level of timber management on these lands is often inadequate, and active wildlife management is virtually nonexistent. The causes of poor timber management on New England's rural forest have been limited landowner interest, low quality growing stock, and poor markets for wood. These woodlots still yield aesthetic and investment benefits to their owners, and provide wildlife habitat.

Forests have always been a key portion of the economy of every Northeastern farm. Farmers burned trees while clearing land and sold the ash as potash. The woodlot supplied fifteen to thirty cords each year of fuelwood for cooking and heating all winter. Merely harvesting and working up such a woodpile was a massive, year-round effort. Some farmers tapped sugarbushes for maple syrup. Many worked as loggers in winter to bring in extra cash. As the principal source of fuel and building material, the farm woodlot has truly been an important adjunct to New England agriculture. In the oak regions, open range grazing of hogs and cattle was important in colonial times, leading to endless quarrels over stray animals damaging crops and constant employment for the town animal wardens and fence viewers. In colonial times, the term "barrens" was applied to any piney woods, because the pines produced no valuable mast for the livestock. Today, as timber supplies shrink elsewhere, the potential exists for farm-owned forests to be more important to the rural economy, especially as global timber supplies change.

State forestry agencies have adopted programs to assist rural landowners in managing their woodlots. Most offer free forestry assistance, help in tree planting, and marketing advice. Where state foresters actually work on the land and supervise timber sales, they are limited to a maximum acreage or number of days of service per landowner. As a result, the average size of treatments accomplished by state programs is extremely small.

The state forestry agencies helping rural landowners are too understaffed to handle all of the region's forestry needs. Few forest owners are even aware of the assistance available to them. Perhaps this is fortunate, or the service foresters' waiting lists would be much longer than they are now.

These programs have promoted better forestry and have helped educate landowners about their woodlots. They are so thinly staffed, however, that they have touched only a tiny fraction of the land needing attention. Most estimates are that it would take a century or more for the state foresters to handle every woodlot needing attention now. As the Report of the 1979 New England Private Non-Industrial Forestry Conference noted, most of the forestry advice in New England is actually dispensed by loggers.

Programs assisting private forest owners are numerous and are organized differently in each state. A recent program, Forest Stewardship, seeks to foster long-range planning for a broad range of multiple-use values. Regionally, stewardship plans were in by 1992 place for 600,000 acres, half in New York (Table 6.3). One of the major programs, the Forestry Incentives Program (FIP) has cost-shared management practices on 272,000 acres of Northeastern forests from 1972–1992 (Table 6.4). The National Conservation Reserve Program has been important in other regions, but only includes about 200,000 acres in the Northeast. Of this, 8,500 acres were planted (Table 6.5). A privately sponsored Tree Farm Program fosters longterm management. This program enrolls 12 million acres in the Northeast, much of which is forest industry land (Table 6.6).

These programs, however, are not without their critics. Points of concern include:

1. Competition with private consultants, removing business that could sustain private consultants (although the rules try to minimize this).

TABLE 6.3. Forest Stewardship Program Accomplishments by State, FY 1995–1996

	1996 Acres	Cumulative Acres
Maine	20,481	203,136
New Hampshire	26,993	300,458
Vermont	30,705	202,717
No. New England	781,179	706,311
Massachusetts	18,951	174,373
Rhode Island	0	10,176
Connecticut	1,109	21,401
So. New England	20,060	205,950
New York	133,493	937,110
New Jersey	4,445	40,596
Pennsylvania	26,185	161,256
Mid-Atlantic	164,123	1,138,962
Total	965,362	2,051,223

Source: USDA-FS, Report of the Forest Service, Fiscal Year, 1996.

2. Small scale of effort and poor education programs, hindering the spread of forestry information.
3. A tendency to spend time heavily on "repeaters" who are already interested in forestry rather than trying to convert new owners to better forestry. Many of these repeaters would probably do the work without subsidy.
4. A tendency to be ruled by the in-box, serving all inquiries rather than setting priorities based on regional needs and real economic payoffs.
5. An "entitlement mentality" and strong control by local committees who are often more interested in pleasing their neighbors than in aggressively pursuing major conservation priorities.

The free gift of forestry practices has had some unintended effects. It has made it harder to convince owners that forestry advice is really worth paying for. As a result there is little follow-through on many management practices. Further, lack of emphasis on follow-through encourages a shallow idea of forestry as the application of isolated practices rather than as sustained longterm planning.

TABLE 6.4. Forestry Incentive Program (FIP) Cumulative Summary of Total Participants and Acres from 1975 through 1992

	FIP Participants	FIP Acres (1,000)	Total Farm & Misc. Pvt. Acres (1,000)	FIP Acres as % of Total Farm & Misc. Pvt.
Maine	1,919	26	8,275	0.3%
New Hampshire	2,104	30	3,353	0.9%
Vermont	1,680	39	3,412	1.1%
N. New Eng.	5,703	94	15,040	0.6%
Massachusetts	2,329	37	2,455	1.5%
Connecticut	445	11	1,530	0.7%
Rhode Island	244	3	286	0.9%
S. New Eng.	3,018	51	4,271	1.2%
New York	4,976	72	13,467	0.5%
Pennsylvania	2,977	42	11,747	0.4%
New Jersey	1,166	13	1,381	1.0%
Mid–Atlantic	9,119	127	26,595	0.5%
Total	17,840	273	45,906	0.6%

Source: USDA ASCS, *Forestry Incentives Program from inception of program through 1992 fiscal year,* Washington, D.C.: USDA, 1993; and Waddell, 1989.

Examples of poor follow-through abound. Plantations have been paved for housing, or forgotten to be overtaken by aspen — a costly way to grow aspen. Blister rust or budworm control is applied to stands that are not being managed anyway. Plantations are not thinned when they should be. Landowner turnover or changes in objectives frustrate state foresters by upending their plans regularly. These constant changes, however, are the basic weak point in any forest management program on a small woodlot, whether managed with public or private assistance.

Changes Needed in Forestry Assistance Programs

What are the biggest needs, then, in promoting better forest and landscape management in the rural forest? They can be briefly listed:

1. *Better information for owners of woodlots.* Improved extension, education, and general information programs about forestry are needed.

TABLE 6.5. Enrollment in Conservation Reserve Program, Northeastern States, June 1992

	Total Acres Enrolled	Total Acres w/ Tree Plantings
NEW ENGLAND		
Maine	38,490	2,569
New Hampshire	0	0
Vermont	193	0
No. New England	38,683	2,569
Massachusetts	32	10
Rhode Island	0	0
Connecticut	10	10
So. New England	42	20
MID-ATLANTIC		
New York	64,498	3,627
New Jersey	723	27
Pennsylvania	101,078	2,242
Subtotal	166,299	5,896
Northeast Total	205,024	8,485
Total U.S.	36,422,733	2,487,767

Source: USDA, RTD Updates, Jan. 1994.

Ultimately, education is the key to better management because there will never be enough foresters and conscientious loggers to do the job. With the high turnover of ownership, the education job is never finished. The university-run cooperative extension services are important, although understaffed. An important contribution is being made by private groups. One example is the American Tree Farm System, which recognizes and publicizes well-managed woodlots.

2. *More private consultants.* The improving markets for wood are helping to make it possible for consulting foresters to earn a living. The region has hundreds of consulting foresters, but most of them earn their living on surveying and real estate work and not primarily on forest management. By introducing many owners to forest management, public forestry programs may have created a more favorable climate for consultants.

TABLE 6.6. Tree Farm Program in the Northeast, 1994

State	Tree Farms	Total Acreage (1,000)
Maine	1,938	8,038
New Hampshire	1,696	966
Vermont	927	514
N. New England	4,561	9,518
Massachusetts	1,365	249
Connecticut	459	136
Rhode Island	204	36
S. New England	2,028	421
New York	2,447	1,385
Pennsylvania	1,612	682
New Jersey	296	86
Mid-Atlantic	4,355	2,153
Total Northeast	10,944	12,092
Total U.S.	72,071	88,824

Source: American Forest and Paper Association.

3. *Improved sense of landowner responsibility.* For serious forestry to occur, landowners must accept their own responsibility for sound forest stewardship. They must recognize that if a logger butchers their lot, it is because they allowed it. It is not the logger's fault. Public programs, handing out free trees, free stand improvement, free fire and insect control, all encourage the view that forestry is really the government's responsibility. This improved sense of landowner responsibility for multiple-use management must be vigorously fostered by all private and public agencies concerned with forestry. One way to achieve this is to promote more active owner decisionmaking, as is being done in Pennsylvania's stewardship program, among others.

4. *Improved logging equipment and systems for light partial cuttings on small areas.*[20] New small machines are becoming available to cut small volumes of small trees economically. In some areas, loggers are buying and using them. These machines avoid the ruts and damage to residual trees caused by large skidders. They are less costly to own and operate. Their

Roadside elms have vanished from the region's landscape, an example of the need for technical assistance on insect and disease issues. *George French photo, courtesy of Maine State Archives.*

improved availability will help sell forest management to owners who do not want large skidders on their land.

Recently, horse loggers, who have survived all along, have received increased attention. Horse logging is a hard way to earn a living but an extremely attractive way for small owners to harvest timber. I do not expect to see a real increase in horse logging, because it is such a strenuous occupation, but horses will continue to be part of the region's logging industry. Improved education and training for loggers will help. There are many conscientious and careful loggers working in Northeastern woods, whose work is a pleasure to see. Foresters would like to see such operators receive more recognition. Several states now have programs for logger certification.

5. *A sound and stable approach to forest taxation.* Forest landowners and foresters understandably spend a lot of time arguing about taxes and advocating that they be kept low. They argue that high annual property taxes and estate taxes eliminate the incentive to hold capital in the form of trees. The recent regional and national conferences on nonindustrial private forestry exemplify this concern with taxes.

Regional documentation on forest tax programs is lacking, but for the four Northern Forest Lands states (Maine, New Hampshire, Vermont, and New York), participation has been summarized. In those states, in 1991–1992, there were 15,955 parcels enrolled and a total of 11.8 million acres of land. While much of this land is industry-owned, these programs are important to many farm and other individual owners as well.[21]

Many of the most contentious issues of taxation involve equity aspects of property or capital gains taxation, and really do not hinge on the incentive effects of the taxes. And it is certainly understandable that owners of timber capital would like to see low taxes. The typical spokesman for small woodland owners is a retired, middle-class person who has managed a forest property for years. Rising land values have raised the annual property taxes. Because of their age, they are concerned with estate taxes; they fear the breakup of the property so diligently assembled.

Yet no evidence has ever been offered that reducing such a landowner's tax problems would induce others to start practicing forestry. Until such evidence is produced, it will be difficult to design forest tax policies that will truly promote forestry. In all too many instances in the past, programs of special tax assessments have subsidized owners who plan no timber management and have helped reduce carrying costs for speculators. And they have reduced cash outflows of those who would have managed their lots anyway.[22]

Still, certain basic desirable features of forest taxes can be identified:

a. Eligibility should hinge on preparation of an acceptable management plan and its careful implementation.
b. Tax payments should be timed to coincide with harvesting so that revenue is available to pay them.
c. To the extent that lowered taxes on forests are below ad valorem values, two problems must be solved:
    - The transfer of tax burden to other taxpayers owning nonforest real estate. This transfer is severe in many rural towns and is a chronic political irritant. Some states make payments to local governments to partially offset these transfers.
    - Means to "recapture," with interest, the value of the tax break when the land is sold for development.
d. Landowners should be able to count on predictable levels of taxation that are stable relative to prices of stumpage. The greatest threat to forestry is not high current taxation but the threat of future confiscatory tax levels.
e. Use-value tax programs need to be kept simple in order to make it straight forward for rural owners to participate, and must minimize perverse incentives. A common temptation is to require owners in such tax programs to meet other social objectives such as keeping land

open for recreation, or meeting certain detailed management standards. The effects of such requirements are usually unpredictable.

Many forest landowners have been caught in a serious tax bind by annual tax bills that have increased faster than timber prices. Annual charges against future crops devour profits quickly. Recent years have seen dramatic increases in assessments and annual tax bills, making owners fearful of still more tax escalation.

Northeastern forests were once valued at zero for local property taxes. They have been, until recent decades, undertaxed in relation to their true economic value. The future holds the fear, for many owners, of dramatic overtaxation as speculative land values reach unheard-of levels. At the same time, however, the owners of rural real estate often hold large, unrealized capital gains from high land prices. The tax problems of rural forestry stem from the fact that many such owners are "land-poor."

6. *Improved emphasis on multiple-use forest values.* Our past programs have been too timber-oriented. Most foresters recognize this and are skilled in harmonizing timber management with wildlife, aesthetic, and other landowner concerns. But the message has still not reached many rural landowners. Until the revival of interest in fuelwood, foresters despaired of ever getting across the message that timber management does not need to eliminate other forest values. The Stewardship Program is making an effort in this direction.

We need to improve our ability to manage sensitively for wood, wildlife, and water, with one eye out for how the land looks from the road or the back porch. Properly educated, most owners will be willing to forego some timber returns to benefit nontimber forest values.

7. *Cooperation among forest landowners must be fostered.* Recent years have seen the emergence across New England of organizations representing wood users, landowners, loggers, and consultants. These groups work at educating members and promoting their interests. It has been customary for foresters to urge the formation of forestry cooperatives. The Northeast has not been especially favorable ground for cooperatives, because of the stoutly independent landowners, past poor wood markets and low prices, and the small size of most of its ownerships. Few owners hold lots large enough to produce substantial annual income, which means that their interest in investing time in a co-op is limited.[23]

Other institutions, such as farm co-ops, soil conservation and resource conservation and development districts, and regional planning agencies, can provide excellent parent groups for local landowner groups. The importance of landowner interest and cooperation cannot be overemphasized. Without landowner commitment to longterm management, the efforts of foresters are wasted. Plantations are lost to brush, are not thinned properly, or succumb to disease. Thinned and improved hardwood stands degenerate through neglect. Soil conservation and stream improvements deteriorate from poor maintenance.

Efforts to maintain a landscape perspective for forestry in the rural forest will require some form of cooperation among owners. Just how this will work remains to be seen. There will likely be no single framework or model for making this happen. A proposal for Landscape Management Areas, raised in Chapter 12 below, offers an outline of an approach.

8. *Existing cost-share and assistance programs for rural landowners should be retained but refocused on cooperators who are most likely to be able to follow through for the longterm.*

Overview

No program of favorable taxation, free forestry services, or forestry television shows will mean a thing for the long-run productivity of the rural forest without landowner commitment to active management in a landscape context.[24] Developing such commitment is the true challenge facing the forestry community. The most promising approach is both to demonstrate the financial advantages and to promote the acceptance of a conservation ethic. Such an ethic argues that sound land stewardship is an obligation of ownership and not just one of a number of options. Good citizens do not strip their woodlots.

To improve forest conditions and implement a landscape management approach in the Rural Forest, it will be necessary to make all of our existing approaches more effective than they have been in the past. Further, no single one of them can carry the load. They all must work together. They will need to be supplemented by new ideas and cooperative efforts in order to ensure that a landscape perspective can be implemented in areas where it is important. Finally, innovative approaches to maintaining and enhancing opportunities for public access are needed.

7 A Place in the Country: The Recreational Forest

THE RECREATIONAL forest includes those areas in which the forest land is principally a green backdrop for resorts, second homes, or lakeside "camps" (Figure 7.1). The ownership pattern has become so altered by subdivision, nonresident ownership, and scattered development as to make future use of the land for commercial timber, public recreation, or other purposes difficult or impossible. Probably some 7 million acres of Northeastern forest land lie in the recreational forest, including forested portions of recreational lots and resorts, and the immediate areas heavily influenced by them. Most of the region's recreational activity occurs elsewhere, in the rural, industrial, and Wild Forests. In the recreational forests, forest ownership and use are strongly influenced by recreation-oriented resort, service, second home, and leisure lot development.[1]

There is no survey information on the numbers of individuals and families that use their own camps or condos, or that visit the many resorts and attractions in the recreation forest. In the 1982–1983 National Outdoor Recreation Survey, the following percentages of Northeastern respondents reported engaging in activities relevant to this forest:[2]

Driving for pleasure	45%
Downhill skiing	9%
Visiting zoos, fairs, or amusement parks	48%
Outdoor concerts, etc.	28%
Camping in developed campgrounds	13%

Outdoor recreation pursuits related to the forests have a major economic impact in the Northeast. Based on work done on a portion of the region, we can estimate that forest-based tourism in the Northeast probably accounts for at least $10–12 billion in direct spending impact. Though only a portion of this sum is retained as income in local communities, it is an indication of the importance of recreation to the region's economy.[3]

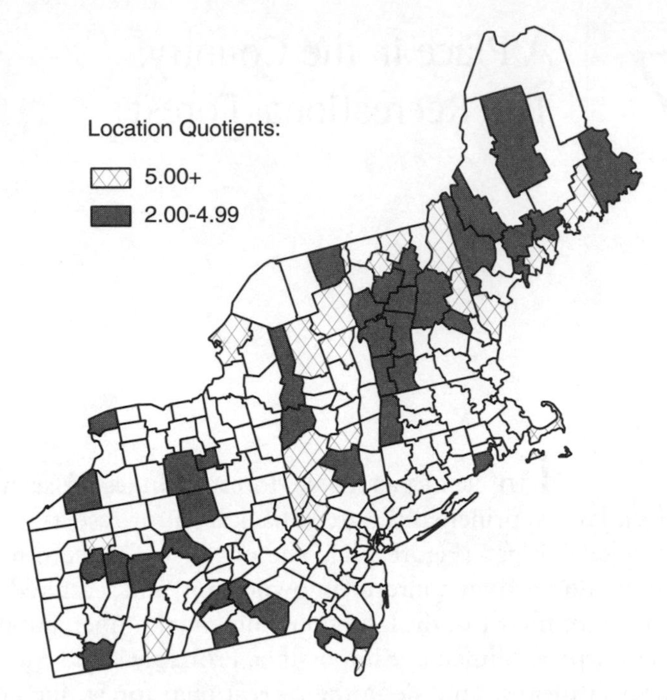

Figure 7.1 Concentrations of Lodging Employment, 1987
Data mapped are location quotients. A county with a location quotient of 5.0 has 5 times as much employment in the lodging industry as the national average.
Source: T. L. Brown and N. A. Connelly, "Assessing changes in tourism in the Northeast," in S. D. Reiling, ed. 1992. Measuring tourism impacts at the community level. *Orono: Maine Agr. Exp. Sta. Misc. Rept. 374 . p. 8.*

Development driven by a mix of recreational demands, suburbanization, and retirement migration can change the landscape quickly. On Cape Cod, intense development pressures caused rapid loss of forest and farmland:

	Percent Change in Forest and Farmland, 1951–1980
Provincetown	–34%
Truro	–39
Wellfleet	–39
Eastham	–76
Orleans	–55
Chatham	–62

Canoeists on the Delaware River, Delaware Water Gap National Recreation Area. *Courtesy of U.S. Department of the Interior.*

In the face of pressures as strong as these, in an area so near to major population centers, this wave of land use change was probably unstoppable.[4]

For generations, families have built remote camps along lakeshores as vacation retreats and for fishing and hunting expeditions. In the twentieth century, families began buying entire farms for rural vacation homes. The overall effects of this activity on the region's rural economy and land use pattern were benign. In the 1960s and 1970s, and later in the 1980s, however, the region endured an explosion in land speculation that resembled such previous bursts as the short-lived nineteenth century sheep boom. A distinctive feature of this new land boom was the sale of land as subdivided lots, usually to distant nonresidents. Many of the sales were made by high-pressure campaigns using television and magazine advertising, touting the beauties of cabin life in the woods. Many played on greed, hawking the high profits to be made from buying now.

The distinctive impact of recreational land booms is well illustrated in the 1970s boom in Vermont, a state that took vigorous public action to control the boom's worst effects.[5]

Changing Land Uses in Vermont

Outdoor recreation is a major Vermont industry, accounting for 15 percent of the gross state product in 1972. The future of this industry is of great economic, social, and environmental importance. In the late 1960s, however, the state's recreation economy began to change in ways that destroyed important social values. In many towns, rural occupations were displaced; land was sold to nonresidents. The future productivity of farm and forest land was seriously impaired by parcel fragmentation and absentee ownership.

The recreation boom repeats a historical pattern of boom and bust in rural land use, as seen in the overextension of lumbering and farming. But the current expansion has created farm and forest land use patterns much less capable of adjustment to meet changing future needs than were previous land use systems. This section was included in *Wildlands and Woodlots;* it still tells a useful story, so I have not updated it to the present.

Agriculture has, until recently, dominated Vermont's rural landscape.[6] Farm numbers and acreage began to decline in Vermont in the late nineteenth century, with abandonments concentrated at first on the poorer soils and higher elevations. The typical Vermont farm during this period included twenty to sixty acres of tillage and pasture, with the remainder in forest.

In 1949, 3.5 million acres, 60 percent of Vermont's area, was in farm ownership. By 1969, farm acreage had fallen to 1.9 million acres (32.6 percent of the state land area) and to 1.5 million by 1992. In 1949, Vermont had 19,000 farms. More than 12,000 disappeared by 1969.[7] These trends reflect the general retreat of marginal farming in Northern New England as well as the new demand for recreational land.

In 1973, at the outset of the recent landboom period, Vermont was 76 percent forested. In contrast, cropland covered only 13 percent of the state's area. The forest has risen in total area, but its ownership composition has changed. Public forest agencies and the forest industry have increased their landholdings, and farmers have reduced their woodland ownership. Other private owners, many nonresidents, increased their acreage by 100 percent in less than two decades. The average parcel size is declining, and turnover in forest landownership is high.[8]

Tourism and recreation are familiar features of the Vermont rural economy. With the advent of the automobile, a significant trade developed based on farmers taking in summer boarders. Improved roads accelerated this process. Recreation brought seasonal jobs, extra income, and improved markets for farm products and property all across Northern New England. This process continued slowly through the 1950s; some areas became concentrations of rural mini-estates owned by the "summer people" from the city.

The town of Stowe, for example, had 88 farms, including 15,000 acres of woodland, pasture, marsh, and tillage in 1940. This number fell to about 30 farm units by 1960, with 7,300 acres of land. During this period, farmland prices roughly tripled, and the brisk market for their land enabled many farmers to leave agriculture. Out-migration, a normal part of Vermont farm life for a century, continued. After 1960, the pace of change accelerated. Only eight farms survived by 1975, covering less than 3,000 acres. Stowe was no longer an agricultural community. Farmland prices rose to nearly $2,000 per acre by the mid-seventies. Woodland prices, which remained stable from 1940 to 1960, thereafter rose dramatically.[9]

From 1940 to 1975, farming declined because of poor land fertility, distance from markets, and low farm incomes. Up to the late 1960s, the recreational land market provided for an orderly succession of land uses that minimized social disruptions. The 1970s land boom, however, involved the active displacement of rural land uses by a new speculative land market. Tillable land, pasture, and forest were all affected.

Recreational forest serves as a green backdrop at saltwater farm, Merrymeeting Bay, Maine coast. *Photo by Chris Ayres.*

Vermont's 1970s Recreation Boom: Trends, Impact, and Community Response

Up to the 1960s, recreational land uses grew slowly, primarily through the sale of farms and their conversion to second homes or full-time residences. But recent land use changes were based on entirely new markets and institutions. The three dominant themes were the growth of ski recreation, the expansion in second homes specifically constructed as leisure units, and the conversion of the land into a commodity, subdivided and sold by experienced marketing firms.[10] These developments increased the economic, social, and environmental impact of recreation on the state.

Skiing is one of Vermont's major industries. In the late 1970s, there were more than 14 million skiers in the country; their numbers have been growing rapidly each year.[11] This growth is expected to continue. In Vermont, the industry has grown steadily. In the 1973–1974 season, out-of-state skiers accounted for 75 percent of the skier visits and 90 percent of skier expenditures.[12] The ski boom brought a shift in the location of the state's recreation industry, away from established lakefront locations and into previously declining mountain villages, although some established resorts such as Stowe shared in the growth.

Based on state surveys, second home numbers in Vermont increased as follows:[13]

	1968	*1973*	*% Change*
Number of vacation homes	22,600	29,600	31%
Vacation home taxes raised (millions)	$4.3	$9.5	121%
Average tax per vacation unit	$191	$322	68%

Second homes are an important part of the tax base in many towns. In 1973, 50 towns derived more than 25 percent of their tax revenues from second homes. A number of towns receive half of their tax revenue from second home property. Many Vermont second homes are owned by nonresidents — 62 percent of them in 1973 (up from 58 percent in 1968). Connecticut and New York residents together owned 32 percent. Second homes owned by out-of-staters, in fact, contributed nearly 7 percent of the total town taxes raised in the state.

The Land Boom in Dover and Wilmington, Vermont

Dover and Wilmington are Southern Vermont towns that illustrate the changes prompted by the new recreation boom. By the mid-seventies, only 18 farms remained in these towns. The recreation expansion based on skiing and the automobile accelerated the removal of land from farming, re-

The recreational forest: ski area, Vermont. *Courtesy of Vermont Department of Tourism and Marketing.*

placing agriculture with an economy dependent on supplying lodging, recreation, and other services to out-of-state residents. In addition, the region has been living off its capital. Local and out-of-state real estate and development firms have been selling land. Many of the customers are nonresidents. In Wardsboro, the town immediately north of Dover, 62 percent of the private land is owned by nonresidents.[14]

According to a survey taken in 1975–1977, after the peak of the land boom, 202,000 acres of forest land in tracts larger than 10 acres were transferred in Vermont, of which 87,000 acres were purchased by nonresidents. This included more than 4,000 acres purchased by owners of record with Canadian or foreign addresses.[15]

Commercial development has been largely oriented toward tourism. Dover and Wilmington in 1973 contained 83 ski lodges and motels, 24 restaurants, and 13 real estate firms. In 1968, the towns contained 609 second homes. In 1973, the number had grown to 1,054. In 1973, second homes accounted for about one-third of total land taxes raised. There are more than 60 subdivisions in the area. Several planned subdivisions have never been offered for sale. Others have been offered and have made negligible sales. In contrast, the large subdivisions, which hired experienced land-marketing firms, have sold relatively well. In many of these, however, few leisure homes have ever been built.

The state's stiff capital gains tax on short-term land turnover was cited as a major reason for a standstill in new subdivisions after 1972.[16] Additional factors are gasoline prices, the effects of environmental legislation, and the prior purchase and preparation for development of most of the area's large available tracts by that time. The slow demand for lots after 1972, and the mid-1970s recession, would have slowed subdivision activity in any case. Since the 1972 peak in land sales, statewide sales of forest tracts larger than ten acres fell significantly, and the acreage involved fell by nearly two-thirds — from 153,410 acres in 1972 to 57,130 acres in 1977.[17]

The 1980s Northeastern Land Boom

The land boom of the 1980s was caused by a brief boom in regional economic growth. It was intensified by the completion of interstate and other major highways that enhanced access to the remote areas of the Northeast. Subdivisions were created in extremely remote areas like Washington County, Maine, and Coos County, New Hampshire, that had been bypassed by earlier recreational lot booms.

Consultants for the Northern Forest Lands Council surveyed parcels 500 acres and larger; the 25 million acre Northern forest lands region of New England and New York was studied. From 1980 to 1991, 532 transactions of that size were found, totalling 7.6 million acres.[18] This total exaggerates the amount of activity, however, as the region's largest industrial ownership, Great Northern Paper Co., was sold twice during the period. Omitting this single property leaves only 2.2 million acres of transfers. The analysts studied the fate of the lands in detail. They found that only about 203,000 acres had been subdivided during these years, from parcels larger than 500 acres initially. Of the acres, 155,000 had already been divided, and 39,000 acres were converted to other uses. These data showed that much of the land remained in longterm forest ownership. While 203,000 acres in 11 years may seem like a small amount of land, it is more significant when it continues decade after decade, and, when it is recognized that the lands involved were probably in scenic, waterfront, and remote areas.

The low rate of land conversion found in the above study is consistent with results of an earlier assessment conducted for the Northern Forest

Lands Study. That survey found that much of the subdividing occurred in parcels initially smaller than 500 acres and often included farmland. Only very small proportions of forest land in 5 sample study areas had been converted to subdivisions, ranging from 1 percent to 10 percent. It was not possible to document wider economic and environmental effects, but the authors strongly suggested that many of these effects would be negative.[19]

Within the Adirondack Park, between 1967 and 1987, the following developments occurred:

- 21,000 new single-family homes;
- 6,500 new vacant lots, including 4,000 on water.

From 1984 to 1989, the number of lots on applications for Adirondack Park Authority (APA) permits tripled, and only about half of the subdivisions were subject to APA review.[20]

Economic and Social Impact

The rapid pace of development, and Vermont's increasing dependence on tourism have prompted heated discussion of the economic and social impact of the new recreation economy. Based on current research, a few tentative generalizations can be made.[21]

Since many ill effects of resort and second home growth are claimed, it is well to note some of the positive effects. Growth in rural towns has enabled some areas to improve the supply of public services, through an increased tax base and growing revenues from nonresidents. In some areas, the amenities associated with skiing and rural life have stimulated settlement by professionals whose services became available to residents and who are often active in civic life as well. These immigrants are often leaders in movements for zoning and planning in their towns. In existing summer resorts, the growth of winter recreation has eliminated winter lulls in activity, leading to less seasonality of employment and more efficient use of social overhead facilities and private capital. Finally, the growing demand for land has helped many a farmer make a comfortable exit from agriculture by selling to a nonresident or developer. The land market has thereby eased the social strains resulting from the steady retreat of agriculture.

Concern has been evident over the effects of the recreation boom on local government costs and revenues. A number of studies suggest that this concern is not warranted.[22] The large direct and indirect additions to the tax base and revenue from nonresident landowners and visitors generally outweigh the costs. Resorts do not place heavy demands on schools. Second homes do not contribute children to the local school system. The distribution of tax benefits, however, is often skewed and inequitable, with a few communities reaping large tax base gains while others receive only higher costs and service needs — the Town Meeting effect in another guise.

The recreation boom, however, has had some serious undesirable effects. Most important are its effects on seasonality of employment in new seasonal resorts, the creation of an uneconomic settlement pattern, the removal of land from productive use, and the creation of an unbalanced economy. Although some communities have reduced employment seasonality through winter recreation, many towns have simply shifted from summer peaks to winter peaks. In Stowe, for example, employment in the first quarter of 1970 averaged 1,097; in the second quarter, the "mud season," employment averaged 622. But employment in many rural occupations was traditionally slack during mud season.[23] Seasonality is a basic characteristic of a resort economy, as are low wages and high employment turnover.

The conversion of farm and forest land to recreational uses proceeded slowly in the first half of the twentieth century. Since it was based on the transfer of farms and large tracts, its impact on settlement patterns and land use was small. But the land boom of recent decades has followed entirely different patterns.[24] Land is cut up into one-acre lots and sold. Roads are cut through wooded hillsides, and condominiums or detached dwellings appear. This process, occurring in a totally unplanned fashion, has created uneconomic sprawl over large areas, resulting in excessive land consumption and high service costs. Additional "shadow conversion" has undoubtedly resulted.

Development has been uneconomic in another, perhaps more durable and important way. The land-marketing industry proceeded with little regard for economically important natural values. As a result, there are countless examples of flood-plain encroachment, water pollution, aesthetic blight, and other lost environmental values. These values are intrinsically important, but they are also the basis of the region's attractiveness to tourists.

The land boom has emphasized trading in small land parcels. From 1968 to 1977, the price of Vermont forested tracts larger than 500 acres rose little, while prices rose significantly for smaller tracts. The average size of parcels sold fell by half in that period — from 80 to 41 acres.[25] The price differentials between land in small and large parcels create irresistible incentives for owners to subdivide. In 1983, 57 percent of Vermont's forest landowners owned less than 20 acres of woodland.[26] Although this is only 4 percent of the state's commercial forest land, it is still more than 150,000 acres of land that will be difficult to manage for tree crops, wildlife, or public recreational uses. Note that this was *before* the land boom of the 1980s.

The land boom has removed large areas of land from commodity-producing and job-producing uses. This reduction has come about temporarily through speculation, in which developers hold land for future sale or development. Frequently, existing productive uses cease for this period. The demand for lots has had another significant effect. High profits to be made by subdividing became an additional incentive for "liquidators" to purchase properties, high-grade them for the most valuable logs, and sell them off in pieces to eager realtors or urbanites. So in these cases, strong lot markets can increase the amount of abusive logging taking place.

The wood processing industry in Vermont has been suffering from declining timber quality, obsolescence, and distance from markets for decades. From 1952 to 1972, almost 300 wood-using plants closed in Vermont.[27] Most employed only a few people. Continued parcel fragmentation, increased absentee ownership, and more rapid ownership turnover make forest harvesting and management less likely on larger areas of land each year. If these trends continue, Vermont's forest resources, which have been growing timber at twice the rate of harvest, will not contribute their full potential toward employment and economic health for rural communities. Especially where there is overdependence on recreation, wood products jobs will continue to be important.

The invisible processes of subdivision and parcel fragmentation, which may not change surface features at all, can permanently remove land from forest, public recreation, or farm uses because of the prohibitive cost of reassembling economically operable units from one-acre lots owned by residents of a dozen distant cities. These land parcels could never, by reason of their shape or size, be economically useful for any other purpose. Subdivision regulations and environmental restrictions have encouraged this perverse process, by setting large minimum lot sizes. Oddly shaped properties created with long, narrow sections to allow minimal road or water access are common in the Vermont landscape.

The removal of land from productive uses for farming and forestry has been widely blamed on the rising taxes per acre that accompany rural recreation developments. This view seems clearly erroneous for most Vermont recreation areas. The land boom does destroy existing productive land uses, but it does so by paying high prices for land and by subdividing it. The "push" effect of rising taxes is much less important than the pull of high demand for land.

Finally, the recreation land boom has created a new and unbalanced economy for many Vermont towns, and, indeed, for the whole state. The region's history of dependence on one-industry towns seems destined to be repeated in the recreation-oriented villages of Vermont. In Vermont, poor snow years or gasoline shortages can create serious hardship. For example, state surveys estimate that skier outlays in Vermont fell from a peak of $54 million in 1971–1972 to $35 million in 1973–1974, due to poor snowfall.

A study of Appalachia's economic development seems relevant for Vermont. The study concluded that because of its seasonality and low-wage workforce, "recreation alone can almost never provide a base for a viable economy."[28] The recreation boom in Vermont has been overextended. Instead of providing a valuable addition to a diversified economy, it has replaced one kind of imbalance with a new imbalance, which may prove much more difficult to correct.

In extreme instances, overdevelopment of recreational subdivisions, resorts, and condo complexes has led to gentrification. High land prices push local residents, and tourism industry workers, out of town and into outlying areas or marginal housing situations. Labor shortages can affect traditional industries. Longtime residents can feel marginalized as wealthy newcomers, controlling large tourism-based businesses, and middle-class retirees begin to assume more and more power in Town Meeting, in state capitals, and behind the scenes.

Response: Social Innovation in Planning

The development brought on by the state's recreational expansion prompted state legislation to establish a planning program. Vermont now possesses a land use control program that incorporates statewide performance standards while assuring a degree of local autonomy and flexibility.[29] This program consists of a water pollution control program (Act 252), a land use regulation system (Act 250), and local and regional planning.

Act 250 created the State Environmental Board and a series of regional environmental commissions. It also established a detailed system of performance standards that must be met by all large developments. Each district commission holds hearings on proposed developments, following examination of the site by the county forester. The commission has the power to reject an application or to authorize the project with modifications. Decisions may be appealed to the State Environmental Board. Among other provisions, the act provides strong protection for all lands above 2,500 feet in elevation. Finally, the Vermont statutes authorize the creation of local planning commissions, boards of adjustment, and regional planning commissions. After the mid-1980s boom had hit its peak, the Vermont Legislature enacted Act 200, which strengthened the plan-

In some towns, skiing, land sales, and the land have coexisted relatively well. Stowe, Vermont. *Courtesy of Vermont Department of Tourism and Marketing.*

ning system and attempted to gain a better representation of fair housing issues in the planning process.[30]

State-level controls have been supplemented by vigorous local activity in many Vermont towns. At a minimum, towns struggled to remain independent of state rules and district environmental commissions. Some went beyond this, however, to install planning and zoning boards, to draft land use plans, and to establish subdivision ordinances and building and plumbing inspection programs. The debates over these innovations often divided residents along factional lines of "old-timers" versus "newcomers," or farmers versus tourists. However, it is now widely accepted that local government cannot ignore land use control and regulation. The state-mandated program of environmental controls is seen by many as a shift of power to Montpelier, but the new programs in fact allow, for the first time, a measure of local public influence on developments affecting a town's future.

The challenge of recreational lot subdividing also prompted major land use enactments in New York's Adirondack Park, in Maine, and in the

A Common Public and Private Stewardship: The Adirondack Park

The open spaces of the Park — its forests, fields, mountains and lakes — are its most important asset, tangible as well as intangible. They assure the quality of the natural and economic systems of the Park; to diminish them means to diminish forestry, agriculture, and much of the opportunity for the pursuit of outdoor recreation.

It is vital to the economy of the state and the Park that these spaces, both publicly-owned (the Forest Preserve) and privately-held (more than half the land within the Park boundary), remain open. It is vital to the quality of life of those who live within the Park and those who visit it. It is vital — it is paramount — in any responsible vision of the Park in the twenty-first century. Yet this open space is by no means guaranteed for future generations.

The open space, today much more so than 20 years ago, is subject to the economic pressures on the lumber and forest products industries, to the tax pressures on estates, to the hunger in the immense, affluent population centers of the Northeast and Canada for a house on the lake, a camp in the woods. These pressures may ebb and flow but they never die, and they will lead inexorably to the fragmentation of privately-held open space.

The Adirondack Park Agency was created two decades ago to cope with this pressure — primarily to save the privately-held forests for both the woodsman's saw and the call of the loon. But the tools it was given cannot prevent fragmentation of the forests. They can only determine the size of the fragments into which the Park's forest land and open space will be sliced.

What is most needed to keep the Adirondack Park a great and true park in the twenty-first century, then, is a strategy to thwart the subdivision of vast tracts into little ones, to keep the forest industry viable and thereby preserve the Park's most fragile asset: the open space of its forests and fields outside the state-owned Forest Preserve. In this report the Commission offers such a strategy, its Open Space Protection Plan.

The plan would keep almost half the Adirondack Park in productive farm and forest use. It would provide tax abatement for owners of such lands, and tax reimbursement to local taxing jurisdictions. It would permit owners of modest tracts to build their dream homes, but would steer the main development pressure into areas already designated as most suitable. It would preserve the appeal of New York's wilderness vacation land.

. . . The Commission urges the addition, over time, of certain parcels to the Forest Preserve for specific purposes: to preserve biological diversity, to round out wilderness areas so all Adirondack ecosystems are represented, to enhance Wild Forest recreation opportunities. The total of such additions, some 650,000 carefully selected acres, would bring the state-owned Forest Preserve to just over 50 percent of the Park.

Source: Commission on the Adirondacks in the twenty-first century, 1990. *The Adirondack Park in the twenty-first century.* State of New York. April. p. 11–12.

New Jersey Pinelands.[31] It also helped motivate the passage of major land acquisition bond issues in most states in the Northeast.

Impacts of Wildland Subdivisions

The 1980s land boom took on a new and different focus than the 1970s boom, which focused on established resort areas. In the 1980s, the most remote forest areas became marketable for the first time. Completion of the Interstate Highway system in the 1970s and 1980s produced a dramatic improvement in the convenience of reaching Eastern Maine, Northern New York, and Western Pennsylvania. This improved access, combined with the remoteness and low cost of forest land in the corners of the region, led to a new development never seen before in the region's wildlands: the recreational lot hawker, subdividing wildland lakeshores and riverfronts in areas reached only by gravel roads. This industry found a great marketing advantage in the hunger of urban Americans for a piece of the outdoors, and in the widespread hope of "making a killing" in land. The gaps and loopholes in subdivision regulations and zoning in these areas simplified the wildland lot-hawking business and increased its profitability. A few aggressive companies were even publicly traded on the stock exchanges.

The recreational forest was spreading into the industrial forest, entering a region where its effects on land use and environmental values were multiplied many-fold in comparison to the traditional recreational areas which were, often as not, abandoned farms.

The first effect of wildland subdivisions is to *privatize* wildness, to convert the enjoyment of vast wild lakes and forest vistas into preserves for the privileged few. There has been a rich tradition in the woods of the Adirondacks, where wealthy families enjoy their "Kingdoms" and hunting clubs their exclusive leases. Similar exclusive arrangements exist throughout the region, from duck hunting leases along Chesapeake Bay to exclusive use arrangements in the Maine Woods. These large-scale preserves have helped maintain wildness and its habitat values, in contrast to the subdivisions. A small recreational subdivision on a remote lake or stream may do far more than privatize the access and the scenery. It can also effectively grant preferred use rights to local fishing and hunting opportunities, foreclosing the option for former visitors from long distances. The occupants of new camps on a remote pond can fish it out in a few weekends when it formerly was a treasured retreat for users who had to hike in before the shoreline was sold and subdivided.

Perhaps more serious than privatization is the fact that wildness itself is inherently fragile. Wildness offers perhaps the ultimate example of shadow conversion. The shadow conversion multiplier for wildness is surely far greater than it is for farming and forestry. When a subdivision occupying 5 percent of a lakeshore is built out with camps, its owners enjoy the view across a wilderness lake. But campers on the opposite shore see the lights well into the evening. Thus have the new owners of 5 percent of the shoreline destroyed the perceived wildness of the remaining 95 percent.

Many of the land sellers took advantage of loopholes in the subdivision laws "big enough to drive a log truck through," that left large lots unregulated. In these instances, faulty public policies contributed to the loss of significant resource and environmental values. In many instances, the wildland subdividers took the opportunity to liquidate merchantable timber in order to maximize profits on their short-term investments.

Reflections

A social tragedy is unfolding in the forests around major recreation areas as well as everywhere the land sales operators have been at work in the remote north woods. The land market, weak or even perverse public policies, and private financial pressures have caused a maximum loss of wildlands values and future productivity, in return for negligible social gains and ongoing public costs.

That recreation was potentially destructive of important values was shrewdly forecast by John Wright, editor of a major survey volume on New England, in a 1933 essay:

> A surging tide of humanity sweeps out of the cities . . . If allowed to take their own course without control these floods are a menace. There is a real danger that they may submerge in a welter of ugliness the little that remains, at least in the environs of the larger centers, of the Old New England of quiet villages and tranquil vistas.[32]

This prophecy is amply fulfilled along countless miles of roadside near Northeastern seacoasts and mountains. In fact, because of sprawling growth around its towns and villages, in 1993 the National Trust for Historic Preservation put the entire state of Vermont on its "Endangered List."

Tourism, of course, has been important in the Northeast for more than a century. Until the 1960s, tourism and recreation played an important role in easing the social transition away from an overextended agriculture. The demand for land and employment opportunities created by tourism assisted individual and community adjustment to a declining farm economy. But the new recreation landsales industry of the 1970s and the 1980s, based on capital-intensive resort development, rapid sale of land

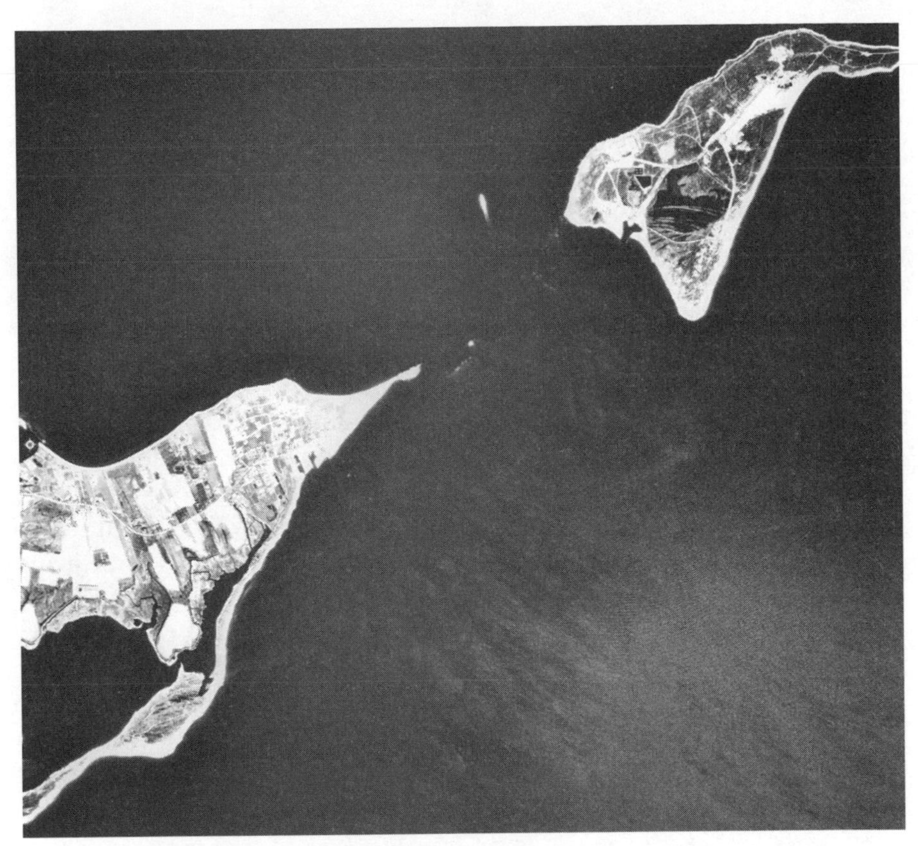

Orient Point State Park and Plum Island, New York (Long Island).
Courtesy of U.S. Geological Survey.

as an investment commodity, and construction of second homes, has created a new form of overextension. The recreation boom has created an economy of new one-industry towns, based on an industry that economists agree to be a poor community economic base. Much research to date has addressed narrow issues of fiscal cost-revenue analysis; little attention has been paid to these broader issues.[33]

There is a full research agenda here for social scientists and planners. What kinds of policies have served well in controlling undesirable side effects of the land boom, and what policies have failed? What policies have made it worse? How these policies work is a case study in Town Meeting government.

The Northeast's rural economy has adjusted to change as past booms subsided. Adjustments have been facilitated by an industrious population, by a flexible system of land use, by a resilient ecosystem, and by the opportunity for out-migration. But the current boom has created a land use system in the recreational forest that is far less flexible. Large acreages are in unsold or partially sold subdivisions, which will probably never be fully "built out." The result is a patchwork of ownerships, consisting of parcels too small for economical farm or forest management. This will preclude their use for producing wood for building products or energy, and food and fodder for humans and livestock. It also may effectively end their availability for hunting, fishing, walking, snowmobiling, and cross-country skiing. From the Eagle's Eye View, the Recreational Forest has left behind many problems.

Nowhere is the cumulative effect of individual subdividing decisions being tracked and evaluated. Studies done for the Northern Forest Lands Study and the NFL Council to assess the situation lacked completeness and regional coverage. They are in any case no substitute for ongoing tracking of land market activity. In our region's Town Meeting culture, how the future use and development of land will evolve in the future remains an open question. Probably, few policy improvements will be made in a decade that is obsessed with "takings" and past excesses of regulation. Ironically, the slower pace of subdividing in the mid-1990s has taken the edge off of the sense of crisis that is usually required to deal with tough policy choices.

The future of the recreational forest is uncertain, but the desire of suburban families for a place in the country, or a condo by the ski area, assures it a continuing significance in the region's life.

8 The Northeast's Wild Forest

SINCE THE DAYS of Thoreau, Northeasterners have appreciated Wild Forest landscapes. They pushed for state forests and parks as early as the 1880s, with the citizen-led effort to create the Adirondack Park.[1] Local citizens pressed for federal acquisition of land in the White Mountain National Forest. Maine contains a splendid example of a privately created wilderness — Baxter State Park. Although little virgin forest remains in the Northeast, the forest has regrown so completely that few hikers realize that many a trailside vista was once farmed or abusively cut, and often burned (Table 8.1). The few remaining scraps of virgin forest are mostly outside the region's large, formally designated wilderness areas.

The boundaries of the Wild Forest are difficult to define. The Wild Forest includes a range of forest lands in public and conservation ownership whose primary objective is the maintenance of natural conditions. There are individual private parcels that fit this definition, but we know little about them. Defining the Wild Forest in strictly ecological terms as vestiges of truly undisturbed forest is not useful.[2] Large acreages of the Wild Forest are devoted to such specialized uses as watershed protection. Perhaps a half million acres of such lands are found in the region. Since some of these lands are available for limited timber cutting, they fit no rigid definition of wilderness. But since they form large green blocks in the midst of cities and suburbs, they fill most of the functions of the Wild Forest.[3] Likewise, the tiny parcels owned as greenspace or preserves by towns and nonprofit groups are so small that they could not accommodate backpacking trips, yet they also are part of the Wild Forest, protected from development, logging, and often from hunting as well.

A listing of major Wild Forest units can be made, however (Table 8.2). This listing includes major federal and state areas that are specifically identified as having a wild status, including research areas. State parks and

TABLE 8.1. Acreages of True Virgin Forest, Selected Northeastern States

	Acres	No.of Units	Limitations
Maine	1,553	15	State lands
Massachusetts	600	28	Berkshires
New York	90,000	10	Adirondacks
Pennsylvania	26,000	n/a	Statewide

Sources: Maine State Planning Office, *Uncut timber stands and unique alpine areas on state lands,* Augusta: Executive Dept., 1986; P. W. Dunwiddie and R.T. Leverett, *Survey of old growth forest in Massachusetts.* Rhodora 98: 419–444, 1996; D. J. Leopold, C. Reschke, and D. S. Smith, Old growth forests of *Adirondack Park, New York,* Nat. Areas J., 8(3): 1866–1879, 1988; and T. L. Smith, *An overview of old growth forests in Pennsylvania,* Nat. Areas J., 9(1): 40–44, 1989.

game lands are not listed, nor are extensive areas of forest land that is unlikely to be managed at all due to slope, swamp, or other limiting physical conditions. All such lands would be considered a part of the Wild Forest.

All told, the region's Wild Forest comes to roughly five million acres of land, mostly forested. This includes more than a million backcountry acres on the National Forests, fish and game lands, state parks, and private nonprofit reservations that will retain a generally unmanaged character. This total comes to about 5 percent of the region's land area. The Wild Forest is a product of the aspirations of the region's citizens for greenspace, for a place to stroll in quiet woods, and for conserving natural processes. The Wild Forest is a hallmark of the region's land use pattern and quality of life. As such it is a resource of national significance. Preserved wild areas are the essential core for sustaining biodiversity of habitat for certain wildlife, and for the wider values of "wildness" associated with the region's more remote forests.

Since most of it is in such tiny units, the Wild Forest is not easily mapped. Except for the Adirondack Parks' Forever Wild lands, several National Forest wildernesses, and Maine's Baxter Park and Allagash Wilderness Waterway, few of the designated units would make more than a hefty dot on a small map of the region (Figure 8.1).

As ecological knowledge has improved, it has become clear that property lines often fit poorly with the needs of resource protection and management. Silt from sloppy logging on side streams can damage the clarity of water and spawning habitat on a wild and scenic river; roads and sub-

TABLE 8.2. Examples of Designated Wild Areas, Northeast, Early 1990s

State	Federal	State or Other	Acres
Maine	White Mountain NF*		12,000
		Baxter Park	203,000
		Allagash WW	100,600
		BPL	44,000
	National Wildlife Refuge		7,392
New Hampshire	White Mountain NF		102,932
Vermont	Green Mountain NF		59,598
Massachusetts	National Wildlife Refuge		2,420
		Quabbin Reservation	50,000
New York		Adirondack Preserve	2,310,000
		Catskill Preserve	275,000
Pennsylvania	Delaware Water Gap NRA (also N.J.)**		69,629
		State Forests, etc.	169,000
	Allegheny NF		8,938
New Jersey	National Wildlife Refuge		10,341
TOTAL			3,424,850

*On the National Forests, additional research natural areas are part of the Wild Forest.
**Includes easements on private land and 7,600 ft. of nonfederal public land.
Sources: Agency reports and documents of various vintage.

divisions can restrict wildlife use of habitat they need outside of game areas, and so on. A new focus on "ecosystem management" is arising, based on the metaphors of the Worm's Eye View and the Eagle's Eye View of the landscape. In this perspective, supporting the broader functions of the Wild Forest will take a whole new approach to land management, planning, and environmental regulation.[4]

Mount Katahdin, centerpiece of Maine's Baxter State Park, part of New England's Wild Forest. *Photo by Chris Ayres.*

Services and Values of the Wild Forest

There is a common view that wildlands placed in reserved areas are "locked up" providing nothing of benefit to society, only quiet preserves for a few hikers. Northeasterners have long known better. When state government action seemed inadequate, private organizations took to the land acquisition business themselves with a vengeance. Much of New England's Wild Forest is a direct result of active private effort, not just of interest groups lobbying government legislatures. In contrast, governmental ownership and policies played a major role in the Mid-Atlantic states.

The citizens who promoted the creation of the region's Wild Forest did so — and still do so — from a variety of motives, from crass to idealistic. In some instances, people sought to create preserves and parks to conserve peace, quiet, and undeveloped views. The most conspicuous example is Acadia National Park, donated to the federal government by wealthy Mount Desert vacationers led by the Rockefeller family. Although in a family tradition of magnificent conservation gifts (Smoky Mountains, Virgin Islands, and Grand Tetons National Park), the gift certainly helped real estate values and kept out development.

Large portions of the Wild Forest were created for utilitarian conservation purposes — to preserve game, fish, and clean water supplies or to conserve channel storage and prevent flood plain encroachment. Protecting water supplies was a major argument for federal acquisition of the White Mountain National Forest and for creating the Adirondack Park. Recreation, birdwatching, and open space values have been high on the list of objectives in virtually every instance. Finally, the pure "preservation" motive, best expressed by Baxter State Park, is seen in dozens of the tiny parcels of woods and marsh held by The Nature Conservancy and other private nonprofit groups. Whatever the motives of the private or governmental groups involved, preserved wildlands serve a broad range of social values, which may be outlined as follows:

1. Scientific
    a. Preserving key ecosystems to sustain biodiversity.
    b. Conserving gene pools and potentially useful organisms.
    c. Providing natural areas for research and monitoring ecosystems.
2. Economic
    a. Backcountry recreation.
    b. Conserving wildlife and fish.
    c. Protecting watersheds and water quality.
    d. Conserving scenic resources to benefit tourism.
    e. Enhancement of nearby real estate values.

Allegany State Park Master Planning Process: Issues and Experience

Setting

Allegany State Park occupies approximately 100 square miles of Cattaraugus County in Southwestern New York. Approximately ninety-five percent (62,600 acres) of the parks 67,000+ acres is state owned. It is estimated that over forty percent of mineral rights beneath the public lands are held by private interests.

The park's location is geologically unique since it was not covered by the last Continental Ice Sheet. Hence, the primary features of the park were not modified by ice erosion or morainal deposition. The region's valleys, wooded slopes and meandering streams create an area of striking beauty.

Significant and Unique Natural Resources

A major rationale for providing natural areas within the park is the contributions of such areas to the diversity not only of the park but of the entire region. The character of the park forest is different than that of the surrounding region due to the relative lack of large scale timber harvesting since the park was created.

. . . the park harbors the largest publicly owned, contiguous block of forest in Western New York and provides important habitat to species of mature forests and to those requiring large home ranges or isolation from human activities. Promotion of the increase of successional and open site species, which have abundant habitat in the region surrounding the park, while reducing habitat of mature forest species could lead to a reduction in the regional diversity of wildlife.

. . . diversity of the park could be increased through active management without adversely affecting regional diversity as long as certain guidelines are followed.

- leaving all existing old growth forests undisturbed;
- allowing forests on slopes greater than 30% to develop into old growth;
- leaving sufficient buffer to prevent edge effects from degrading conditions in the interiors of the mature stands;
- allowing old growth stands to function as natural systems;
- elimination of exotic species;
- reduction of the deer herd; and
- designation of preserves in the following areas: The Big Basin area (entire), southwestern oak and mixed mesophytic forests and the area lying south of Quaker Run from Mt. Tuscarora to Bear Bog.

Designation of some areas of Allegany State Park as natural areas is consistent with the current management of most New York State Parks. Since these areas are not actively managed for wildlife or timber, they are, in effect, natural areas, although they may not be formally designated as such. The only forest management activities occurring on a regular basis in most state parks are fire control, hazardous tree removal, scenic vista maintenance, insect and disease control and mowing (e.g., along road shoulders). Thus, the forests within state parks are basically allowed to undergo natural succession. . . .

Forest Management

. . . The draft plan calls for placing 70% of the park in some type of natural area designation. A forest inventory conducted in 1974 recognized five forest types. . . .

. . . The most obvious change in the forest at Allegany State Park is the loss through natural succession of early successional forest. Approximately 17 percent of the land evaluated in 1937 was Aspen type. . . .

The need to manage the forest resource at Allegany State Park is centered about two basic concerns, 1) wildlife habitat and management, and 2) long term perpetuation of the forest and its composition. . . .

Wildlife Management

It is estimated that 361 species may occur on Allegany State Park. Information on their abundance and distribution is limited. The total number of species by major groups are as follows:

Estimated number of species in Allegany State Park by major groups

Mammals	49	
Birds	242	(107 residents, 135 migrants)
Reptiles	16	
Amphibians	24	
Fish	30	

. . . A decision to do no wildlife habitat management at Allegany could lead to a reduced number of wildlife species. The management decision to create and maintain all forest growth stages over the entire park may maximize wildlife abundance and diversity, but would significantly reduce the opportunity to provide large contiguous blocks of mature and old growth forest conditions which have regional significance and could also lead to a reduced number of wildlife species.

Source: New York Office of Parks, Recreation and Historic Preservation, 1993. *Allegany State Park master planning process: issues and experience.* In USDA-FS, NEFES, Proceedings of the 1993 Northeastern Recreation Research Symposium, Saratoga Springs, New York, April 18–20, 1993. Gen. Tech. Rept. NE-185. p. 13, 16–17, 19–20.

f. Avoiding costs of development (services, pollution, congestion).
g. Promoting a balanced land use pattern.
3. Cultural
a. Conserving a Cultural Heritage.
b. Preserving aesthetic values.
c. Providing educational opportunities.
4. Ethical
a. Providing for biodiversity and preservation of natural processes (existence value).
b. Providing scope for individual freedom.
c. Social value of exercising restraint.
d. Bequest motives . . . bequeathing wilderness to the future.

Most of the values benefit individuals who never visit the areas involved. It is also true that parcels of well-managed forest can provide most of these services — it is a matter of degree and of which services hold priority.

One value of Wild Forests not often discussed is the promotion of a balanced land use pattern. It is not easy to define a balanced land use pattern, but it is easy to cite examples of imbalance. Every reader probably has a favorite example, but I think long stretches of U.S. Highway 1 will do well — uninterrupted stretches of shabby commercial strip development. Furthermore, I assert that the large acreages devoted to development each year should be balanced by proper commitments of land for open space, recreation, water supply protection, and aesthetic relief from visual blight.

If we can afford to pave and develop millions of acres in the next few decades, can we not afford to preserve for the future a corresponding undeveloped acreage? Much of this should be in working farms and carefully managed forests and not necessarily in classic wilderness. An example would be the debate over New York's Allegany State Park (box), in which a professionally prepared plan made a case for a measure of active management. Under heavy political pressure, the Governor vetoed the plan in favor of a preservation option. Choices of this sort will arise with increasing frequency. In my view, opportunities to use light-on-the-land management should not be regularly and lightly tossed aside.

An important shift is occurring in the balance between these values. New York's Adirondack Park and New Hampshire's White Mountain National Forest were originally conceived by people with a strong appreciation for the values of nature. Yet their primary public policy justifications were instrumental ones — they focused on the use values of tourism, game, and water supplies and the need to prevent destructive forest fires.

Even today, much of the research and management attention is devoted to use values. Existence value refers to the fact that people want to know that some wild land survives, even if they never see it.[5] Bequest values refer to the importance of wildlands as a Bequest to future generations. It is becoming ever clearer that the existence values and Bequest values of wild areas are more important to most people than the use values or instrumental values. One would suspect that this is so since relatively few people actually visit these places, yet a concern for their preservation is widely shared.

Recreation in the Wild Forest

The best publicized use of the Wild Forest is recreation: backpacking, mountain climbing, cross-country skiing, snowshoeing, and fishing. In the scattered patches of Wild Forest near the cities, day walks, picnics, and nature walks are popular. The region's organized hiking community takes national leadership in the efforts of the various trail and outing clubs, organized under the Appalachian Trail Conference, to maintain hiking trails. The volunteer efforts of these groups give the trails their distinctive character.[6]

Nationally, use of backcountry areas rose rapidly, at a rate of 7 percent per year from 1960 to the late 1970s, then paused for much of the 1980s. National surveys estimate that 5 to 10 percent of the adult population participate in backpacking and in "dispersed camping." The USDA-FS expects the use of wilderness areas to grow.[7]

Campers enjoy *Twilight in the Adirondacks,* Sanford Robinson Gifford (1864). *Courtesy of The Adirondack Museum.*

Data for Northeastern wild areas roughly reflect these trends.[8] Figures collected by the Appalachian Mountain Club and the USDA-FS show that use levels peaked earlier in major public wild areas. In a few instances, this may reflect crowding and the imposition of rationing or reservation systems at some areas. Or it may represent a market saturation for back-country recreation, together with the aging of the baby boom generation. Walking and related activities popular in the wild patches are enjoyed by 50 to 60 percent of the population.

According to the 1982–1983 National Outdoor Recreation Survey, Northeasterners participated in a number of major activities relevant to the Wild Forest:

Canoeing/Kayaking	11%
Backpacking	3%
Primitive Camping	10%
Birdwatching/Nature Study	15%
Cross Country Skiing	5%

These activities attract a smaller proportion of the population than do many others, but on the region's population base, the number of participants is large.[9]

The heavy use pressures on the Wild Forest create overcrowding and reduce the enjoyment of users. Trails are beaten down, fuelwood disappears, and litter proliferates. Tenting sites are increasingly threadbare. Forest and park managers have resorted to limits on party sizes; first-come/first-serve reservation systems; carry-in/carry-out programs; and intensified maintenance, management, and patrolling.[10]

The cost of these management programs is becoming an important issue. Costs directly attributable to backcountry recreation management on major public areas are large. The users, who benefit the most — a small minority of citizens — and whose activities occasion these outlays, contribute little. Costs per user-day of $3 to $6 are common. Since recreation budgets in public agencies are facing budget cuts, the issue of higher user contribution to management cost will surely concern managers and recreationists in the future.[11]

Adirondack Wilderness

The concept of wilderness in the Northeast was pioneered by the advocates of the Adirondack Park. In 1827, Governor DeWitt Clinton told the New York legislature that future generations would regret the squander-

ing of the forests. In 1864, George Perkins Marsh advocated a major reserve:

> It has been often proposed that the State should declare the remaining forest the inalienable property of the commonwealth, but I believe the motive of the suggestion has originated rather in poetical than in economical views of the subject. Both these classes of considerations have a real worth. It is desirable that some large and easily accessible region of American soil should remain, as far as possible, in its primitive condition, at once a museum for the instruction of the student, a garden for the recreation of the lover of nature, and an asylum where indigenous tree, and humble plant that loves the shade, and fish and fowl and four-footed beast, may dwell and perpetuate their kind, in the enjoyment of such imperfect protection as the laws of a people jealous of restraint can afford them. The immediate loss to the public treasury from the adoption of this policy would be inconsiderable, for these lands are sold at low rates. The forest alone, economically managed, would, without injury, and even with benefit to its permanence and growth, soon yield a regular income larger than the present value of the fee.
>
> The collateral advantages of the preservation of these forests would be far greater. Nature threw up those mountains and clothed them with lofty woods, that they might serve as a reservoir to supply with perennial waters the thousand rivers and rills that are fed by the rains and snows of the Adirondacks, and as a screen for the fertile plains of the central counties against the chilling blasts of the north wind, which meet no other barrier in their sweep from the Arctic pole.[12]

By the late 1870s, New York had sold all but some 700,000 acres of state lands. The fires, erosion, and bare soil left by logging had enraged citizens and worried businesses who depended upon tourism and on the flows of water in the Hudson and the Erie Canal. A group of early conservationists began a crusade to preserve the region for the future and to re-establish a large domain of publicly owned lands as its core. Since the brief period of private ownership had yielded such destructive results, Park advocates sought to have these public lands maintained "Forever Wild." Since they were almost entirely cutovers, often burned, there was no noticeable sacrifice of current timber values or employment involved in this decision.

As early as 1872, a state commission was established to look into establishing a park. In Sargent's report in the 1880 Census, his correspondent in New York estimated that the remaining virgin growth in the Adirondacks covered about 1.6 million acres.[13] In 1885, a Forest Commission was created to administer the Park and acquire lands. The Forever Wild character of state-owned lands, and the "Blue Line" surrounding the entire Park were made a part of the State Constitution by public referendum. By 1908, state ownership had increased to 1.6 million acres. By the 1980s, the Forest Preserve acreage in the Adirondack and Catskill Parks exceeded 2.5 million. These acreages are not contiguous, and from time to time land exchanges and additional purchases are conducted to block up existing units, protect important areas, and render public access and management more effective.

A century later, the Forest Preserve lands of the Adirondacks contain stands that have had a century or more to re-grow after logging. These lands provide watershed protection, recreation, wildlife habitat, and other values of incalculable importance to the region. Another important contribution of the Park has been the state's leadership in being one of the first states in the nation to embark on a program of conservation land acquisition.

Late in the nineteenth century debates over the Adirondacks, a movement arose to preserve parts of the Catskills. By then, the area had been so heavily cutover that scientist Charles Sprague Sargent felt that there was not enough forest left to warrant a preserve there. The Catskill Preserve was created, with its core consisting largely of cutover lands left to the state for nonpayment of taxes.[14]

Baxter State Park

The region's most splendid example of private initiative is Baxter Park, which protects Mount Katahdin, the northern terminus of the Appalachian Trail. The park was purchased over a period of years by the late Governor Percival P. Baxter and donated to the state. In making his donations, the governor asked the legislature to accept the deeds by resolves, to give them the effect of law. He donated a substantial endowment for management expenses and specified that the park be administered by an Authority, consisting of the commissioner of Inland Fisheries & Wildlife, the director of the Bureau of Forestry, and the attorney general. These officials provide broad policy guidance to the staff. Strictly speaking, the 200,000 acres of the park are not state land, but are held in trust by the state for the people of Maine.

The authority has recently engaged in complex management planning for the park. It has at times been embroiled in litigation over proposals to salvage blown-down spruce trees and to spray for budworm control, and over fire control procedures. The fire controversy erupted after an intense 3,500-acre fire in August 1977, which was contained by conventional

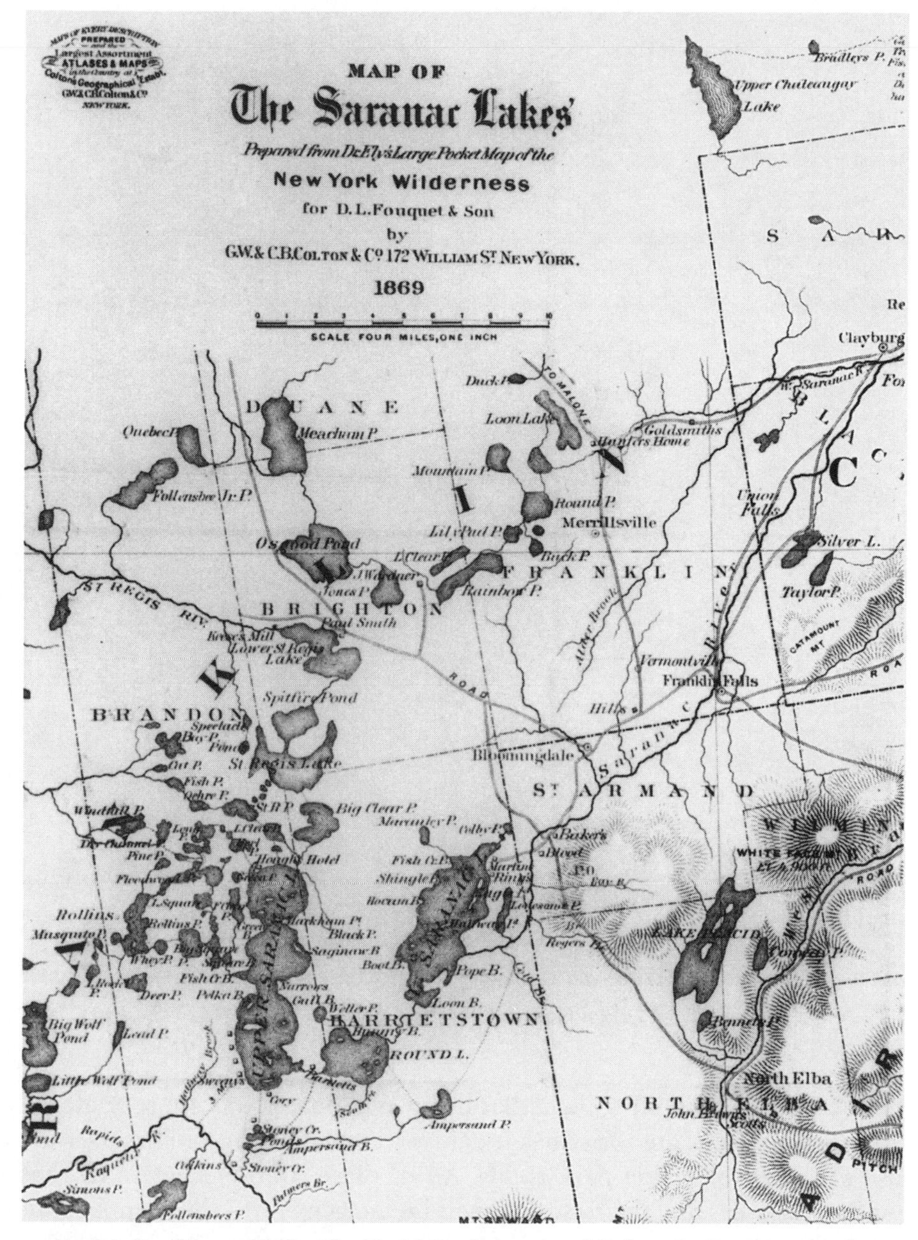

Eagle's Eye View of Lake Placid, 1869. Courtesy of Colton & Co., New York.
Reprinted by permission of Forest History Society.

bulldozed firelines. That fire started in blown-down timber and damaged a sizable acreage of nearby paper company land, consuming a nice crop of postcut regeneration. Critics questioned whether the fire should have

Eagle's Eye View of Lake Placid, 1985. *Courtesy of U.S. Geological Survey.*

been controlled, and if so, whether heavy equipment was appropriate. In addition, the park authority has fought with users and neighbors over the use of snowmobiles on park roads. All of these controversies show that wilderness areas and their users demand management — they cannot simply be left alone.

Governor Baxter specified that the park be managed in its natural state, as a "sanctuary for birds and beasts." He modified this mandate in one corner, where hunting was permitted, and in another, where "scientific forestry" was to be practiced.[15] Baxter Park remains New England's largest dedicated wilderness, and it may always be. Its size dwarfs the entire acreage of federal wilderness in New England. It sprang from the

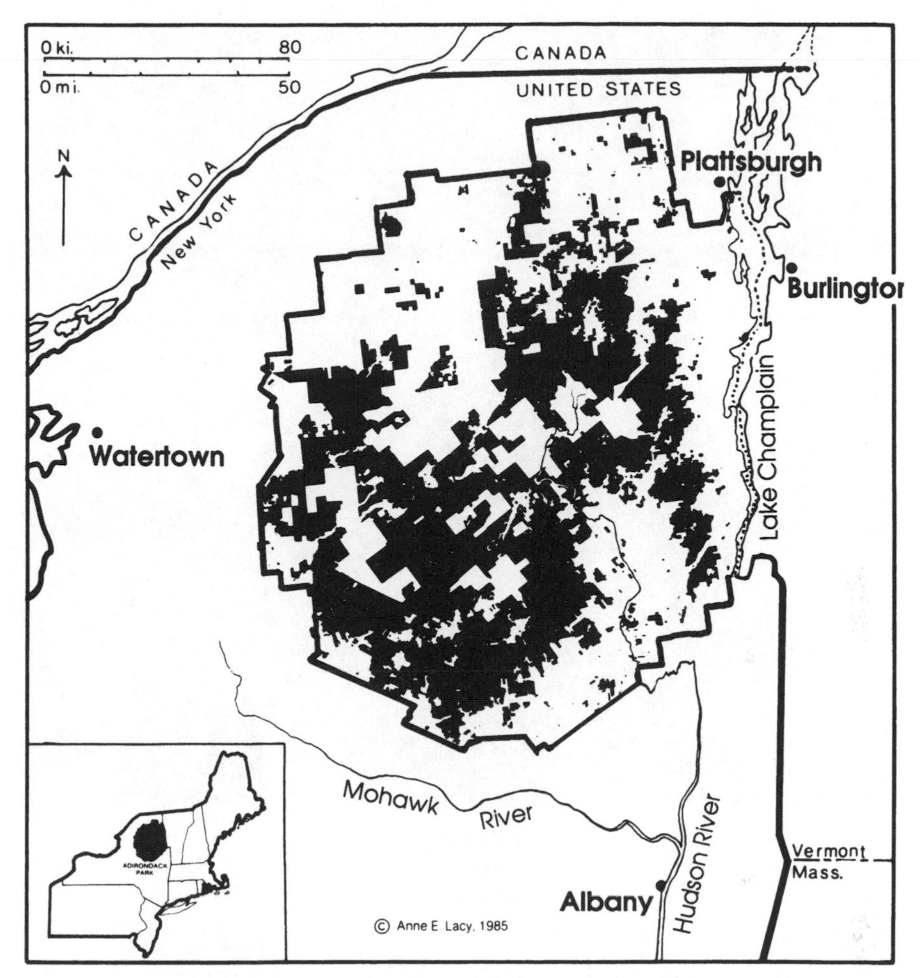

The Adirondack Park, which is a patchwork of public (black) and private lands, contains six million acres within its "blueline" of which 58 percent is privately owned. *Courtesy of Adirondack Council.*

conviction of one man, and not from a loud controversy, as did other New England wild areas. There was no "Save Mount Katahdin" lobby.

Federal Wilderness

The best-known elements of the Wild Forest are the federal wilderness units, which totaled some 200,000 acres in 1996. These units owe their origin to the establishment of major federal reservations — National Forests and wildlife refuges. These were intended to protect watersheds and timber supply or to shelter wildlife.

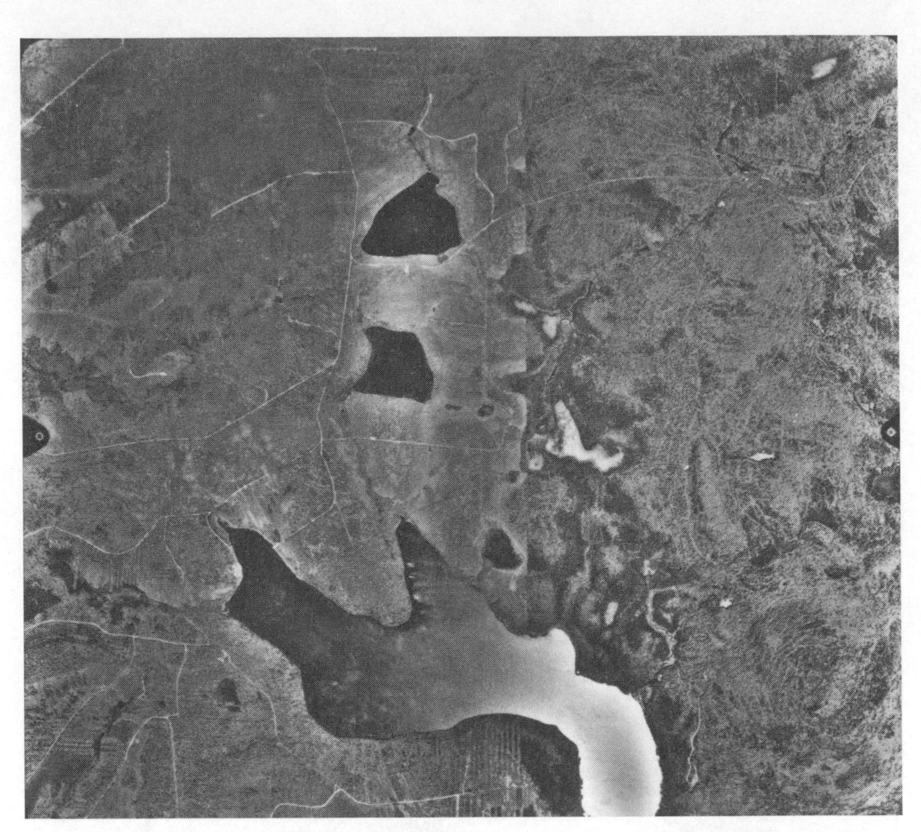

Boundary line of Baxter State Park shows clearly on this air photo of Nesowadnehunk Lake. Trails from old horse logging remain visible within the Park. *Courtesy of U.S. Geological Survey.*

In 1964, following a seven-year legislative controversy, Congress passed the Wilderness Act, which states that land preservation was to stand on an equal footing as a potential use of federal lands. Congressional designation as a wilderness was to protect an area strictly from development. In the Northeast, the areas in wildlife refuges proposed by the U.S. Fish and Wildlife Service were not controversial because their designation threatened no locally important resource uses. The Great Gulf Primitive Area was formally designated, but since it was already managed as wilderness there was little dispute over its official dedication. The other National Forests were a different matter.

For several years after passage of the Wilderness Act, matters were quiet, reflecting a general lack of regional interest. In the early 1970s, hikers and wilderness advocates in the region gained in numbers and organizational strength. They were enraged when a nationwide Forest Service wilderness review, released in 1972, gave little attention to Eastern wilder-

ness, and when ongoing National Forest land use plans allocated insufficient land — in their view — to wilderness.[16] To get around the agency's unwillingness, wilderness groups went to Congress with an Eastern wilderness bill. They shrewdly lined up Senator Henry Jackson of Washington, chairman of the Senate Interior Committee, and Senator George Aiken of Vermont, as sponsors. After a complex legislative fight, the Eastern Wilderness Areas Act of 1974 emerged.[17] All of the designated areas were tiny by western standards; all were second growth on previously logged areas. And they did not satisfy the wilderness advocates, who worked still harder for more.

The next opportunity came in 1977–1978, when the Forest Service conducted a new wilderness review (RARE-II). This time, the review included substantial emphasis on Eastern areas. Its recommendations emerged from complex deliberations and were transmitted by President Jimmy Carter to the Congress in spring 1979. The RARE-II evaluation identified six additional areas in the Green Mountain National Forest, totaling 55,720 acres, as roadless. In its conclusions, however, no Vermont areas were recommended as wilderness. In the White Mountains, thirteen areas totaling 262,257 acres were studied, and three units of about 160,000 acres were recommended for wilderness. In Pennsylvania's Allegheny National Forest several roadless areas were identified, but only about 9,000 acres became designated wilderness. Years will pass before this struggle over National Forest wilderness allocation is settled. As of late 1996, only 184,000 of the 1.6 million acres of National Forest land in the region was dedicated wilderness in 13 separate units. Roading and logging are not planned or are tightly limited on a much larger proportion of the area. In Winter 1998, another major review of the future of non-wilderness roadless areas was announced by the Forest Service.

State and Local Wild Areas

State and local governments manage thousands of acres of forest that qualify for inclusion in the Wild Forest, as noted in Table 8.2 above. The largest example is the 2.5 million acre New York State Forest Preserve in the Adirondacks and the Catskills. Others range from the 14,000 acres of natural areas managed by Vermont's Department of Forests and Parks to the 170,000 acres of designated wild lands on Pennsylvania's State lands, and the Preserve lands in New Jersey's Pinelands. They would include Maine's Bigelow Preserve, although timber will be harvested there. Much of the acreage in state parks and municipal watershed lands could be considered Wild Forest. Future land use planning on state and municipal lands may result in more areas being so designated. For example, Biosphere Reserve status has been proposed for watershed protection lands at Quabbin Reservoir.[18]

Wild and Scenic Rivers

Rivers have formed the central axis of the Northeast's economic history. They have floated logs, powered mills, and carried cargoes, leading to the location of much of the region's manufacturing and settlement along their banks. Most of the rivers work hard — they are everywhere dammed for power and navigation. They carry huge waste outflows from cities and factories. A vigorous tradition of canoe travel in the northwoods has attracted a rapidly growing following and has spawned imitators in calmer waters, even along the coastal islands where sea-kayaking has become popular.

Surprisingly, however, an active interest in stream fishing and canoeing did not translate itself into a strong citizen movement for river conservation. Perhaps people had written off the poisoned and dammed streams and did not perceive threats to the great northwoods streams — the Allagash, St. Croix, Machias, Penobscot. So, while citizen conservationists labored for the state, national, and community forest preserves, they ignored the rivers.

In the early 1960s, this changed. A new assertiveness for conservation was felt nationally, expressed by Rachel Carson's *Silent Spring,* Stuart Udall's *The Quiet Crisis,* and the spectacular sales of Aldo Leopold's reissued *Sand County Almanac.* The catalog of national conservation legislation of the decade is notable. This new movement was also expressed at state levels.

In particular, a movement, led by the Maine Natural Resources Council, emerged to seek protection of Maine's historic Allagash, a major nineteenth century logging river.[19] Its peace and beauty seemed threatened by a rise in timber harvests and new truck logging roads that were supplanting the river log drives. Unlike many controversial wild rivers, the Allagash was not subject to any immediate threat from a dam. In 1964, following a massive public controversy and threats of federal designation, the Maine legislature created the Allagash Wilderness Waterway, from Telos Lake to below Allagash Falls. A state bond issue was used to buy a narrow protective corridor, and private logging was to be regulated in a one-half-mile corridor to allow logging consistent with preserving the scenery.

The Allagash expresses the essence of the mystique of northwoods wilderness canoeing. It is not pristine — it is dammed in several places and crossed by logging roads. But it retains its wild atmosphere. Some ca-

The Bristol Cliffs Wilderness, Green Mountain National Forest, Vermont. *Courtesy of USDA-FS.*

noeists complain that, since its designation, the use level has risen and diminished the wilderness experience. The Allagash is managed by the Bureau of Parks and Lands, an agency of the Maine Department of Conservation.

In 1968, the U.S. Congress passed the Wild and Scenic Rivers Act,[20] which designated a series of instant wild rivers on federal lands. It included Maine's Allagash and Wisconsin's Wolf by state request and set a ten-year schedule for federal studies of 27 streams, among them Maine's Penobscot.[21] After publication of the U.S. Bureau of Outdoor Recreation's wild river recommendation in 1976, the state engaged in lengthy negotiations with Great Northern Paper Company, the principal landowner, to develop a protection program. In 1981, complex legal documents and management plans were adopted by the Maine Land Use Regulation Commission, which incorporate the plans into its zoning rules for the riverway. The company donated a conservation easement to the state.

There have been many innovative efforts to conserve river corridors. One is the Saco River Corridor Association, which administers a zoning program protecting the Saco, a well-used canoe route. This program has been highly successful because of intense local interest and involvement. The Saco, Allagash, and Penobscot efforts all involve private lands, as do the other study streams. Other examples are the Mianus River Gorge Association's work on a tiny creek in New York and Connecticut, and the Coe-Pingree-Brown Company ownership in the Thirteen Mile Woods along the Androscoggin River.[22]

Federal agencies have made their contributions as well. In the Boston area, the U.S. Army Corps of Engineers proposed the purchase of marshlands and easements along the Charles River to preserve channel storage for flood control. Along the Delaware, the Delaware Water Gap National Recreation Area has protected an extensive zone of scenic hills along the river, protecting more than 70 miles of the river. The New England River Basins Commission backed a plan, after years of debate, for reducing flood damage in Massachusetts and Connecticut along the Connecticut River by floodplain management.[23]

Because much of the region is privately owned, these programs will be difficult and costly. Massachusetts has pursued the issue vigorously. Its Wild Rivers program reviewed 180 streams and considered 663 miles for a statewide system. The system will include five categories of streams, including recreational urban rivers.[24] Under 1972 legislation, New York has established extensive Wild Scenic, and Recreational Rivers primarily in and around the Adirondack Park.[25] As of 1994, the state's system included 1,314 miles of rivers, including 4 on Long Island. The New Hampshire Office of Comprehensive Planning, through a consultant, identified 67 streams of wild, recreational, or scenic importance and outlined a protection program. In Maine, the state passed a Rivers Bill in 1983 to protect important streams.[26]

A 1979 federal survey of potential wild and scenic rivers identified about 50 river segments that met the criteria of the 1968 act.[27] This study is to be supplemented by an additional survey of rivers of cultural or recreational significance. When that inventory is completed, river conservationists will have a challenging agenda of work before them.

In Maine, the extraordinary recreational, fishery, and aesthetic quality of the state's wildland lakes drew special planning attention from the Land Use Regulation Commission in a 1989 report. The Commission has developed a Lakes Concept Plan process to attempt to guide landowners to use more responsible development patterns. A private owner, Lowell Associates, submitted a pioneering, low-density scheme for a property around Attean Pond in Western Maine, which was authorized by the Commission.[28]

High energy costs of the 1970s and early 1980s raised interest in small hydroelectric dams, many of them rebuilds or expansions of existing dams. Restoring such dams generated conflict among power developers, fish and game interests, river recreationists, and lakefront cottagers. While

lower oil prices have moderated this concern, the controversy will certainly be back in time.

More serious, however, is the booming market for rural real estate and the dramatic cleanup of many badly polluted streams. While choked with logs and stinking from pulpmill effluent, these streams offered few attractions. Now being cleaned up, and with fishing improving, they display miles and miles of high-value, developable real estate.

The gains in water and fishery quality on thousands of miles of Northeastern streams are a source of great satisfaction to all concerned with the outdoors. But those gains are likely to be funnelled into private pockets unless vigorous public and private effort is exerted to protect riverside corridors and maintain and expand public access points. In many more instances, the taxpayers whose dollars paid for pollution cleanup will go to these streams and find "No Trespassing" signs. During the land boom of the late 1980s, this fear was realized in all too many cases.

Along the West Branch, Penobscot River in Maine, these lands are protected from development by a conservation easement totalling 8,000 acres. But they could not be protected against loss to the spruce budworm, so trees behind these spruce were salvaged. *Photo by author.*

Recreation on Private Lands

In the Northeast, provision of backcountry recreation has necessarily emphasized private lands to an unusual degree compared to other regions. Put-in points for rafting and canoe trips, campsites, and trails rely heavily on private land, even in cases where visits have public units as a destination. The North Maine Woods management program on the St. John River, and Great Northern's limited access area on the Debsconeag Lakes are examples. In the Mid-Atlantic states, hunting leases on private lands are managed by private groups and public agencies. The development of the Appalachian Trail, and of snowmobile trail systems, are prime examples of using private and public lands to meet recreation needs. Experience has demonstrated that by and large, these arrangements can work. To meet future needs, we will have to get even better at improvising public-private-nonprofit management for backcountry recreation.

Outlook

The Northeast has essentially no wilderness, if by wilderness we mean vast tracts of truly untouched virgin forest. Aldo Leopold once wrote that a wilderness ought to be large enough to absorb a two-week pack trip. By that definition, none remains here. Probably the longest wilderness trip in the region is the ten-day canoe voyage on the St. John — almost entirely private land. Because of their small size, few of these areas are complete natural habitat units for wide-ranging wildlife species. But few residents would guess that the region today has 4 to 5 million acres of land in its Wild Forest, where active land management is banned or restricted in favor of natural processes. Little of this wildness is in the well-known federal areas — most is on state land, and some is privately owned or was conveyed to public agencies by active private groups. Half of the region's total is in the Adirondack Park.

The public values attached to wild areas are shifting toward a heavier emphasis on biodiversity, existence and habitat values in contrast to a traditional interest in recreation, scenery, or scientific values. One implication is that there will be much more public concern for attributes of wildness outside of the identified units of the Wild Forest itself.

The Northeast expresses a particular pragmatic branch of wilderness politics. In the late 1970s, the state of California sued the federal government to compel it to designate more wilderness. In New England, some local political leaders have threatened dire consequences if the Forest Service designates any. The wildernesses of the Adirondack Park owe their creation to an extraordinary coalition of political forces in the late 19th cen-

Municipal Watershed, Scituate Reservoir, Providence Water Supply Board, Rhode Island. Such lands are an important part of Southern New England's Wild Forest. *Courtesy of Rhode Island Division of Forest Environment.*

tury. Maine's largest wilderness, Baxter Park, was created by the vision and generosity of a single man, not to a popular movement with signs, brochures, petitions, and lobbyists. Maine's 1976 referendum creating the 35,000-acre Bigelow Preserve is the exception, not the rule — and timber cutting and snowmobile use are allowed there. Such a wilderness could not pass muster in California.[29]

Northeasterners support wildland preservation, but only to a limited degree compared with other regions of the country. Most of them are prepared to allow the logger and builder a place in the landscapes where they hike and camp on weekends. Local groups of snowmobilers in particular often oppose wilderness since it bars use of vehicles. Political tensions over designating Wild Forest and wild rivers become very complex.

The region is developing its own approach to the wilderness question, and it is doing it quietly and slowly. Major scenic jewels of the landscape have been preserved for nature's slow work. More acres will be devoted to such uses in the future, but only slowly and mostly by strictly private initiatives. As an example, in the 1970s, the private, nonprofit Nature Conservancy acquired more land in Maine than did the Maine Departments of Conservation and Inland Fisheries & Wildlife put together. As in the Appalachian Trail tradition of volunteerism, this arrangement is consistent with the region's values and ways of doing things.[30]

Virgin hemlock stand, Colebrook, Connecticut, 1912.
Courtesy of G. E. Nichols Collection, Yale University.

I believe that a sensible conservation program for the region would attempt to increase the acreage in the Wild Forest by 50 percent by the year 2020 — from 5 to 7.5 million acres. This would still be only 7.5 percent of the region's land.[31] The effort should focus on bolstering existing large remote areas, especially those with key wildlife values, but would also involve private groups acquiring small, key parcels. An enlarged Wild Forest would be a prize Bequest for this generation to pass to the future. Since the bulldozers and skidders will be busy, time is not on our side. Many acres would be donated and acquired by bargain sale, while others would fetch high prices. From the Eagle's Eye View, the region would still be dominated by nonwilderness landscapes.

If the forecasters are right, perhaps six million more people will live in the region by the year 2025. Providing houses, streets, and shopping centers for these people and for urban emigrants will easily consume another one to two million acres of land and compromise recreation and traditional rural uses on much more. There will surely be more leisure lot subdivision land booms. It is time to plan for preserving an acre for each acre that will be developed. The price will be cheap, compared with the benefits that will be enjoyed by our grandchildren.

Adding acres to the region's public estate will not be the best solution to conservation problems in every area. Also, outright acquisition may not be cost effective or politically feasible. A large role will remain for carefully designed programs of development right acquisition, regulation, and other measures aimed at preserving the elusive character of wildness that is so valued in the balance of the region's working landscape. An example is the Forest Legacy program, a federally funded program acquiring easements in several states.

Thoughtful scientists and citizens realize that the region's heritage of wildness, its wildlife habitat values, and its traditions of public recreational uses cannot be preserved by designated wilderness areas alone. For this reason, the Wild Forest must be seen as an essential element — but only one element — in a broader effort designed to retain wildness, habitat, other natural values, and public access over larger landscapes. More sensitive management of large public forest units and industrial properties can provide many of the values of the Wild Forest over a much wider area.

Major proposals for expanding the region's Wild Forest include proposed additions of some 600,000 acres to the Adirondacks' Forever Wild Lands, and an ambitious Maine Woods Reserve of 2.7 million acres proposed by The Wilderness Society. The Reserve is not intended to be entirely publicly owned, and the details of its management have not been developed. Even more ambitious proposals, for a 3.2 million acre Maine Woods National Park, and a 5 million acre Thoreau National Reserve, have been offered.[32] A campaign has been launched to convert the White Mountain National Forest to a National Park. The future prospects for these proposals are uncertain. But clearly, protecting the Northeast's remaining wildness will depend on developing more innovative ways to secure the protection of wildlands for the future. This generation's Bequest of wildness to the future is being shaped now.

9 Forest Landownership

An important perspective on forest land is achieved by simply asking, "Who owns it?" The ownership of the land will largely determine its uses and its future contribution to the region's economy and landscape. The land use and ownership changes of the 1980's economic boom drew public and legislative attention to the issue of forest landownership in the region with an intensity not seen since the early 1900's conservation movement. This chapter reviews the history of forest landownership in the region and the basic facts about current ownership patterns and issues.

Early Forest Landowners

The Northeast's history of forest landownership is a rich and colorful one, reaching back to original Indian occupancy and royal claims based on discovery and conquest. Some Indian land claims are now being reasserted, but space prevents proper treatment of that story.[1] In the original passage of the land from Indian ownership to royal claims to private hands, injustice, official corruption, timber stealing, and speculative binges leading to fortunes and bankruptcy were common. Because of chaotic, overlapping royal land grants and poor surveying, land titles were uncertain in frontier areas for generations, where many titles were obtained by squatting.

At times, speculators and royal governors amassed large holdings, as in the two-million-acre Maine Purchase of William Bingham, and the New Hampshire holdings of Benning Wentworth and his family, and some of the New York patroonships along the Hudson. These empires collapsed under inadequate markets for wilderness land or dissolved in periodic financial panics. Their names remain on our maps, but their original owners are long forgotten.

In colonial times, English monarchs made extensive land grants to fa-

vored royal associates and relatives. The Great Proprietors, like John Mason and Sir Ferdinando Gorges, the Duke of York, and the Holland Land Company controlled vast unsurveyed frontier domains, hoping to convert them into thriving communities that would enrich their owners through trade and land sales.[2] New England land passed into private ownership over the period 1620–1878. This is an astoundingly long period, spanning the entire history of the nation until the very peak of agricultural growth in Southern New England. During this time, New England remained a major factor in national manufacturing, shipbuilding, finance, and trade. By contrast, most of the land sold or granted nationally by the federal government was disposed of in less than a century after the Louisiana Purchase.

Land distribution in Southern New England was virtually complete by the Revolution. The bulk of the land there was owned by farmers; little was retained for public purposes. These colonies did not go through a long stage of massive private landholdings because, by and large, towns were sold or granted one by one for settlement, not in large blocks. They were not held for long periods in speculative hands, but for the most part were quickly settled.

In 1664, Charles II granted to his brother, the Duke of York, the lands that later became New York, New Jersey, Pennsylvania, and Maine east of the Kennebec. The Duke at the stroke of a pen became one of the world's largest landowners. The Duke's naval expedition then induced the Dutch at New Amsterdam to surrender; the city that was later to dominate the region then took its new name. New Jersey was quickly granted to two of the Duke's associates. After that time, large sales were made. The Macomb Purchase of 1791 covered 3.7 million acres, and the Holland purchase of 1792 covered 3.3 million acres.[3] The vast Massachusetts Purchase of 6 million acres was resold in 1788–1793 in large pieces.

William Penn's wealthy father had lent money to the Duke of York. In discharge of these debts, Penn obtained the entire area of Pennsylvania as a proprietary colony in 1681. Penn developed a plan for settlement and land distribution, emphasizing making land widely available. Yet, in Pennsylvania as early as 1726, there were already 100,000 squatters. Early Swedish settlers along the Delaware River are said to have brought the log cabin to America, and later German immigrants established distinctive farming and architectural patterns in the region.[4] But in 1779, the Commonwealth assumed ownership of 22 million acres, and in succeeding years, several very large land sales were made, including 1.5 million acres to the Holland Land Company.

The large ownerships of the Hudson Valley slowed settlement in Eastern New York until after the Revolution. The Adirondacks and regions beyond the crest of the Alleghenies were affected by the ban on settlement in the Proclamation of 1763, so these regions saw only occasional traders, squatters, and hunters before the Revolution. The opening of the Erie Canal in 1825 and the growing trade on the upper Ohio River were major forces promoting growth beyond the Alleghenies. The farmlands of Western New York and Ohio were far more attractive than the stony soils of the Adirondacks. So even after the Civil War, New York still owned extensive lands in the Adirondacks, which later became the nucleus of the Adirondack Park.[5]

Current Patterns of Forest Landownership

The best statistics on forest ownership are USDA-FS estimates, based on available surveys and data sources.[6] At times, the figures reflect differing definitions or administrative choices about land use rather than actual forested area. Since the most convenient figures are for commercial forest land area (now called "Timberland"), they will be used here. Assessing the forest this way emphasizes economic values, not ecological ones. Commercial forest (timberland) is land that is available for timber harvesting and is also capable of growing more than twenty cubic feet of wood per acre per year. This is about one-fourth of a cord, or a pile of four-foot sticks two feet high and four feet long. Commercial forest land does not include about half a million acres of forested land not available for timber cutting because of wilderness designations, and almost a million acres that is forested but of extremely low timber productivity, such as spruce and tamarack bogs.

Northeastern forests are 88 percent private, with similar acreages in farm holdings and in industrial ownership (Tables 9.1, 9.2). Nonprofit groups hold about 1 percent of the region's forest. The public forest estate is about 12 percent of the total, including roughly equal amounts of federal and state land. The county and municipal share is roughly equivalent to that of the nonprofit sector. About two-thirds of the publicly owned forest is in the Mid-Atlantic states.

Few large land holdings have endured. In Connecticut, for example, the largest parcels held by private individuals are roughly 5,000 acres. The larger state forests and the 20,000 to 30,000 acre ownerships of several major water companies were laboriously assembled by buying up one farm after another. The basic unit of trading in land in settled areas of the region has traditionally been the individual farm or "lot," usually surveyed but often laid out by "metes and bounds." The typical sizes of these farms varied, but a lot "100-acres more or less" was common in Northern New England. Farms were as large as 400 acres in Lancaster County, Pennsylvania before the Revolution. Today the state average is 154 acres. Many of these lots acquired the names of their early owners and are so indicated even on tax records.

Many small woodlots participate in the American Tree Farm program, which recognizes more than 40,000 landowners for sound forest management. *Courtesy of American Forest Institute.*

TABLE 9.1. Ownership of Commercial Forest Land, 1992 (Percent)

Owner	Northern New England	Southern New England	Mid-Atlantic	Total Northeast
Private				
Nonfarm	49%	71%	65%	62%
Farm	10%	14%	15%	12%
Forest Industry	35%	1%	5%	15%
Total Private	93%	86%	86%	89%
Public				
Federal	3%	1%	2%	3%
State, County, & municipal	3%	13%	13%	8%
Total Public	7%	14%	14%	11%
All Owners	100%	100%	100%	100%

Source: Powell, et al., 1993.

The decline of farming after the mid-nineteenth century left land in the hands of farmers. But from the mid-twentieth century, farmers sold land rapidly — first to "summer people," then to tract developers, investors, and recreational subdividers. Improved roads and widespread automobile ownership have promoted rural subdivision and recreational land purchases. On the fringes of the industrial forest, paper companies have been active buyers of farms.

Farmers sold three-fourths of the land they owned in 1880 in the succeeding century. Regionwide, from 1952 to 1992, much of the forest land remaining in farmers' hands was sold — a total of 7.5 million acres, or 12 percent of the region's forest. The land sold, however, retains a profound imprint in its parcel size, land use, and forest condition from its period in farm ownership. Farmers owned most of Connecticut's forest in the nineteenth century, but only 7 percent of it in 1977 — though this was out of a forest area double what it had been in the nineteenth century. With their strong farm sectors and fertile soils, Pennsylvania and New York farming has fared better; their farmers hold three-fourths of the Northeast's farm woodland.

The region's privately owned land is held by more than 1.8 million ownership units (Table 9.3). Individuals accounted for 95 percent of the ownership units, but they controlled only 64 percent of the acreage. Corporations, on the other hand, were only 2 percent of the ownership units, but held 15 million acres, or 25 percent of the privately owned acreage.

The extreme skewness of ownership by size has many implications for policy. For example, 61 percent of the private ownerships are 9 acres or less, but these units own only 5 percent of the forest land (Table 9.4). On the other hand, there are about 1,400 units with 1,000 acres or more; they hold a total of about 13 million acres or about 13 percent of the private forest land in the region. So a modest number of ownership units owns the bulk of the region's forest acreage. Yet policies for land use, access, and taxation can profoundly affect the 1.1 million owners of tiny woodlots as well. This table shows why it is essential to understand the size distribution of ownerships and not to treat forest owners in terms of averages.

Ownership turnover is an important aspect of forest ownership, especially in the smaller size classes. While this is a well-recognized situation, it often obscures the fact that there has been a core of fairly stable longterm ownership as well. Thus, the population of owners divides into different groupings based on length of tenure (Figure 9.1).

Motivations for forest ownership are similarly skewed. Because of the large area in corporate hands, timbergrowing is the leading motive by acres, though it is held by a small portion of the ownership units (Figure 9.2). When number of owners is considered, the primary motivations are

TABLE 9.2. Forest Landownership Trends and Patterns, 1952 and 1997 (commercial forest land = timberland)

State and Year	All Owners	All Public	Total Private	Forest Industry	Non-industrial Private
Maine					
1997	16,952	630	16,322	7,298	9,024
1952	16,609	182	16,427	6,617	9,810
% Change 52–97	2.1%	246.2%	–0.6%	10.3%	–8.0%
Percentage 1997	100.0%	3.7%	96.3%	43.1%	53.2%
New Hampshire					
1997	4,551	792	3,759	513	3,246
1952	4,819	682	4,137	771	3,366
% Change 52–97	–5.6%	16.1%	–9.1%	–33.5%	–3.6%
Percentage 1997	100.0%	17.4%	82.6%	11.3%	71.3%
Vermont					
1997	4,461	593	3,869	227	3,642
1952	3,845	297	3,549	528	3,021
% Change 52–97	16.0%	99.7%	9.0%	–57.0%	20.6%
Percentage 1997	100.0%	13.3%	86.7%	5.1%	81.6%
N. New England Total					
1997	25,964	2,015	23,950	8,038	15,912
1952	25,273	1,161	24,113	7,916	16,197
% Change 52-97	2.7%	73.6%	–0.7%	1.5%	–1.8%
Percentage 1997	100.0%	4.6%	95.4%	31.3%	61.3%
Connecticut					
1997	1,815	249	1,565	0	1,565
1952	1,973	159	1,818	3	1,815
% Change 52-97	–8.0%	56.6%	–13.9%	–100.0%	13.8%
Percentage 1997	100.0%	13.7%	86.2%	0.0%	86.2%
Massachusetts					
1997	2,965	480	2,486	71	2,415
1952	3,259	399	2,860	259	2,601
% Change 52-97	–9.0%	20.3%	–13.1%	–72.6%	7.2%
Percentage 1997	100.0%	16.2%	83.8%	2.4%	81.5%

Rhode Island					
1997	356	69	287	0	287
1952	430	26	404	0	404
% Change 52-97	–17.2%	165.4%	–29.0%	0.0%	–29.0%
Percentage 1997	100.0%	19.4%	80.6%	0.0%	80.6%
S. New England Total					
1997	5,136	798	4,338	71	6,333
1952	5,662	584	5,082	262	4,820
% Change 52-97	–9.3%	36.6%	–14.6%	–72.9%	31.4%
Percentage 1997	100.0%	10.3%	89.8%	4.6%	123.3%
New Jersey					
1997	1,864	500	1,364	0	1,364
1952	2,050	181	1,869	4	1,865
% Change 52-97	–9.1%	176.2%	–27.0%	–100.0%	–26.9%
Percentage 1997	100.0%	26.8%	73.2%	0.0%	73.2%
New York					
1997	15,406	1,153	14,252	1,220	13,032
1952	11,952	895	11,057	1,172	9,885
% Change 52-97	28.9%	28.8%	28.9%	4.1%	31.8%
Percentage 1997	100.0%	7.5%	92.5%	7.9%	84.6%
Pennsylvania					
1997	15,853	3,518	12,334	613	11,721
1952	14,574	3,229	11,345	442	10,903
% Change 52-97	8.8%	9.0%	8.7%	38.7%	7.5%
Percentage 1997	100.0%	22.2%	77.8%	3.9%	73.9%
Mid Atlantic Total					
1997	33,458	4,847	28,611	1,648	25,678
1952	28,576	4,305	24,271	1,618	22,653
% Change 52-97	17.1%	12.6%	17.9%	1.9%	13.4%
Percentage 1997	100.0%	15.1%	84.9%	5.7%	76.7%
NORTHEAST TOTAL					
1997	64,558	7,660	56,899	9,757	47,394
1952	59,511	6,050	53,466	9,796	43,670
% Change 52-97	8.5%	26.6%	6.4%	–0.4%	8.5%
Percentage 1997	100.0%	10.2%	89.8%	16.5%	73.4%

Source: U.S. Forest Service, RPA Website

TABLE 9.3. Ownership of Northeastern Private Forest Land by Form of Ownership, 1993

	No. Ownership Units			
	Individual	Corporate	Other	Total
New England	737,200	13,600	10,500	761,300
Mid-Atlantic	1,009,600	19,000	49,400	1,078,000
Total	1,746,800	32,600	59,900	1,839,300
	Acreage Owned (1000 A)			
	Individual	Corporate	Other	Total
New England	15,991	10,816	2,809	29,617
Mid-Atlantic	20,918	4,205	3,153	28,276
Total	36,909	15,021	5,962	57,893

Source: T. W. Birch, *The private forest landowners of the Northern U.S.,* 1994. USDA-FS, NEFES, Res. Bulletin. NE-136 (1996).

"part of residence," "esthetic enjoyment," and "recreation." Taking these three categories together, they account for more acreage than does timber production alone. This information can be further broken down by class of owner and other categories to gain a much more refined picture of the region's landowners.

TABLE 9.4. Owners and Acres by Size Class, Northeast, 1993

Size Class	# Owners	Percent of Total	1,000 Acres	Percent of Total
1–9	1,103,200	59.97%	3,031	5.24%
10–49	492,500	26.77%	10,468	18.08%
50–99	130,900	7.12%	8,684	15.00%
100–499	105,100	5.71%	15,512	26.80%
500–999	5,100	0.28%	2,935	5.07%
1000+	2,806	0.15%	17,261	29.82%
Total	1,839,606	100.00%	57,891	100.00%

Source: T. W. Birch, 1996, op. cit.

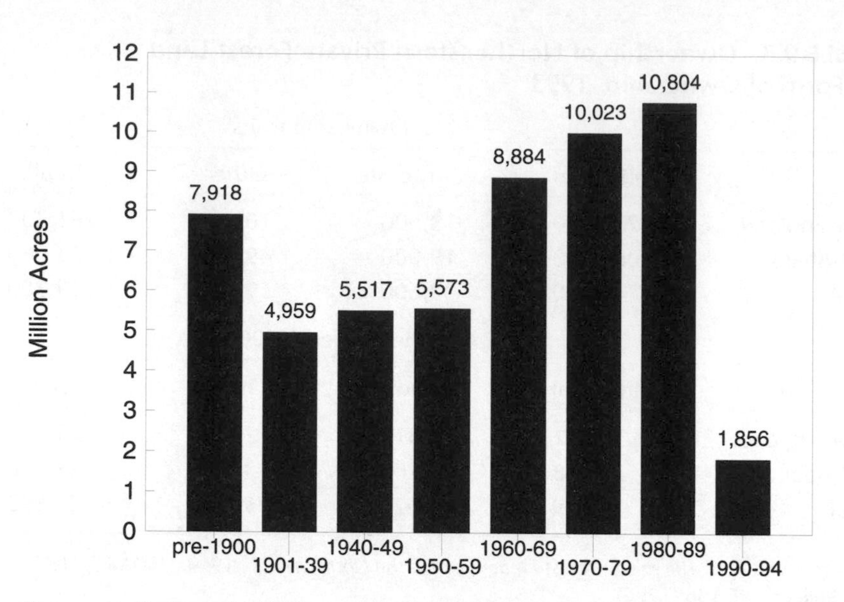

Figure 9.1 Estimated Number of Forest Land Acres by Date of Acquisition, 1993
Source: T. W. Birch, op. cit.

Land, Farm, and Timber Booms

From colonial times to recent decades, the boom-and-bust cycles of the Northeast have left behind growing forests. The booms up to the 1840s were speculative. They rearranged ownership of the land but not its timber stands. The mast trade and shipbuilding booms resulted from political and market trends. They were not halted by timber exhaustion but by changes in shipping that made the products obsolete. With that bust went the employment base of many coastal towns, especially in Maine.

It has been written that if the country had been settled from west to east, much of the Northeast would remain a wilderness. This may be so. But in fact, millions of acres of forest were cleared, mostly in a brief period from 1790 to 1880. Many of the hill farms scratched out of the forest simply could not provide reasonable livelihoods and could not compete with midwestern farms. In the vast contraction of settlement and farming after 1880, forests reclaimed millions of acres, as Torbert put it, in "a rising sea of woodlands." Though soils may have been degraded during this interlude, it was the impersonal market for crops and labor that released these lands to forests. A brief sheep boom from the 1840s to the 1880s was but an episode in this longer expansion and retreat of farming.

The northwoods saw rapid lumber expansion in the 1840s through the 1890s. This boom was ended partly by the exhaustion of large pine and spruce. Competition from midwestern pineries, made accessible by

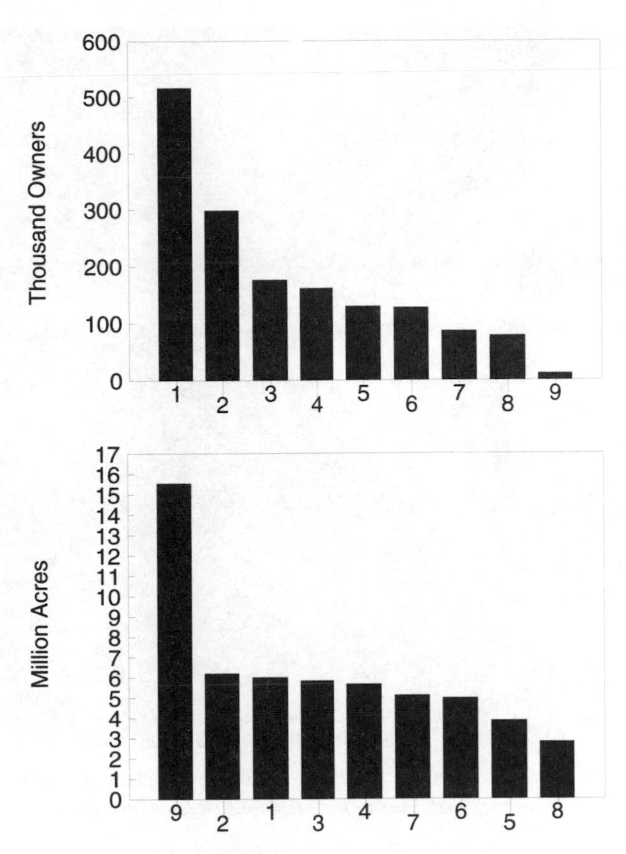

Figure 9.2 Estimated Number of Ownership Units and Acres of Northeastern Forest Land, by Primary Reason for Owning Forest Land, 1993

1. Part of Residence
2. Esthetic Enjoyment
3. Recreation
4. Other and No Answer
5. Part of Farm
6. Farm and Domestic Use
7. Land Investment
8. Estate
9. Timber Production

Source: T. W. Birch, op. cit.

the Erie Canal, and later from cheap West Coast lumber, in 1919 was also important. Exhaustion of the forest itself was not the sole cause. From the 1890s to the 1920s, the region was swept by the "boxboard boom." Hundreds of tiny mills cut low-grade boards and box shook from old field pines, pushing the region's lumber output to its all-time peak in 1909.

In the north, a pulpmill boom followed the era of the sawmill business, making the Northeast the source of a large share of the nation's pulp and paper by World War I. Paper companies bought up the old lumber holdings, and collected farm woodlots. In 1890, according to geographer Henry Gannett, lumber firms owned about 4.5 million acres in the region,

The land market has a voracious, if cyclical, appetite for small forest and lakeshore parcels. *Courtesy of Maine Department of Conservation.*

of which 2.1 million were in Maine.[7] This was 32 percent of the recorded industry ownership nationwide.

These lumber and paper booms again reflect outside market forces, relative costs, and corporate investment calculations. They also reflect changing perceptions of timber supply conditions. The energy crisis of the 1970s spurred a dramatic fuelwood boom that brought the use of wood for energy back up to a par with the paper industry's consumption. The falling oil prices after 1985 caused many consumers to turn away from wood fuel, however.

Development now underway is of a different character. The speculative land booms of recent decades significantly reduced the timber production potential of a large area of forests. This was not a biological change but simply the result of parcel fragmentation and absentee ownership, together with scattered residential and commercial development creating local vested interests opposed to public access or logging.

Suburban development after World War II was much more wasteful of land than previous development. Farms and woodlands vanished before the bulldozers with alarming speed. In Connecticut, for example by the early 1970s, landclearing was the principal cause of timber removals. This was also true in Southern New Hampshire and New Jersey in the late 1980s boom. The consumption and fragmentation of land and spiraling land prices have threatened much of the region's agriculture with extinction, but have left behind a substantial suburban forest. Continued wasteful sprawl, however, could eliminate that forest in a few decades. The commitments of forest to houses, pavements, and tiny isolated parcels are largely permanent in terms of forest values of wood, wildlife, watersheds, and amenity. The suburban sprawl of the post-1950 era will not be a transitory feature of the region's landscape.

During the 1980s, an extraordinary land boom swept over the Northeast. The causes of the boom were numerous, and need not detain us here. The extraordinary nature of the boom could be described with a basketful of statistics, but we can use two as proxies for the entire boom. First, at a time when farmland prices nationwide were falling, Northeastern farmland prices boomed (Figure 9.3). This price trend was caused not by the profitability of growing grain, milk, or vegetables, but of growing houses. Even more extraordinary was the nearly three-fold boom in private housing starts in just four years (Figure 9.4). This private housing boom was reflected in wildland subdividing in Maine and New York (Figs. 9.5 and 9.6). In the crash from 1986 to 1991, all five of New Hampshire's largest banks failed — an extreme example of the financial excesses that helped fuel the boom.

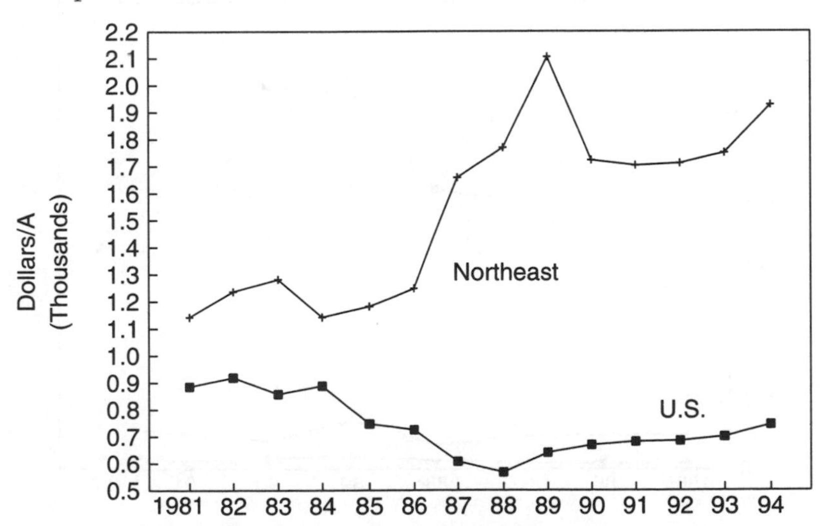

Figure 9.3 Northeast's 1980's Land Boom: Farmland Values, 1981–1997
Source: USDA Agricultural Land Values, AREI Updates, Dec. 1996, No. 15, and USDA, NASS, Agr. Land Values and Cash Rents, Oct. 1, 1997.

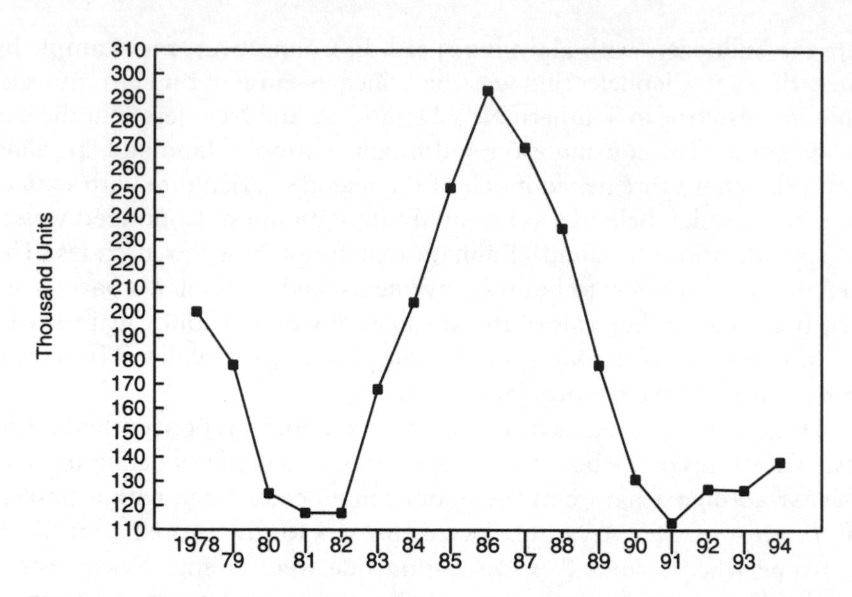

Figure 9.4 Northeast's 1980's Land Boom: Private Housing Starts 1978–1994
Source: U.S. Dept. of Commerce

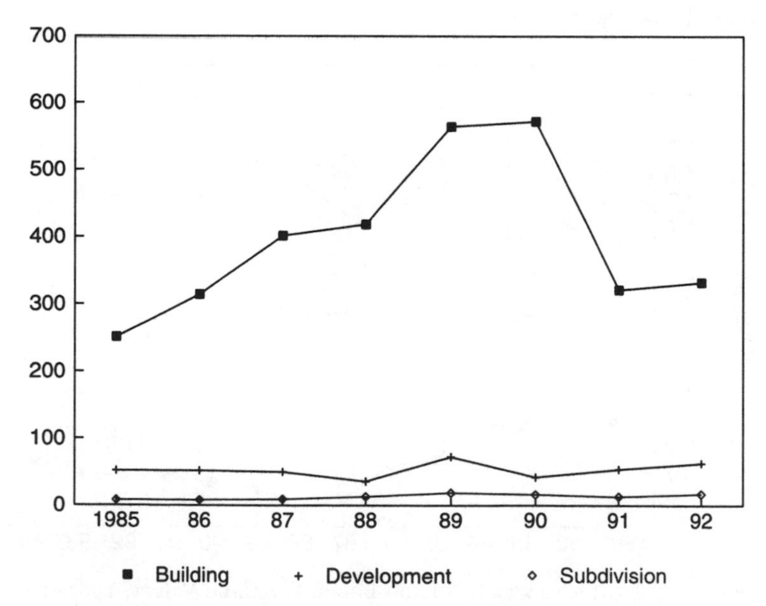

Figure 9.5 New Permit Actions by the Maine Land Use Regulation Commission
Source: Land Use Regulation Commision.

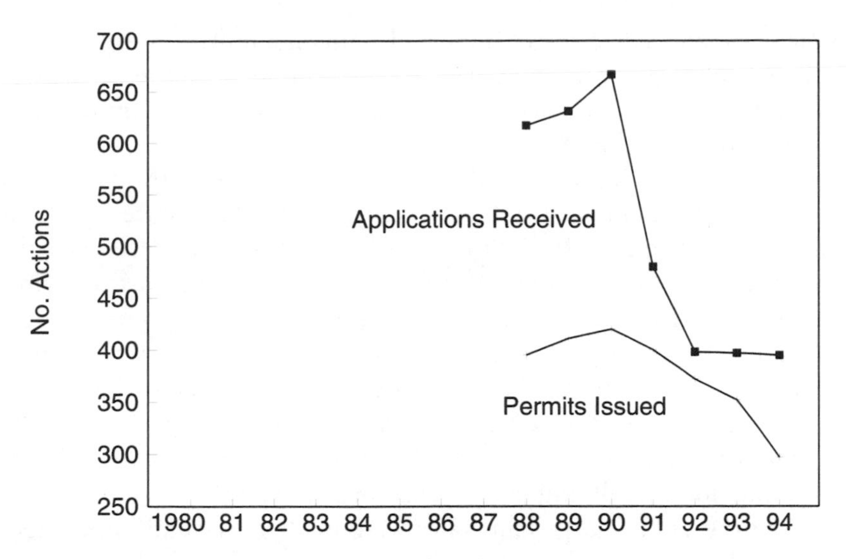

Figure 9.6 Adirondack Park Agency Permit Activity, 1980–1994
Source: Adirondack Park Agency.

As this discussion shows, stability in land use and ownership is not characteristic of the region's history. Instead, waves of land use change sweep the region over long periods. This will continue to occur, modifying the region's forests and their uses.

Industrial Ownership in the Forest

Wood-using companies, especially those in the capital-intensive pulp and paper industry, have found it desirable to supply a certain proportion of their wood needs from their own land. Over the years, they have built an impressive empire in Northern New England and New York, where valuable species of trees can grow well and where large acreage blocks were available. Today, no major firm acquires all of its wood needs from its own land. They buy wood from the large nonindustrial owners and individual truckloads from farmers and weekend woodcutters. They also sell large volumes to others who process species that their own mills do not use, whose mills are more favorably located, or who can offer a better price.

The forest industries now own 17 percent of the Northeast's commercial forest, or 11.4 million acres. Most of the industry ownership is in Maine, where 48 percent of the forest is owned by large pulp, paper, and lumber companies. An additional 20 percent of Maine's forest is owned by large landholders who do not own mills but are in the business of selling pulpwood and sawlogs to Maine and Canadian mills. Several of these massive holdings, of 150,000 acres and more (equal to all the public forest in Connecticut), are owned by direct descendants of the families who bought the towns in the 1840s.[8]

In Maine, forest industries added 1.4 million acres to their holdings from 1952 to 1992. In the three Northern New England states, perhaps three dozen firms control 10 million acres of forest — an empire twice the size of Southern New England's entire forest area. In these vast domains, local government is often absent, and population is low — only a few thousand people live year-round in the entire ownership. Several owners dominate the scene. The three largest industrial owners control 4 million acres. The largest is the Great Northern Paper Company, which owns just over 2 million acres. By comparison, Glatfelter's 107,000 acres in Pennsylvania may look small, but it is a substantial holding.

During the 1980s, large land sales were made by several large paper companies. In addition, several corporate takeovers changed ownership of some of the largest paper companies in the region.[9] According to a federally-sponsored study, between 1980 and 1991, 7.6 million acres of land in parcels above 500 acres changed hands in Northern New England and New York. Yet only a small acreage, 203,000 acres was subdivided. In Maine alone, a single landsales firm sold 52,000 acres in eight years, 80 percent to nonresidents. Some major tracts in New Hampshire, New York, and Maine were publicly acquired.[10]

In Maine and Northern New Hampshire, a special form of landownership persists. This is "common and undivided" ownership, a survival from two colonial practices. The colonies sold or granted towns to groups of buyers, each of whom exercised one "right" in the town. Many Southern New England towns were sold to groups of thirty to fifty owners. Some promptly sold their rights, seeking greener pastures. Some accumulated rights in groups of towns. Many settled the lands they acquired. Where districts were later settled, the rights were located on the ground by various methods, usually in such a way as to assure equitable distribution of a village lot, and some marsh, woods, and good tillable land to each right-holder. In some instances, the holding of rights in common and undivided form ("unlocated") persisted.[11] In Maine, the state itself held — and still does to this day — some of its reserved lands in undivided ownership.

Another source of this ownership pattern was the leading role of shipowners and merchants in early land investment and speculation. These men, like David Pingree of Salem, Massachusetts, were accustomed to owning undivided shares in individual ships. They naturally applied the same concept to joint ventures in land. This helped amass capital and

spread risk, and allowed owners to take in partners readily by selling undivided interests. Common tenancy also minimized the cost of subdividing distant and poorly surveyed parcels. Division of estates among heirs was also simplified. As a result of such repeated divisions, some towns are now owned in minute fractional interests (Table 9.5). The majority owners or their representatives normally make management decisions and send cotenants their share of net revenues.

In parts of Western Pennsylvania, development of coal and oil extraction led to the prevalence of severed ownership of mineral and surface rights, with attendant economic, political, legal, and environmental conflicts. In the 1930s, it was estimated that as much as 6.6 million acres of Pennsylvania forest land was owned by mining firms.[12] In many cases, the publicly owned lands in that state are underlain by privately held mineral rights. Fully 93 percent of the Allegheny National Forest is underlain by mineral rights that are privately held; only 7 percent of subsurface rights are government owned.

An emerging presence in ownership of managed timberland is the institutional investor. Several major institutions now manage timberland for institutional clients such as pension funds. Until recently, their presence in the Northeast was limited. By 1998, John Hancock had acquired about 450,000 acres in Maine, the largest such acquisition in the region up to that time. Whether more institutional acquisitions will occur remains to

TABLE 9.5. Common and Undivided Interests in the Northwest Quarter, Township 11, Range 4, WELS, Aroostook County, Maine, 1981

Owners	Percentage Interest
Irving Pulp and Paper Co., Ltd.	66/180
Prentiss and Carlisle Co., Inc.	4/180
McCrillis Timberlands, Inc.*	1/180
State of Maine, Bureau of Public Lands	23/180
E. G. Dunn Heirs	33/180
Fernald Group, E. G. Dunn Heirs	23/180
Marjorie D. Fernald	20/180
Peter Dunn Heirs	10/180

*Managed by Prentiss and Carlisle. This 1/180 interest is equivalent to about 30 acres.

Note: This table includes common and undivided interests in the quarter town, and in the timber and grass on the public lot. T11R4 lies southwest of Presque Isle, Maine. The northwest quarter is northwest of Squapan Lake.

Source: Maine Bureau of Public Lands.

be seen. The pension fund type of investor has a time horizon that would fit well with managed timberland as a portfolio asset. But these investors are very cautious about adding new asset classes to their portfolios, so any trend in their activity would be slow and deliberate. At present, these investors have concentrated their activity in the west and south. This story has major potential impacts on forest ownership and use but cannot be further developed here.[13]

The Public Forest Estate

The Northeast has a rich and diverse public forest estate. Most of this estate was acquired by purchase or tax default in the twentieth century — after the public values of forests for recreation, watershed protection, wildlife habitat, and timbergrowing became appreciated. Citizen interest in protecting especially scenic areas and important watersheds like the White Mountains, the Adirondacks, and the Poconos was high.

A portion of New England's public estate traces back to reservations from land sales made to provide for local schools and churches. The "public lots" in Maine's unorganized towns and Vermont's "glebe lands" are the notable examples.[14] Sizable portions of the public estate were acquired by donations made by wealthy individuals — the grandest examples being Baxter State Park (Governor Percival Baxter) and Acadia National Park (John D. Rockefeller and others). In Southern New England, public parks, forests, and water supply lands were pieced together from abandoned farms, brushlands, and cutover. Eager foresters planted pines on every bare acre they could, and many such plantations have already yielded one or more harvests of wood.

The White Mountain, Green Mountain and Finger Lakes, and Allegheny National Forests were acquired under the 1911 Weeks Act, which authorized federal purchases of land for National Forest in the East. Much of the land so acquired had only been in private hands for several decades and had been skinned by lumber companies; it was often tax delinquent. Today, these 4 forests account for most of the 1.8 million acres of federal forest land.

Slight additions to federal ownership have been offset by significant allocations of land to wilderness and limited use, so that federal commercial forest acreage fell by 23 percent from 1952 to 1987. The federal agencies — especially the USDA-FS, which administers the four National Forests (Table 9.6) — owned about 2.7 percent of the region's commercial forest in the late 1970s. As further wilderness allocations are made in these forests, federal commercial forest will decline somewhat further, al-

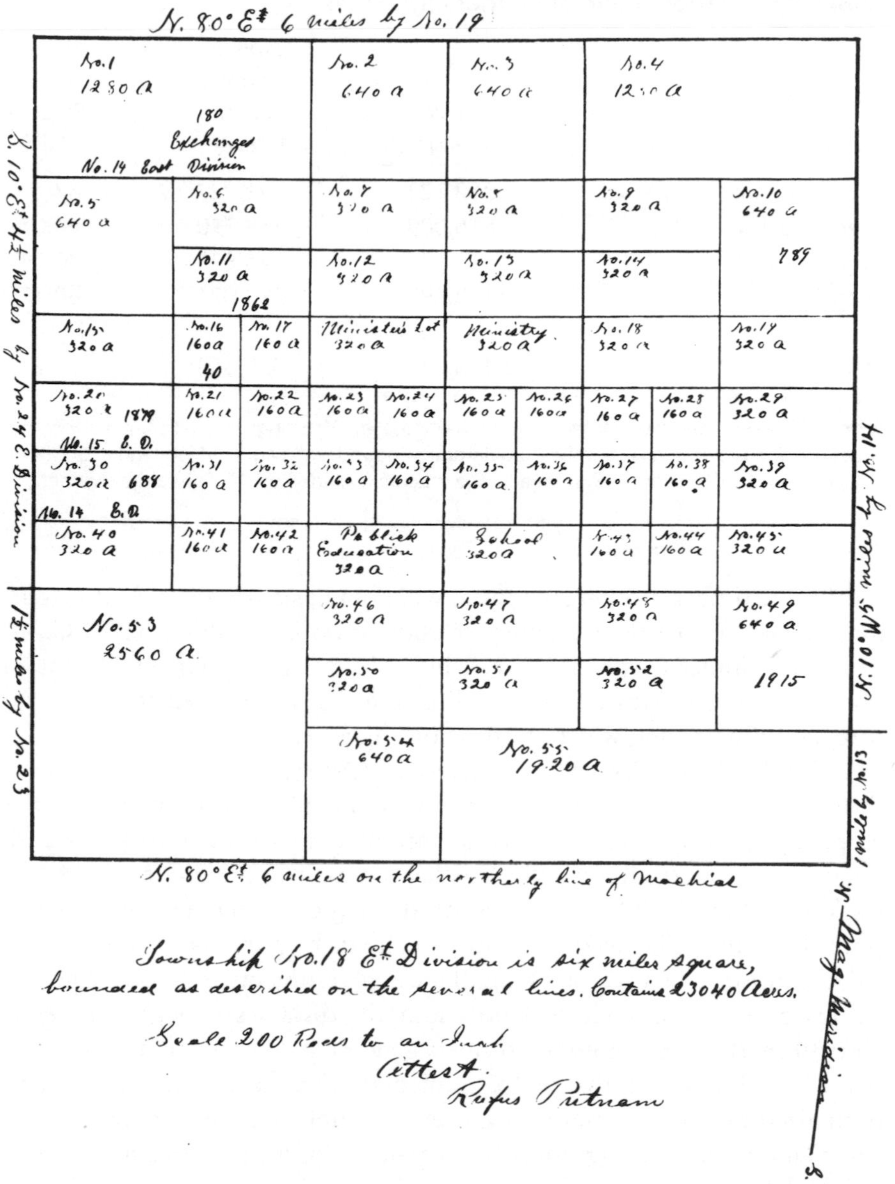

N. 80° Et 6 miles by No. 19

N. 80° Et 6 miles on the northerly line of Machias

Township No. 18 Et Division is six miles square, bounded as described on the several lines. Contains 23040 Acres.

Scale 200 Rods to an Inch

Attest.
Rufus Putnam

Forest land as real estate. Original survey of T18, E. D., Eastern Maine, for sale to absentee investors by the Commonwealth of Massachusetts in the late 1780s. No indication of any natural feature is given. *Courtesy of Maine State Archives.*

TABLE 9.6. National Forest Ownership, 1996

National Forest	Gross Acreage	National Forest System Acreage	Percent NFS
White Mountain, NH & ME	852,123	741,927	87.1%
Green Mountain, VT	815,000	358,790	44.0%
Fingerlakes, NY	15,825	15,825	100.0%
Allegheny, PA	742,693	513,021	69.1%
Total	2,425,641	1,629,563	67.2%

Source: USDA-FS, *Land areas of the National Forest System.* Washington, FS-383, Jan. 1997. Gross acreage is the area within the "proclamation boundary," within which land acquisition is authorized; National Forest System acreage is actually owned by the federal government.

though the timber growing potential will not fall significantly. At this writing, budget proposals are under consideration in Washington, D.C., to halt most timber cutting on the White Mountain and Green Mountain National Forests. Pressure in this direction will likely intensify in the future, as an increasing wave of litigation suggests.

New York and Pennsylvania were the first states in the nation to authorize public expenditures for conservation land acquisition. Yet despite its pioneering role in forestry, conservation, and parks, and its longstanding conservation traditions, the Northeast remains a region of private forest ownership. Only 12 percent of the region's commercial forest is publicly owned. The need to acquire the lands by purchase has been a decided hindrance in a region of small local governments jealous of their local tax base and opposed to landholding by distant senior governments. Until recently, overabundant timber, general public access to privately owned land, and the absence of immediate threats to greenspaces have minimized public perception of a need for public landownership. Except for wildlife and hunting, and the occasional park, extraordinary scenic point, or boat ramp, much of the region's dispersed recreation activity occurs on private land. The size and controversy associated with The Adirondacks, Baxter Park, and the National Forests often obscure this.

Northeastern states and counties have added to their landholdings, through bond issues, federal grants for parks and wildlife habitat, donations, and other measures. The net increase in commercial forest owned by states, counties, and municipalities was 400,000 acres from 1952 to 1992, an increase half the size of the White Mountain National Forest. At

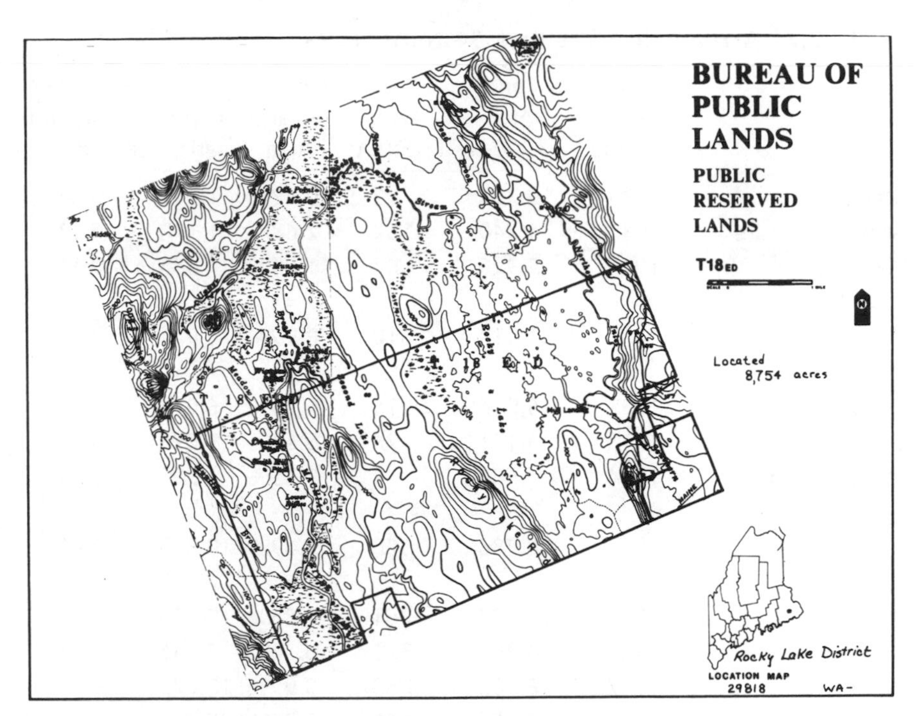

Forest land as public estate: managed by the Maine Bureau of Parks and Lands, T18 E. D.'s extensive areas of lake and marsh are shown on this agency map. *Courtesy of Maine Bureau of Parks and Lands.*

times in the past, especially during the Great Depression, states and counties acquired land through tax delinquency or after abandonment. In New York, an active program acquired farms and planted them for state forest. In Pennsylvania, large state forests, parks, and game management areas were acquired during this century. As a result, these two states contain the bulk of the region's nonfederal public forest land.

The state park and the wildlife conservation movements have promoted the assembly of an important conservation land system (Tables 9.7 and 9.8). Many of these areas are small — they averaged only 990 acres in New England in the 1980s — but they protect important habitats and access opportunities. This acquisition was funded in part by federal funding programs. These lands are managed for a diversity of specific purposes, but they all preserve open space, protect public access, and enhance game populations. State and federal fish and game lands are often actively managed to improve habitats.

Pennsylvania has the largest system of game lands, equal in size to three times the forest acreage of Rhode Island, and larger than the public landownership of Maine. On these lands, the agency harvested products

TABLE 9.7. **Acreage in State Park Systems, 1995**

	Year System Established	Land Area (1000 A)	Park Acres (1000 A)	Percent of Land Area	Population	Acres per Thousand Population
Maine	1935	19,753	92	0.47%	1,241	0.07
New Hampshire	1935	5,740	154	2.68%	1,148	0.13
Vermont	1936	5,920	64	1.08%	585	0.11
N. New England		31,413	310	0.99%	2,974	0.10
Massachusetts	1939	5,016	315	6.28%	6,074	0.05
Connecticut	1913	3,101	176	5.68%	3,275	0.05
Rhode Island	1904	669	9	1.35%	990	0.01
S. New England		8,786	500	5.69%	10,339	0.05
New York	1924	30,223	261	0.86%	18,136	0.01
New Jersey	1905	4,748	321	6.76%	7,945	0.04
Pennsylvania	1927	28,685	283	0.99%	12,072	0.02
Mid-Atlantic		63,656	865	1.36%	38,153	0.02
NORTHEAST Total		103,855	1,675	1.61%	51,466	0.03

Note: Definitions of what is a state park vary from state to state. For example, Maine's 200,000 acre Baxter State Park is administered separately from its State Parks and is not included in this summary.

Source: P. Myers, *State parks in a new era,* Washington, D.C.: Cons. Foundation, 1989, vol. 1, p. 31 and vol. 3, p. 34 and Statistical Abstract of the U.S.

worth $11 million on 9,656 acres in 1993–1994. This management activity improved habitat and access, provided raw material for local industries, and produced revenues equal to 20 percent of the agency's entire budget.[15] A small area of designated wilderness exists on the federal wildlife refuge lands (Table 9.9).

The largest conservation land unit in the Northeast is the Adirondack Park, a 6 million acre region encompassed by the "Blue Line." The Park contains both private and public lands. The state-owned lands constitute the Forest Preserve, established in the New York State Constitution in 1895. These lands are to be managed in a "forever wild" condition and represent an extraordinary heritage of wilderness, lakes, and mountains. The Park is home to 130,000 residents. The use of private land within the

TABLE 9.8. Acreage in State Fish and Game Systems, 1980s

State	Acreage of Land
Connecticut	19,728
Maine	35,222
Massachusetts	50,961
New Hampshire	20,614
Rhode Island	28,163
Vermont	52,195
New York	144,000
Pennsylvania	1,229,300
New Jersey	210,000
Northeast Total	1,790,183

Source: Various agency reports.

TABLE 9.9. Acreage Owned or Controlled by U.S. Fish and Wildlife Service, Sept. 30, 1995

	Acres Under Agreement Easement or Lease	Total Acres Owned or Controlled	Designated Wilderness
Maine	622	46,737	7,392
New Hampshire	35	5,907	—
Vermont	86	6,513	—
N. New England	743	59,157	7,392
Massachusetts	697	12,923	2,420
Connecticut	0	713	—
Rhode Island	20	1,523	—
S. New England	717	15,159	2,420
New York	1,868	26,832	—
Pennsylvania	45	10,051	—
New Jersey	2,414	61,567	10,341
Mid-Atlantic	4,327	98,450	10,341
Northeast Total	5,787	172,766	20,153

Source: USDI, Annual report of lands under control of the USFWS as of Sept. 30, 1995, Washington, D.C.: USFWS, Dir. of Realty, tables 2 and 5.

Blue Line is regulated by the Adirondack Park Agency. A recent commission reporting to Governor Cuomo listed 245 recommendations for securing the future wilderness values and economic opportunities for the region.

Nonindustrial Private Owners

About 37 million acres of Northeastern forest is owned by individuals like you and me — who hold small parcels as parts of a residence or recreation property, as a farm bought for a vacation retreat, or as an investment (Table 9.10). While these owners are highly varied and may live in other states and regions, they have been of considerable interest to foresters. The owners of many small tracts seem uninterested in investing in woodgrowing, in selling timber, in allowing public recreation, in managing for wildlife, or even tolerating hunters. In some areas, however, owner preferences are changing. Because these owners hold such a vast acreage, they control the key to industrial and fuelwood production in most of the region, outside the industrial forest.

Because of busy real estate markets, a predominance of speculative motives, and our national mobility, ownership of land is highly unstable. Compared with the Dunn or Pingree heirs of Maine, who still hold lands acquired soon after 1840, the average period of tenure in most of the Northeast is short. A national landownership survey for 1993 provides a convenient overview of regional ownership. Only a modest percentage of the private nonindustrial forest had been acquired prior to 1940.

Occupational differences within this owner group are important. According to Birch's data, the leading groups in 1993 were:

Owners Group	No. of Owners (thousand)
White Collar	670.4
Retired	496.5
Blue Collar	263.4
Farmers	89.0

Retirees are the second-largest owner group. It was once thought that this group would have little interest in forest management, but this is not true. Many leading tree farmers are retired, and many forest owners have strong Bequest motives. They view their forest as an important asset to pass on to their children.

Rapid ownership turnover may be of little consequence in itself for aesthetic, investment, and some wildlife values of forest land. Such high turnover, however, does not help in establishing longterm plans for timber

and wildlife management and reduces the chance that owners will apply management treatments with deferred returns.

The size distribution of forest parcels by owners and by acreage has potent public policy and resource management implications. There are roughly 1.8 million owners in this region with parcels larger than one acre. The total number of owners will affect public access for recreation, opportunities for wildlife management, and programs designed to promote or regulate wood production. As the number of owners increases, the public efforts to communicate information about forest values and management are stretched thinner and thinner.

We could consider a size of 100 acres to indicate potential for serious future timber management, though some foresters would use 25 to 50 acres. At 100 acres, there are only about 90,000 nonindustrial owners, plus a few dozen industrial and land management concerns, who really count in the region's industrial wood supply (see Table 9.10). This is not to say that timber cutting and some management will not occur on the smaller tracts, only to emphasize how a small proportion of the owners control the best management opportunities. Parcels in this size class present the most attractive opportunities for wildlife management and are most likely to remain available for public recreation.

Most of the region's owners control a small total acreage of land, hold their properties for brief periods, and own too little land to be interested in planned timber management for financial gain. These tiny tracts are

TABLE 9.10. Individuals: Estimated Number of Private Ownership Units and Acres of Forest Land Owned, by Size Class and Subregion, Mid-Atlantic and New England Regions, 1993

	OWNERS		ACRES OWNED	
Acres	Mid-Atlantic Percent	New England Percent	Mid-Atlantic Percent	New England Percent
1–9	58.4	67.1	9.1	6.5
10–49	30.6	19.5	31.3	18.7
50–99	6.5	7.1	20.4	21.9
100–499	4.4	6.0	32.5	36.9
500–999	0.1	0.3	3.7	6.8
1000+	0.0	0.1	3.0	9.2
Total	100.0	100.0	100.0	100.0

Source: T. W. Birch, 1996, op. cit.

hard to harvest because they usually cannot supply enough wood to make a job economically attractive. In most areas, logging jobs on ten to twenty acres have received few bids unless the wood was standing next to a mill or was all veneer logs. Better timber markets and the fuelwood market are making small sales more common.

Private Ownership and Public Uses

The trend toward parcellation and absentee ownership continues to threaten the traditional rights of access to rural and forest land. The extent of posting varies across the region (Figure 9.7). In New York, posting is high because leasing of lands to private clubs is widespread there. In the other states, posting is far lower on average, but in the urbanized areas it is high.

During the land booms of the 1970s and 1980s, a less visible trend became apparent: the widespread "privatization" of spaces formerly used by large numbers of people. Land values became so high that low-intensity uses along lakeshores and in the wildlands could not support the inflated values of the underlying land. So sporting camps were condo-ized, each cottage sold to an individual family. Campgrounds and youth camps were sold or subdivided. Regionwide, the cumulative total effect has not been measured. But the result has surely been to convert facilities and spaces formerly serving many people each summer to havens for only a few.

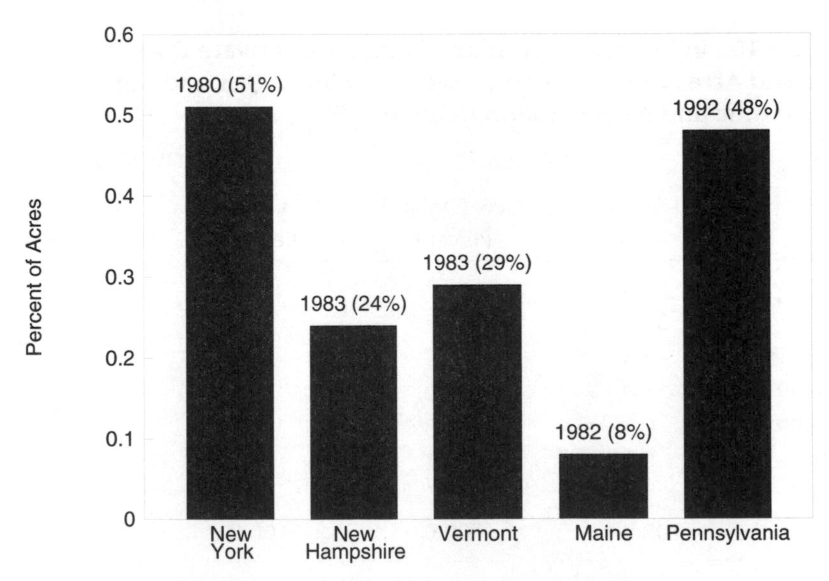

Figure 9.7 Forest Lands Posted, Northeastern States
Source: USFS Res. Bulletins.

While for the most part these places were already in private ownership, their use has become much more exclusive.

Even as parcellation and "privatization" are shrinking public opportunities to walk, hunt, pick mushrooms, or fish on private lands, vigorous efforts are underway to preserve public access. In Pennsylvania, the state leases 2.5 million acres from 21,000 landowners for public hunting. In Maine, a private association called North Maine Woods manages roads, gates, and campsites in about 2.5 million acres of private land, including the campsites along the St. John River. The North Maine Woods system has succeeded in maintaining a rough balance of timber and recreational uses in a portion of Maine's wildlands.[16]

In other areas, cooperative programs maintain trails on private land for cross-country skiing and snowmobiling. Efforts by fish and game agencies and landowner groups try to encourage owners to leave land open to public access. As ownerships fragment, however, the task grows more difficult each year.

Nongovernmental, Nonprofit Organizations

The Northeast has fostered a vigorous and diverse community of nonprofit groups that own land for education, research, conservation, and preservation. More than 2 million acres of Northeastern land, mostly forest, have been preserved by these groups, and the region leads all other regions in total area protected by state, local, and regional groups.[17] The region's long history of scientific research and its strong private universities have provided some of the largest and the best-known research properties, such as the Harvard Forest, several Yale forests, and the Pack Forest of the College of Environmental Science and Forestry at Syracuse. The region was an early home of the conservation movement: the Audubon societies, the Nature Conservancy, and hiking groups play strong roles as major landholders. State-level and regional conservation groups, like the Massachusetts Trustees of Reservations, the Society for the Protection of New Hampshire Forests, and the Western Pennsylvania Conservancy, operate major preserve systems (Table 9.11). Also, they often transfer acquired lands to public agencies for management.

Landownership is also pursued by a bewildering variety of local land trusts, Audubon chapters, hunting clubs, and community recreation groups, such as the Boy Scouts, churches, and YMCAs. In the aggregate these groups must be important landholders though no detailed figures are available. In the region's pattern of forest use, they are important because they preserve shorelines and other significant lands from development. In some cases, the lands may be available for timber harvesting.

TABLE 9.11. Land Protected by Local, State, and Regional Land Trusts in the Northeast as of 1994

	Number of Land Trusts	Acres Protected*
Connecticut	112	42,575
Delaware	3	33,846
Maine	76	94,125
Maryland	36	64,949
Massachusetts	122	160,782
New Hampshire**	24	100,000
New Jersey	36	65,789
New York	69	125,248
Pennsylvania	55	326,836
Rhode Island	29	9,999
Vermont	28	91,155
Northeast Total	590	1,202,400
U.S. Total	1,145	3,144,339

*Acres reported by location of land trust (not necessarily by location of acreage). Some acreage may be protected by more than one land trust.

**The figure published for New Hampshire is clearly in error. It is replaced here by an estimate of 100,000 acres; U.S. total adjusted accordingly. Land trusts protected 14 million acres nationally as of 1994.

Source: USDA, AREI Updates, No. 13 (1995).

A special institution that apparently developed in the region is the land conservation trust.[18] These trusts are usually based in a single town. They acquire small but sensitive habitats like bogs, salt marshes, and woodlands for preservation and recreation. They use intensive fund-raising methods but rely heavily on outright land donations. They have been able to acquire and manage a multitude of tiny properties. Being community-based, they are able to identify and acquire small tracts without the resentment that often accompanies state and federal land purchases. The lack of county-level or regional governments has made government an especially blunt instrument for identifying and meeting local land conservation needs. Communities have faced intense suburban growth, farmland conversion, and loss of natural amenity.

In Southern New England, the nonprofit sector rivals state government as a landowner, measured by acreage. In general, the private groups

H. J. Crawford Reserve, a tract of 11,300 acres owned by the Western Pennsylvania Conservancy. Under strict covenants, the Conservancy sold timber rights on this Reserve to a private firm. *Source: Western Pennsylvania Conservancy.*

have devoted more energy to expanding ownership and developing effective public use and interpretation programs than have state agencies.

Although traditionally tax-exempt as charitable and educational institutions, nonprofit landholders are under increasing pressure as taxes rise, especially in suburban and coastal areas. Revenue-strapped towns refuse to allow tax exemptions for nonprofit owners. Some groups must pay property taxes on some of their lands, which dilutes their fund-raising efforts. An increasingly skeptical public attitude toward tax exemptions must be countered by aggressive efforts to demonstrate to citizens and tax payers the public benefits of conserving open space. The Northeast's Town Meeting culture encounters difficulties when conservation and fiscal needs clash.

Foreign Ownership

The 1980s saw increased activity by foreign investors in many U.S. real estate markets. Publicity reached a peak when Rockefeller Center was purchased by a Japanese company. The sales of the Diamond International lands in New York, New Hampshire, and Maine by a British financier

Coastal islands have been a focus for public and private conservation group land acquisition programs. *Photo by Will E. Richard, Outdoor Ventures North, Inc., Georgetown, Maine.*

who controlled the company, were widely noted in the press as a source of concern about who really controlled the region's forests.

Foreign individuals and foreign-controlled corporations have owned forest land in the Northeast for many years. Today, foreign owners hold about 3.5 million acres of forest in the region (Table 9.12). Forests comprise a large portion of the total land owned by foreign entities in the region. In 1990, for example, foreign owners added 530,000 acres to their ownerships in Maine alone, in 46 separate transactions. In the same year, foreign interests in the region sold about 91,000 acres of forest and other nonagricultural lands.[19] In 1994, the S. D. Warren Division of Scott Paper Company, with about 930,000 acres of Maine lands, was purchased by a South African paper company. In 1998, J. D. Irving, Ltd. of New Brunswick purchased 1 million acres of northern Maine lands, becoming the state's largest single landowner. While foreign owners may have many motivations for investing in U.S. real estate, there is little reason to believe that their resource management policies are much different from those of U.S. investors. In some instances, however, the longer time horizon and stewardship ethic of Europeans has produced a higher standard of forest practice.

Policy Themes

Public perceptions about forest ownership and management continue to change. The tension between rights of private owners and needs of the public is never clearly or finally resolved. In the Northeast, environmental

TABLE 9.12. Foreign Ownership: Forest Land and all Agricultural and Forestland, Northeast, December 31, 1997 (1,000 Acres)

	Forest*	All Agricultural and Forest Land
Maine	2,969.4	3,037.2
New Hampshire	16.6	18.9
Vermont	53.2	84.5
Massachusetts	1.6	2.6
Rhode Island	—	—
Connecticut	0.1	1.2
Subtotal	3,040.9	3,144.4
New York	230.5	286.4
Pennsylvania	49.7	103.7
New Jersey	5.6	23.2
Subtotal	285.8	413.3
REGION TOTAL	3,326.7	3,557.7

*And other nonagricultural.

Note: Sappi, Ltd. purchase of S. D. Warren Maine lands (930,000 acres) is included in these tables; that land was almost entirely sold in late 1998. Acquisition of one million acres in Maine by J. D. Irving, Ltd. (of New Brunswick), to close in 1999, not included.

Source: C. H. Barnard and J. Stokes, 1998. *Foreign Ownership of Agricultural Land through Dec. 31, 1997.* USDA-ERS, Stat. Bull. 943.

groups are advocating some combination of public ownership and regulation as a means of preserving wild and undeveloped landscapes. It would seem that in urban areas, there is considerable support for this. On the other hand, a major political backlash in the wake of the 1992 elections has challenged past policies at their very roots.

In places like the Pine Barrens, the Adirondacks, and Maine's Unorganized Territories, zoning systems already in place are being used to protect more subtle values than ever before. Even in these areas, some conservationists are advocating additional public ownership. Examples are the proposed 3.2 million acre Maine Woods National Park around Baxter Park in Maine (Figure 9.8), and the 650,000 acres of additions to the Adirondack Park proposed by the Commission on the Adirondacks in the Twenty-first Century.

Spurred by public concerns and by active conservation groups, a special federal appropriation was obtained for a unique interstate group, the Northern Forest Lands Council. It was to address issues in the northerly

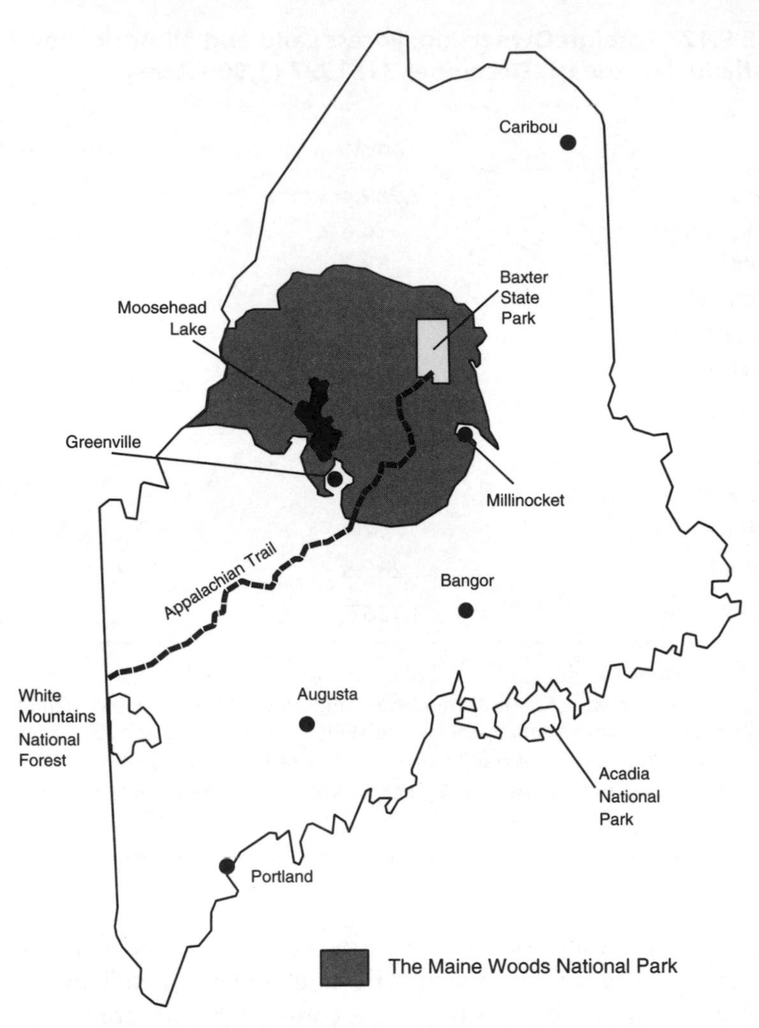

Figure 9.8 Proposed Maine Woods National Park
(3.2 million acres)
Source: Restore North, 1994.

portion of New York and Northern New England. This council was a task force of state agency and private representatives to guide an assessment of the region's forest problems. In a major report, the Council recommended a package of measures and further studies, which were completed in 1994.[20] Following this model, an assessment of the "Highlands" in New Jersey and nearby New York was conducted in 1992.[21] Also in New York State, the Tug Hill Commission and the Catskill Park are attempting to deal with changing land uses and conservation issues.

In addition to these debates focused largely on particular areas, the coming century will see further tensions over state and federal regulation of private land uses. Interest in water quality and fish habitat can only grow; with this may come revisions in the federal/state approach to protecting streams from sedimentation by logging operations. New regulations applicable to coastal areas are being adopted, and a major debate continues over regulating wetlands, many of which are forested. The land boom of the 1980s threatened many traditional forest values and drew public and professional attention to the importance of landownership. Finally, many municipalities and states are implementing various forms of regulations over forest practices. After the 1994 elections, in Washington and in state legislatures, there occurred a dramatic backlash against these and other regulations. It is too soon to tell what the outcome will be. Legislatures, landowners, and the public will further challenge the basic balance in our society between protecting legitimate public concerns while respecting the rights of private owners.

Recent trends in rural landownership have radically increased the number of owners — to almost two million. The median size of a forest parcel is now about ten acres outside the industrial forest. Despite the trend toward parcelization, fully three-fourths of the region's forest is still owned by relatively few owners of 100 acres and larger. The opportunities for serious longterm forest management, for any purpose, are greatest on tracts above this size. The good news for timber and wildlife management, and for public access, is that this ownership category accounts for most of the region's forest, and involves a fairly small number of landowners.

The future use and values of this forest will be determined by the incremental decisions of these two million individuals, families, and companies. These decisions will be swayed, for good or ill, by a host of local, state, and federal regulatory programs aimed at a variety of purposes.

The concept of "ownership" is growing ever more blurry. Private landholdings can be "owned" in a multitude of different components:

- Subsurface minerals;
- Powerline rights of way;
- Fishing access easements;
- Snowmobile trail right of way;
- Maple sugaring rights;
- Conservation easements;
- Timber cutting rights;
- Cottage site leases;
- Hunting rights.

New York-New Jersey Highlands Regional Study

Recreation

- 15 percent of the Highlands is owned by Federal, state, or county agencies. Of this 147,800 acres, 18 percent is in Federal ownership, 63 percent in state parks and forests, 9 percent in state wildlife management areas, and 10 percent in county parkland. Not included in this accounting are municipal parks.
- There are several significant tracts of privately held land that have historically been made available to the public. Notable are: Schunemunk Mountain, NY (6,000 acres), Black Rock Forest, NY (4,000 acres), Sterling Forest, NY (17,530 acres) and Pequannock Watershed, NJ (35,000 acres).
- There are over 48,000 acres of lands and waters owned and managed by water purveyors and although a percentage of these areas are not physically accessible, these lands provide scenic and wildlife values. . . .

Timber

The Timber component of the Highlands forest resources is underutilized. The current harvest level is 10 percent of the annual growth rate, indicating a potential for an increase in timber harvesting without depreciating current stock. *(continued)*

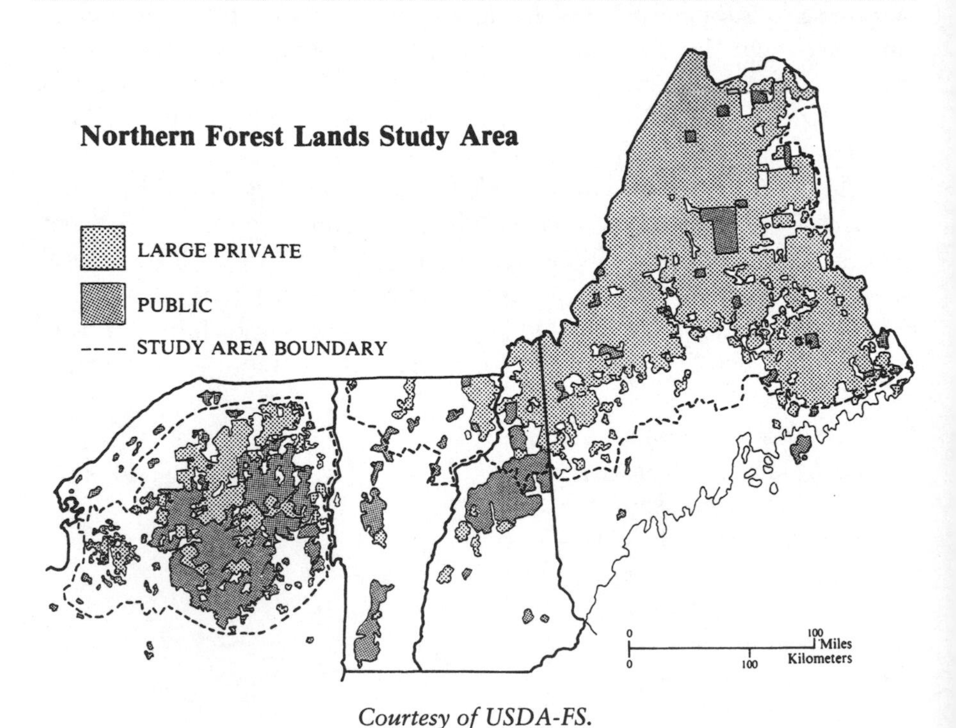

Courtesy of USDA-FS.

The economic value of the sawlog and veneer harvest is over $3.2 million. Another $8 million dollars of cordwood is also harvested.

Most of the Highlands timberland is broken into small tracts of land — 85 percent are less than 19 acres. Surveys have found that these landowners value the forest more for its scenic value then for timber production. Large contiguous tracts of timberland which could be efficiently and economically managed are rare. . . .

- 48 percent of the Highlands is forested. 41 percent is considered to be timberland, forested land both capable of sustaining good tree growth and not legally or administratively reserved from timber harvesting.
- 89 percent of the timberland is in the central hardwood forest type, with oak species comprising 40 percent of the growing stock.
- The Highlands forest is mature and will need selective harvesting to maintain its current level of productivity. 61 percent of the timberland areas can be classified as having too many trees per acre causing overcrowding and a loss of vigor and health.
- 75 percent of the timberland in the Highlands is privately owned, mostly in small lots. Over 85 percent of the Highlands tracts are less than 19 acres; 70 percent are tracts of less than nine acres.
- 72 percent of New Jersey landowners mentioned aesthetics, enjoyment, or increased property value as the primary reason for timber ownership. No owner listed income from timber as a primary benefit of landownership. Only one percent listed income from timber as a secondary benefit.
- 6 percent of New York State land and 24 percent of New Jersey timberland was placed in preferential assessment programs, which give landowners a reduced tax rate in exchange for their promise not to develop the land.
- 42 percent of the towns within the Highlands have enacted forestry management laws that place restrictions on harvesting.

Source: USDA-FS, NEFES, n.d. New York-New Jersey Highlands Regional Study. p. 23–24.

All of these rights may be held, on a single piece of land, by different individuals, groups, or even units of government. Clearly, what it means to "own" land is not, and has never been, a simple matter.

For the coming century, the several traditional sharp lines — such as public-private, plantation-wilderness, park-forest-game area — will be increasingly irrelevant to the real problems:

- Public concerns now address the appearance, uses, and habitat *values of the landscape* as a whole and not just tiny points. The ecosystem does not stop at property lines — or state lines;
- Interminglings of common and undivided tenures, multiple parcels resulting from past farm ownership and failed subdivisions, and severed

George Washington State Management Area, Rhode Island. Killing the overtopping hardwoods has released these pines. *Courtesy of Rhode Island Division of Forest Environment.*

mineral estates all create obstacles to solving problems by outright fee simple ownership.

- Ways to keep farms and forests working are important to rural areas.
- Private ownership rights will demand more serious and thoughtful consideration.
- Public funds for acquisition will continue to be limited, despite recent improvements in the fiscal position of federal and state governments.

This means that innovative concepts of shared ownership and "the Commons" will be needed to forge new tools for the future.[22]

In late 1998, two dramatic land purchases were announced by leading conservation groups. The Washington-based Conservation Fund assembled a financial package to acquire the entire 330,000 acre ownership being sold by Champion International in New York, New Hampshire, and Vermont. At about the same time, the Maine Chapter of The Nature Conservancy announced its purchase of 185,000 acres of northern Maine timberlands from International Paper Company. In both instances, the

The Northern Forest Lands: A Way to the Future

The conditions which up to now have conserved the Northern Forest can no longer insure its perpetuation. In our discussions time and again we faced a fundamental conflict — between market-driven efficiency that encourages maximum consumption of resources with the least amount of effort in the shortest time, and society's responsibility to provide future generations with the same benefits we enjoy today. We believe that until the roots of this conflict are addressed and the economic rules changed so that markets reward longterm sustainability and recognize the worth of well-functioning natural systems, existing market forces will continue to encourage shorter-term exploitation instead of longterm conservation of the Northern Forest.

This report does not address all aspects of this conflict. Instead, we have chosen to concentrate on feasible, effective steps toward changing a range of public policies and trends that now inhibit conservation of the region's forest resources. They include:

- increased polarization among forest user groups.
- rising property taxes, causing loss of land from natural resource uses.
- pressure for development of high-value areas near shorelines and scenic places.
- jobs lost to competition from other regions and countries.
- incomplete knowledge of land management techniques to maintain or enhance biological diversity.
- lack of funding and clear priority-setting for public land and easement acquisition.
- insufficient attention to and funding for public land management.
- fear of losing public recreational opportunities and access to private lands.
- loss of respect for the traditions of private ownership and uses of private land.
- failure to consider forest land as a whole, as an integrated landscape.

We believe that to ignore what the Council has discovered about the forces for undesirable change and take no action would be to guarantee an uncertain future for the Northern Forest, one that could lead to break-up of large undeveloped tracts of forest land, a steadily weakening economy, and continuing pressure on finite natural resources. For these reasons, the Council believes people must act in a careful way to shape change, to conserve the important public and private values of the forest. We see a Northern Forest with extensive forests rich in natural resources and values cherished by residents and visitors; timber, fiber, and wood for forest products and energy supporting successful businesses and providing stable jobs for residents; lakes, ponds, rivers, and streams unspoiled by pollution or crowding human development; viable communities in which people can live, work, and raise families; forest tracts large enough for wide-ranging wildlife, protected and managed in ways that sustain the diversity of plant and animal species; a culture deriving its identity from the environment in which it has evolved. . . .

Source: Northern Forest Lands Council, 1994. *Finding common ground: conserving the northern forest.* 1994. p. 11-13.

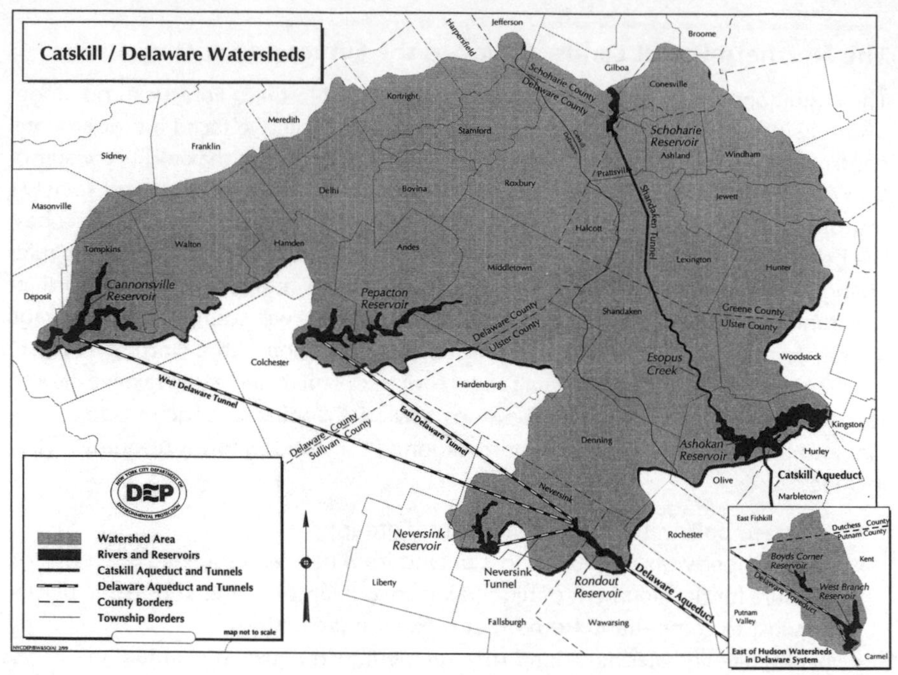

Protecting water supplies has been a traditional motive for forest ownership by public agencies and water districts. The New York water supply system in the Catskills is the focus of a major effort. *Courtesy of New York City Department of Environmental Protection.*

plan was to retain significant acreages in forest management while placing in public ownership tracts of high ecological or recreational value. Then in 1999, the Pingree Heirs, whose lands are managed by Seven Islands Land Company, agreed to sell a conservation easement on 755,000 acres of their Maine holdings to the New England Forestry Foundation. Whether these dramatic developments presage yet a new trend in landownership is too early to tell.

In our region's Town Meeting culture, and at a time of political backlash against government in general, how the region's needs for forest conservation will be met is increasingly uncertain. It may be a time to develop new ideas based on regional traditions, rather than continuing the policy gridlock that increasingly paralyzes federal policy.

10 The Northeast's Timber Budget

A PERSPECTIVE ON the forest that is perhaps unique to foresters is that of the timber budget. Using statistical samples, foresters try to estimate the total volume of trees, the growth, harvest, and losses to mortality. From such estimates, they can see how the forest is changing, determine if the current cut is sustainable, assess management measures, and make recommendations about industrial developments. As for a woodlot, also for a region. This chapter reviews the history of the Northeast's timber budget and the current condition of the forest.

Colonial Times to 1920

Until the mid-twentieth century, statisticians, foresters, and industrialists were unable to calculate the balance of the region's timber budget. Data on timber inventories did not exist. Instead, people could see direct and dramatic evidence of the local timber situation, in cutover lands and higher prices for wood products. In the earliest Massachusetts Bay settlements, fuelwood, and building material scarcity made some locations less desirable for settlement, as William Wood commented in his *New England's Prospect,* published in the 1630s. In Boston as early as 1637–1638, fuelwood was seriously short. Later in the century, Boston was regularly importing fuelwood and the shipbuilding towns of Cape Cod were importing ship timber. By 1690, Rhode Island colonists were relying heavily on construction and ship timber brought in from elsewhere.[1] Winter fuelwood shortages were commonly felt by the urban poor. Town officials took steps to supply fuelwood for the indigents as early as 1635.

So even before the Revolution, the timber budgets of large coastal cities were clearly evident to all, if not in quantitative terms of acres, cords, growth, and cut, at least in practical everyday business terms. Local overseers of the poor distributed fuelwood to the needy, a tradition

Lumber Schooners at Evening on Penobscot Bay, Fitz Hugh Lane, 1863 illustrates Economic Interdependence. *Courtesy of National Gallery of Art, Washington. Gift of Mr. and Mrs. Francis W. Hatch, Sr., © 1995 Board of Trustees.*

that survives in rural areas. During the great Ice Storm of 1998, our church purchased a load of wood, leaving it in the parking lot for neighbors who had run out of wood. Schoolteachers and ministers were paid partly in fuelwood, and widows were provided with fuelwood in the wills of deceased husbands — customs that may yet re-emerge.

By the 1660s, His Majesty the King was being squeezed by the Baltic timber merchants. The Lords of the Admirality worried about the ease with which an opponent could cut off shipments from that region by blocking the Danish straits. This situation stimulated their interest in North American sources of masts and tar. The Broad Arrow Acts, passed at the beginning of 1691, reserved the large pine, 24 inches in diameter and up, for the king, on all unappropriated lands in New England. This policy produced a colorful history of pine-rustling, sporadic arrests by timber agents, stealing by the agents themselves, and general irritation among rural colonials against the distant monarch's policies. It also produced large amounts of 23 inch paneling in colonial houses.[2]

The mast trade revealed the changing geography of a specialized resource — large straight pine and spruce trees — under a primitive transportation system. The mast traders moved slowly up the Delaware, Hudson, Susquehanna, Connecticut, and north from the Merrimack and its tributaries to the Piscataqua and the Maine coastal streams. From stream after stream, the best mast pines close to water were cut, skidded

Snub rope and bridle chains: getting out the masts. *Courtesy of S. F. Manning.*

Hovels for draft animals, colonial mast trade. *Courtesy of S. F. Manning.*

Mast depot for loading masts into specially designed ships.
Courtesy of S. F. Manning.

to water by oxen, and floated out. Through the late 19th century, spars were an important product on the Susquehanna.

During the late nineteenth century, farmland clearing reached its zenith, railroad and industrial use of fuelwood achieved enormous proportions, the wooden shipbuilding industry peaked and declined, and the center of the nation's lumber industry moved westward. It was also evident in the import of timber for ships. By 1850, local wood no longer met the needs, for example, of the Essex and Gloucester yards.[3] The six-masters built on the Maine Coast in the 1890s were planked with southern pine and live oak, and their masts were hewn from douglas fir from Oregon. Because of its rising population and industry, the Northeast ceased to be self-sufficient in lumber after 1880–1890.[4] Before the Civil War, New York was importing lumber. In the mid-nineteenth century, Connecticut probably consumed five to ten times as much wood as it did a century later. This volume was harvested from half as much forest land. The statistics of lumber production told the story clearly (Figure 10.1).

The regional statistics concealed local declines, mirrored in falling employment in major lumber towns like Bangor, the large milltowns along the Connecticut River, Glens Falls, and Williamsport. From the 1890s to the 1920s, a brief boxboard boom throughout Southern New England

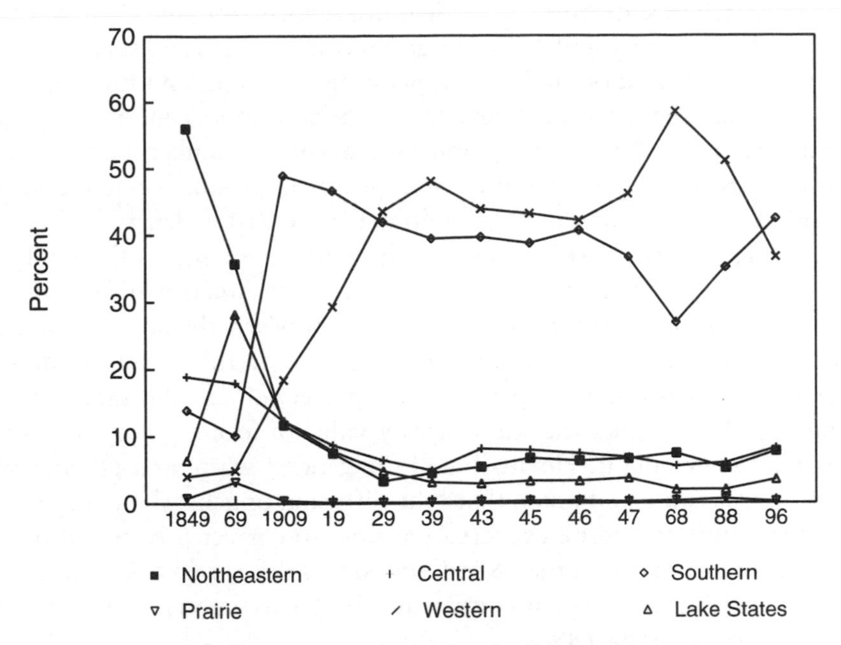

Figure 10.1 Lumber Production by Geographic Region
Source: H. B. Steer, 1947 , and U.S. Bureau of Census, Lumber Production and Mill Stocks.

and New York boosted production there. In local areas, later peaks occurred when hardwoods were cut.

Even by the 1840s, loggers were cutting pines from stands grown on farmlands that had been abandoned before the Revolution. They were still clearing land, but at a diminishing pace. Sizable areas, as on coastal islands, Cape Cod, and the major farming areas of the Mid-Atlantic states had been denuded of forest cover by generations of fuelwood cutting and sheep grazing.

In 1846, George B. Emerson wrote his "Report on the Trees and Shrubs Growing Naturally in the Forests of Massachusetts." He urged that a comprehensive forest survey be conducted and listed the familiar values lost by forest devastation — their aesthetic value, shade, healthfulness, and usefulness for building materials. He noted that in the coastal towns, most wood was already imported. He estimated the fuelwood consumption of the average family at 13 to 14 cords per year, at a price of four dollars per cord.[5]

Early in the nineteenth century, merchants accepted fuelwood in payment for debts. During the Civil War, urban fuelwood prices rose, and cordwood was sent by water to Boston, New York, and other coastal cities. A search of state board of agriculture reports from the Civil War to the turn of the century will yield a large harvest of essays and speeches decrying forest destruction and urging planting, research, and fire control. In the late nineteenth century, then, citizens became more concerned about the region's timber budget. They noted the decline of lumber output in its traditional regions, the expanding acreages of unproductive abandoned pasture, the losses to fire, and the continually increasing demand.

In 1864, George Perkins Marsh, who had experience in the lumber business, decried the destruction of forests in North America and the damage to rivers from log driving. He argued, "I doubt whether any one of the American States, except perhaps Oregon, has, at this moment, more woodland than it ought to permanently preserve."[6] In the same paragraph, he admitted that the scant market value of forest land gave slight incentive for owning it. He attributed the general waste and plunder of forests to this cause and noted that "the diffusion of general intelligence" on forest values would be expected, in time, to correct private mismanagement. In his report on the 1880 Census, Charles Sprague Sargent provided vivid portrayals, often contributed by local experts, of the forest conditions of the early 1880s.[7]

A few hardy workers began to examine their local timber budgets in a more rigorous and quantitative way. In 1875, the New Hampshire Forest Commission reported on the loss of timber volume in wasteful cutting, and the prospect of imminent exhaustion of the forests around the White Mountains.[8] In the late 1890s, pioneer forester Austin Cary conducted a detailed personal field survey of the timber in Maine woods.[9] Since river log driving was the only transportation, he reckoned the timber situation by river drainages — the Androscoggin, Kennebec, Penobscot. He showed the trend in log drives and assessed remaining timber stands. In many areas, he found that previous cuttings, taking only large trees, had left behind well-stocked stands for new harvests. In a 1900 Census Report, geographer Henry Gannett noted that previous estimates of the region's inventory had been far too low. A thorough assessment for Maine in 1904 argued that even with the major boost in paper production, Maine's spruce resources could meet existing demand indefinitely, but the 1909 peak in lumber, the growth of pulp and paper, and the budworm outbreak, changed all that. By 1919, the forest commissioner of Maine reckoned the remaining stand of timber in Maine. He estimated that the timber cut exceeded the growth by tenfold.[10]

The New Jersey State Geologists' Report of 1895 reviewed conditions in that state, finding them as dispiriting as everywhere else. In New York, successive commissions reviewed forest conditions, finding wasteful cutting, timber theft, and destructive fires. A 1916 summary for Pennsylvania was entitled "Forest Devastation in Pennsylvania."[11]

In his remarks to President Roosevelt's 1908 White House Governor's Conference, James S. Whipple, Commissioner of Forests, Fish, and Game outlined the situation as it appeared at the time:

> . . . We have more than 50,000 square miles of area, 32,129,920 acres. It is a rolling, hilly country, with broad valleys and great upland plateaus in the Adirondacks and Catskills and the foot-hills of the Alleghanies. From these uplands most of our rivers run. We now have standing about 41,000,000,000 feet of timber, board measure. Last year we cut off more than 1,300,000,000 feet. About 27% of our total area is covered with forests, good, bad, and indifferent. The State owns, of this woodland, more than 1,600,000 acres. That, under the Constitution, can not be cut. That amount must be deducted in the calculation. A simple mathematical calculation indicates that, at the rate we are now going, in about twenty-two years we will not have a sawing-stick left standing, except on State land.[12]

In Connecticut, foresters conducted a rough survey of the state's forest in 1913, yielding the first statistical picture of forest area and condition in

Sawmill, North Colebrook, Connecticut, 1912, just after lumber output in New England passed its historic peak. *Courtesy of G. E. Nichols Collection, Yale University.*

that state. The first survey of Massachusetts, reported in 1929, took state foresters 14 years to complete![13]

A report on Rhode Island, issued by the state's Agricultural Experiment Station in 1902, listed a significant increase in forest area from 1887 (24 percent of the state) to 1900 (40 percent); the forests were mostly overcut sprout stands. Thirty-three tiny sawmills cut a meager 18 million board feet per year of mostly pine lumber.[14]

Regional Surveys, 1920–1976

The first statistical overview of the region's timber budget was prepared in 1920 by the USDA-FS. The national picture at the time was grim: 326 million acres of cutover forest; 8 to 10 million acres of forest fires per year; rising lumber prices; companies overcutting because of overextended financing; burdensome taxation. Comprehensive forest policy and agencies to implement it were in early stages.[15]

The foresters made their best estimate of regional forest conditions. In using these statistics and comparing them with later surveys, it must be remembered that they are based on limited data, a good deal of judgment, and on prices and marketing standards of the times. They cannot be compared with more recent estimates prepared using statistical sampling, field

Cutover land, New York, late nineteenth century. *Courtesy of New York State Conservation Department, Division of Conservation Education.*

plot measurements, aerial photos, and broader standards of commercial usefulness. For example, different species are now cut, and smaller trees are now used. Since it was on these estimates, however, that legislators, governors, administrators, and industrialists made policy, the view of the time is significant even if statistically weak by current standards.

The region's timber budget in 1920 was badly in deficit. Annual growth was far exceeded by total drain, and the region depended heavily on imported wood and wood products. Regional lumber output had already fallen dramatically from its prewar peak. Just after World War I, the region that had led the nation a century earlier (and that now has 13 percent of the nation's commercial forest) relied heavily on imported lumber, pulpwood, and paper products. The timber supply outlook for the region's paper industry, said the analysts, "holds no particular promise." Some mills were estimated to have only ten years supply of pulpwood. It appeared that without better forestry and major pulpwood imports, the Northeastern pulp industry would be "a thing of the past within 30 years." Interestingly, though a massive spruce budworm outbreak had just severely damaged Maine's forests, the 1920 Capper Report did not mention it. Perhaps the most extreme conditions prevailed in Pennsylvania, where the latest wave of cutting had occurred. By the 1920s, one expert estimated that 5 million acres of forest land was "barren or unproductive, while many more acres are poorly stocked with trees."[16]

The 1940s saw further reviews. The first, covering only New England, was prepared by Henry Baldwin for the National Resources Planning Board in 1942.[17] The report gave considerable attention to the effects of the 1938 hurricane. Its timber data were based on judgment and limited data, since detailed field surveys of New England's forests had yet to be conducted. Of an estimated annual drain from 1932 to 1942 of 570 million cubic feet, fuelwood comprised 42 percent — a high percentage due largely to the collapse of other markets for wood in those years. In the early 1940s, the drain still exceeded the growth of sawtimber. The pulpwood harvest now eclipsed the lumber harvest, and losses to mortality were on about the same scale as lumber drain. The forests at that time were young, recovering from budworm and heavy cutting, and regrowing rapidly in old pastures. Baldwin called for detailed field surveys of forest conditions, such as were under way elsewhere.

After World War II, the USDA-FS conducted its Reappraisal, a major review of the nation's timber budget.[18] The estimates showed continued overcutting of sawtimber trees (hardwoods larger than 11′; softwoods larger than 9″ that are suitable for lumber) in New England while sawtimber growth had begun to exceed drain in the Mid-Atlantic. For all growing stock (trees larger than 5″), however, the balance had turned into surplus — the growth/cut ratio was 1.35. By 1944, fuelwood drain was only 25 percent of the total drain, nearly as great as lumber but in excess of that for pulp. Mortality, a major component of drain, continued to take an amount of timber roughly equivalent to the harvest for lumber. On these estimates, then, the period of depletion was over, as the forests began recovering from overcutting and fire.

In 1947, Dr. V. L. Harper, director of the Northeastern Forest Experiment Station, depicted a grim softwood timber supply situation for a paper industry audience.[19] He noted the heavy dependence on imported pulpwood, the looming budworm epidemic, the high mortality, and the waste in cutting. There was a rough balance in growth and cut, he said, but timber quality was poor and going steadily downward. He urged his listeners to support more use of hardwoods, better harvesting practices, and protection from fire and insects.

Turnaround at Mid-Century

In the 39 years after 1952, the 13-state Northeastern region's timber budget changed markedly. Measured in terms of *sawtimber,* or trees suited for lumber or veneer, the region's inventory increased from 125 billion to 303 billion board feet. Growth of sawtimber doubled, while removals increased by 60 percent.

Growth of *growing stock* increased by 54 percent, to 3 billion cubic feet per year (Table 10.1). Total growing stock removals were also rising, with hardwood removals increasing faster. The region's timber budget showed a growth/cut ratio in 1991 of 2.3 cubic feet grown for each cubic foot cut — the best since foresters began compiling such budgets in 1920 and a healthy physical surplus by any standard. These figures are the first to have been based on two or more detailed field surveys of each state and hence are far more reliable than earlier data. These increases occurred with minimal active forest management.

Despite rising harvests, the total growing stock volume in the region's forests rose dramatically after 1952 (Figs. 10.2 and 10.3). This was primarily due to the increase in volumes per acre, and secondarily to increases in forest acres.

In local areas, however, inventories are declining; in Maine growing stock volumes fell by 18 percent for softwoods while increasing 14 percent for hardwoods from 1982–1995.[20]

Baldwin estimated that 38 percent of New England's forests were under twenty years old in the early 1940s. The highest proportions of stands younger than twenty years were in Massachusetts, with 58 percent, and

TABLE 10.1. Growth and Removals of Growing Stock on Commercial Forest Land, 13 Northeastern States, 1952–1991 (Million cubic feet)

	1952	1970	1991	% Change 1952–1991
Net Annual Growth				
Softwood	652.6	901.7	713.8	+9%
Hardwoods	1,358.0	1,972.9	2,378.8	+75%
Total	2,010.6	2,874.6	3,092.6	+54%
Removals				
Softwood	n/a	412.9	499.5	n/a
Hardwoods	n/a	737.7	819.6	n/a
Total	n/a	1,150.7	1,319.1	n/a
Growth/Removals Ratio				
Softwoods	n/a	2.2	1.4	
Hardwoods	n/a	2.7	2.9	
Total	n/a	2.5	2.3	

Source: Powell, et al., 1993, tables 33, 34. The data are not readily available for our 9-state region, but this summary provides a reasonable picture.

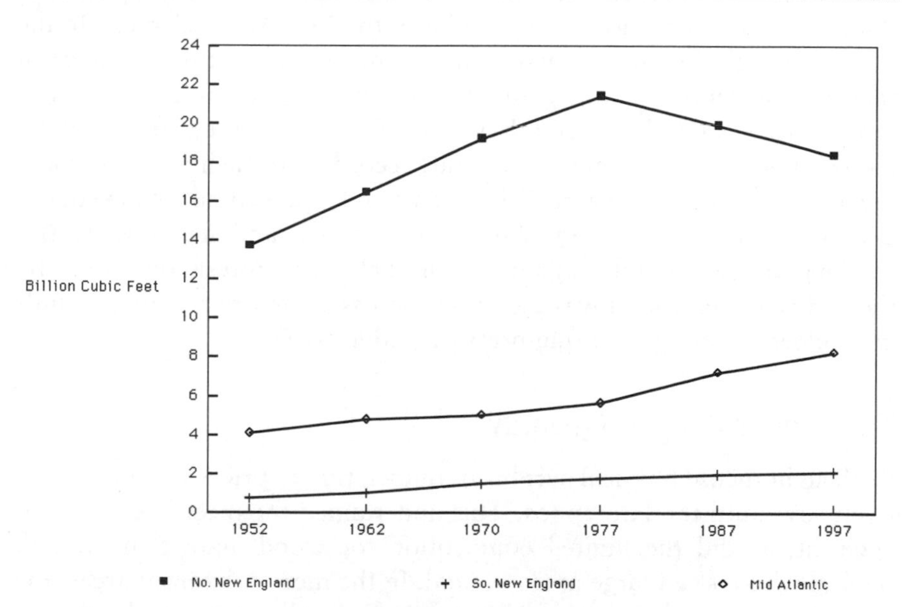

Figure 10.2 Softwood Growing Stock, 9 Northeastern States, 1952–1997
Source: Powell, et al., 1993, Table 11; and USDA-FS RPA Website.

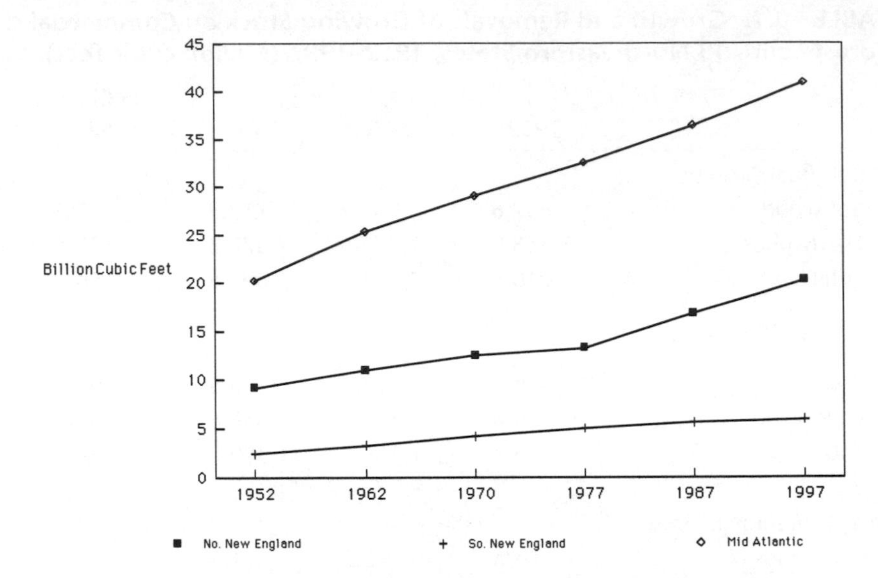

Figure 10.3 Hardwood Growing Stock, 9 Northeastern States, 1952–1997
Source: Powell et al., 1993 , Table 11; and USDA-FS RPA Website.

Rhode Island, with 79 percent. Although about 2 million acres were tallied in 1941 as cutover, nonrestocked, those lands have regrown better than expected. Conditions were similar in the Mid-Atlantic states. In the 1950s through the 1970s, these young stands were entering the stage of their lives when they would gain merchantable volume rapidly. They will now mature, and their growth will decline unless they are carefully tended. Forest surveys rate forest lands according to their level of stocking, or wood volume per acre. Whether a stand is judged fully stocked, understocked, or overstocked depends on its age and its volume. But stocking can be a useful rough indicator of changing forest conditions. By the 1980s, overstocking was increasing across the entire region as stands grew older under light cutting pressure (Table 10.2).

The Timber Budget: Quantity

With an immense physical surplus of timber, timber prices remained relatively low until the late 1970s. This undermined the incentive for management, as did the limited competition for wood among major mill owners who owned large areas of land. In the more urbanized areas, the paper industry gave up pulpmaking and subsisted on imported pulp, recycled stock, and rags. Lumber production limped along. In fact, the poor marketability of forest land persisted until the 1970s, in some remote ar-

TABLE 10.2. Percent of Northeastern Forest Acreage Overstocked, Recent Forest Surveys

State and Recent CFL (Acres)	Overstocked Acres (Year)	Overstocked Acres (Year)	Increase in Overstocked Acres	Latest Year Percent Overstocked
Maine 17,060	4,372.00 (1971)	8,248.00 (1982)	3,876.0	48%
New Hampshire 4,812	774.50 (1973)	1,680.40 (1983)	905.9	35%
Vermont 4,422	252.60 (1973)	878.10 (1983)	625.5	20%
S. New England 5,078	1,374.50 (1972)	2,323.90 (1985)	949.4	46%
New Jersey 1,864	n/a	338.20 (1987)	n/a	18%
Pennsylvania 15,873	n/a	4,237.10 (1989)	n/a	27%
New York 15,406	2,121.30 (1980)	4,589.70 (1993)	2,468.4	30%

Note: Stocking class is shown for growing stock trees only, except for Southern New England, which considered all live trees. For the most recent round of surveys, definitions of stocking levels have changed, so new estimates will not be comparable with earlier ones.
Source: State resource bulletins.

eas — land could be bought for less than the value of the timber standing on it. Shrewd operators can still do this today in some rural areas.

The trends from 1952 to 1997 show an impressive increase in total wood volume in physical terms. But there are reasons to doubt that this situation will continue for very long. First, the regional timber land base continues to erode slowly through land conversion to other uses, to parcel fragmentation, and to incremental effects of regulations. Second, timber harvests are rising with increased mill capacity. Also, the spruce-fir resource has endured a spruce budworm outbreak that reduced growth dramatically and killed a large number of trees. Finally, all across the

Northeast, unmanaged stands are growing older and are often overstocked. They are, over widespread areas, at an age and in a condition in which growth rates naturally decelerate. For most of this century, forest owners faced their most difficult problems in marketing wood, not in growing it. High-grade logs could always find buyers, but many stands could not be managed without markets for lower-grade wood, and those markets were skimpy and erratic.

The principal reason for the resurgence of Northeastern forests was that from World War I to the 1970s, local mills could not compete with construction lumber cut from the forests of the South, the Pacific Coast, and Canada. Paper made from chipped sawmill wastes in Oregon and Georgia was cheap and produced in growing quantities. During this period, the region's residents bought much of their wood products needs from other regions.

As would be expected in a maturing, largely unmanaged forest, mortality is high relative to product use. In Vermont and Southern New England, Pennsylvania, and New Jersey, where pulpwood markets are limited, pulpwood harvests have been less than the average annual mortality. Even in Maine, by the early 1970s, pulpwood harvests were only slightly greater than mortality. Markets were so poor that much of the wood harvested in landclearing — a volume that in the early 1970s and 1980s exceeded harvests for industrial uses in Connecticut and New Jersey — was simply piled and burned or buried. In the 1990s, this wood is being recycled into products like landscaping mulch. The White Mountain, Allegheny, and Green Mountain National Forests have until recently been unable to sell the wood volumes that need to be harvested to maintain timber productivity.

In view of expected increases in wood consumption, a massive forestry job is needed even in the face of the statistics of abundance. With the increase in absentee ownership, ownership of farm woodlots as green backdrops for second homes, and fragmentation of parcels, the difficulty of buying wood has increased. Many owners are unwilling to let loggers on their land, having seen a neighbor's lot rutted by sloppy skidding, a few spindly survivors tottering over a field of stumps and slash. On the other hand, the increased use of fuelwood in the 1980s enhanced landowner willingness to cut trees — at least for a time. Higher wood demand and prices have already stimulated higher harvests in most areas.[21] Economic supply refers to the willingness of landowners to sell trees for harvest. At present, the regional economic supply of wood is less than measured growth of wood. How rapidly this is changing, nobody knows.[22]

The products harvested from Northeastern forests are equivalent to about 25 million cords of wood. They were worth, delivered at roadside

and to local mills, several hundred million dollars to those selling, cutting, and hauling them. This sum does not count the value added in converting the raw logs to products. The intensity of harvesting varies widely across the region, as shown for pulpwood in Figure 10.4.

The Quality Problem

The region's total *inventory* of wood has been rising. But what about the *quality* of that wood? Poor tree quality is a serious problem in Northeastern forests. Much of the spruce-fir forest grows on wet, shallow soils that make trees prone to butt-rot and windthrow. The white pine resource is badly damaged by repeated weevil feeding, which reduces lumber value; many trees are crippled by blister rust. Hardwood stands everywhere suffer from centuries of abuse and neglect. Past high-grading has removed high-quality trees. In most areas, high-grading continues. Sprout regrowth, woodland grazing, and fires have promoted butt scars, root damage, and rot. Diseases and insects, like the beech scale, reduce quality. In 1920, the Forest Service estimated that fully 60 percent of the volume in New England's forest was "fit only for firewood," although utilization standards have changed since then. In New York, 62 percent of the forest

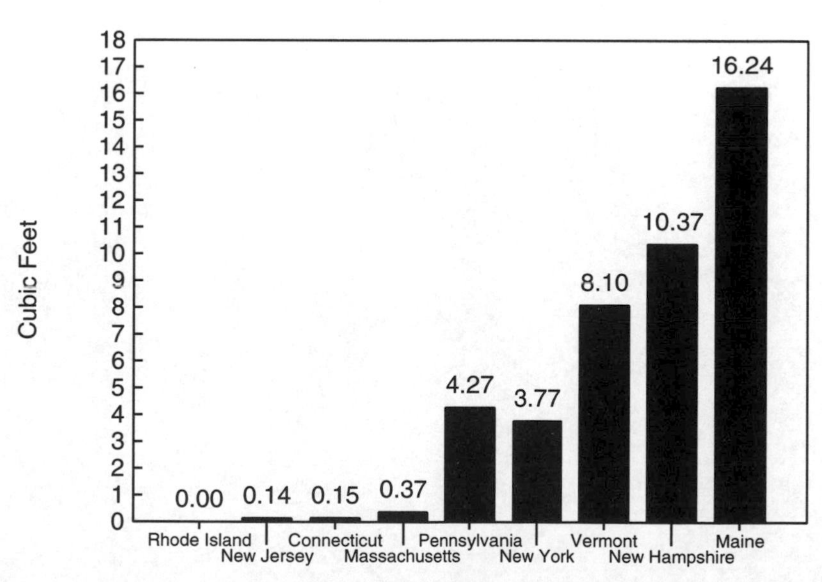

Figure 10.4 Northeastern States Pulpwood Harvest Per Acre of Commercial Forest Land, 1994–1996
Source: R. H. Widmann, and E. H. Wharton, Pulpwood Production in the Northeast. *1994, 1995, and 1996. USDA-FS NEFES, Res. Bull. NE-140 (1998), p. 26.*

area was judged fit only for fuel or "acid wood," and in Pennsylvania only 50 percent of the forest growing stock was felt to be useful, even for mine pit props. Fires in the Keystone State ravaged half a million acres per year, out of 12 million forested acres, and "nearly one seventh of the entire state, once richly wooded, is said to be practically barren."

The forest surveys of the 1970s showed that the tree quality problem was serious. In Southern New England, the 1972 surveys found that an average acre carried 605 trees.[23] Of these, 205 were rough, or less often, rotten. Fully 34 percent of all stands there were understocked with commercially valuable trees by reason of their burden of cull trees.

In Maine, the quality of the sawlog inventory declined significantly from 1959 to 1972. The average grade of hardwood sawtimber fell, and the sawtimber volume in the top log grade fell by a striking 35 percent. Hardwood sawlog quality in Vermont and New Hampshire likewise declined. In Vermont, the forest is seriously burdened by low-quality trees. As forester Neal Kingsley wrote, "Almost 30 percent of all the live trees in the State over 5 inches in diameter at breast height, 4.5 feet above the ground are considered rough and/or rotten." In more recent surveys, volumes of higher value logs rose, but mostly because of increases in size. In Pennsylvania in 1987, 14 percent of the trees were culls, and in New York the situation is similar, but in New Jersey only 10 percent of the trees were culls.[24]

Carefully thinned stands will produce higher quality wood for the future, and may aid in retaining larger trees on the landscape. *Photo by Robert Seymour.*

Quality trees yield lumber valuable for furniture and cabinets. *Courtesy of Steve Lawser, Wood Components Manufacturers Association.*

The hardwood forests of this region owe their origin to the accidents of ecology and human disturbance. Indeed, considerable acreages of hardwood type are occupying lands well suited to growing high value softwoods. Considering their origins, general lack of conscious tending, and the buffeting of occasional insect outbreaks and storms, they could hardly be expected to be in exemplary condition.

Conscious longterm forest management has been applied only to limited acreages. Sloppy or indifferent cutting has been prompted by poor markets, by landowner indifference, or by submission to financial pressures. Except in limited areas, poor markets for hardwood pulpwood and firewood still make serious, intensive management nearly impossible without making large investments per acre. Few individuals or companies have been willing to make these investments. Public agencies have displayed longterm and visible leadership in hardwood silviculture only in exceptional instances. But the various landowner assistance programs have helped promote intelligent timber stand improvement activity on a limited area.

On the face of it, federal survey data on log quality are reassuring. The surveys so far show total inventories of high grade logs to be increasing in the Northeast. Whether this can continue as cutting pressure rises on the large-diameter, quality logs of just a few premium species, must be doubted. Considering the poor quality of management that generally prevails, it is hard to believe that many landowners are practicing true quality renewal (as contrasted with volume renewal) except in limited areas. If actual log quality is indeed increasing, it is generally fortuitous and not due to conscious landowner actions.

Hardwood timber volumes have risen dramatically. But talking to log buyers regionwide would not give you that impression. Why do logs seem harder to find if timber is coming out of our ears? First, the supply of desirable stems that are also in high volume stands is small. So a good deal of the reported volume of high grade wood is scattered here and there and is in inoperable stands. To some extent, desirable stems are being mismarketed for lower grade uses in some areas. And competition for the best logs is increasing. Finally, at any given time, owners are only willing to sell

Wholetree chipping equipment at work in the Maine Woods.
Courtesy of Maine Department of Conservation.

a fraction of the total inventory. So we can see why logs may be harder to get even if volumes of quality material are increasing.

Poor tree quality exacts heavy economic costs. Low-grade trees are useful for fuel, but provide poor material to support industry. Their presence reduces stumpage values and fills growing space that could be occupied by high-quality trees or by young regeneration. Many Northeastern woodlots could be improved dramatically if one-third or more of their poorest quality trees could be made to vanish, or be promptly sold for fuel or pulp.

Of course, rough and rotten trees are valuable as den and nesting trees for birds and wildlife. Oaks and beech provide mast crops for deer, squirrels, and bear. For these reasons, no forester would strip all the low-grade trees from a woodlot. Many large public organizations have guidelines for retaining den trees in harvesting operations.[25] Additionally, it is increasingly recognized that forests need a certain amount of "down woody debris" — basically rotting logs — on the forest floor as habitat for forest floor creatures and to complete nutrient cycles.

The question of quality of forest practice has risen onto the political agenda in a number of states. A state-appointed committee in New Hampshire studied the issue intensively in 1997; a Vermont "Heavy Cutting" law designed to prevent excessive cutting was enacted in that year. Quality of cutting practice was hotly debated during the Maine Clearcutting Referendum. Field studies are under way at this writing in Pennsylvania and New York to initially assess the quality of cutting practices there. As the importance of this issue is realized, we can expect it to receive even more policy attention.[26]

Poor tree quality, and widespread cutting that is degrading stand quality further, are major reasons why foresters, landowners, and citizens concerned with wood supply cannot be complacent about the future of the Northeast's forest.

Third-Party Certification

A new development in addressing the issue of quality of management practice is the concept of Forest Certification or "green certification." This is an application of a broader concept of environmental certification that is being applied to products as diverse as printing paper, plastic containers, and furniture. In countries like Germany, governmentally sponsored certification programs rate products according to strict environmental criteria. These certifications serve as Good Housekeeping Seals of Approval, or the Underwriters Laboratories safety certifications. They provide information to consumers about how products are produced. "Green certi-

Keuka Highlands: State Forests Unit Management Plan

The legal mandate enabling the Department of Environmental Conservation to manage state forests for multiple use is located in Article 9, Title 5, of the Environmental Conservation Law. Under this law, state forest lands shall be forever devoted to "reforestation and the establishment and maintenance thereon of forests for watershed protection, the production of timber, and for recreation and kindred purposes." . . .

The first goal is to ensure ecological diversity and continuous flow of forest products. In order to meet this goal, two basic silvicultural systems will be employed:

1. Uneven-age Management — This is a system of management where trees of different ages or sizes are maintained throughout a forest stand at all times.
2. Even-age Management — This is a system of management where all of the trees in a forest stand are maintained at the same age.

Uneven-age management will be used where it is necessary to maintain large trees at all times for aesthetics and/or where forest stand structure shows a minimum of three broad size classes.

Even-age management will be used when it is not important to maintain large trees on the site at all times and where an even-age stand already exists. Even-age management will be used where deer browsing creates a regeneration problem.

The Keuka Highland Management Unit is comprised mostly of even-aged forest stands. Because of the predominant even-aged stand structure, the even-age management system will be the system most commonly applied.

A second timber management goal is to create a better balance between the age classes of the forest stands.

Visually and ecologically sensitive areas will be managed as natural areas with harvesting limits applied. This comprises 1% of the entire area.

A third goal is to maintain a conifer component. The conifer component will be perpetuated by natural regeneration and planting, if necessary.

To ensure that the goal of sustained yield and ecological diversity is achieved, this plan strives to more evenly balance the age classes of the forest stands and also to perpetuate a portion of the unit in conifer stands.

Management strategies for forest stands reflect the multiple objectives for the unit. For example, because the unit lacks old growth forests, the rotation has been lengthened on natural forests. Because the unit also lacks seedling-sapling forests, some plantation stands have been placed on a shorter rotation. The ultimate result of the various management strategies is a more diverse, better balanced ecosystem.

Some stands fall into more than one management objective during the ten year planning period. For example, stand B-2 is currently a conifer plantation. During the next ten years, it will be converted to a natural hardwood stand by natural regeneration. Consequently stand B-2 is included in each of the first three timber management objectives.

The objectives for timber management in the Keuka Highlands FMU are to:

1. Manage 2,795 acres of natural hardwood and natural conifer on a 150-year rotation.
2. Manage 778 acres of plantation on a 50-year rotation. These stands will be converted to natural stands after they have been harvested.
3. Manage 940 acres as conifer stands. These will be perpetuated by natural regeneration and if necessary, planting.
4. Manage 56 acres of natural hardwoods on a 60-year rotation.
5. Manage 24 acres in a grass/shrub condition.
6. Harvesting limits will be strictly applied on 42 acres of natural hardwoods.
7. Manage 448 acres of forest land on an uneven age basis and 3181 acres on an even age basis.
8. Conduct forest inventory on a 15-year cycle.

Source: New York State Department of Environmental Conservation, 1993. Keuka Highlands: state forests unit management plan. January. p. 24, 26–28.

fication" of forest management programs and wood products is in a very early stage in the U.S. and Canada. The concept is that producers voluntarily submit their management programs and products for certification. Consumers concerned about environmental quality can then purchase certified products, knowing that the environmental claims made by producers are validated by credible third parties.

National programs of management standards are being developed by the American Forest and Paper Association and by the Canadian Pulp and Paper Association. In addition, several private and nonprofit standards-writing groups are developing certification programs. An international group, the Forest Stewardship Council, is attempting to provide a common base of standards so that products entering into world trade can carry valid standards with them to their end users. This is a dynamic process at an early stage of development.[27]

Two examples from this region will illustrate the process. These were both certified by private certifiers. The firms used independent experts and peer review panels to rate management practices against its own list of standards. One was the Kane Hardwood company of Pennsylvania, the other is the public Quabbin Reservoir Watershed in Massachusetts. Also, a leading Maine firm, Seven Islands Land Company, manages 900,000 acres of certified Maine timberland, and a number of wood products from that property now carry labels that follow them to the ultimate consumer.

Whether the various forms of green certification will promote the

Green Certification: Pennsylvania Industrial Timberland

Kane Hardwood, a division of the Collins Pine Company, has recently been certified "well managed." Certification was conducted by Scientific Certification Systems (SCS), which previously certified the Collins-Almanor Forest in Northern California, managed by Kane's parent company.

Headquartered in Northwestern Pennsylvania, Kane manages 133 separate parcels of land covering 116,000 acres in the Allegheny Plateau, one of the few large blocks of contiguous forest in the Northeastern United States. There are 14 commercial tree species in the Kane forest. Predominant hardwoods include black cherry, red oak, hard maple, soft maple, ash and white oak. Also present are Eastern white pine and hemlock. The Kane forests are mainly even aged, having been established after extensive clearcuts around the end of the last century. A smaller component is comprised of 65- to 75-year-old stands that were established after wood harvests in the 1920s and 30s.

The SCS team of two professional foresters and a wildlife biologist evaluated Kane's performance in three vital areas: timber-resource sustainability; ecosystem maintenance; and a variety of financial and socio-economic considerations. The team reported that the properties under Kane's management are "notable for the manner in which industrial timber production has been harmonized with the longterm stability of the timber resources and consideration of the nontimber components of the forest."

Certified veneers are being manufactured by the Freeman Corporation of Winchester, Kentucky. Freeman slices about 20 million sq. ft. each year of certified cherry, oak and maple veneers from the Kane properties.

Source: Understory, Spring 1995 (Journal of Woodworkers' Alliance for Rainforest Protection).

wider use of environmentally and silviculturally sound forestry remains to be seen. As useful as certification may turn out to be, it seems likely to be applicable only to a modest area of forest and a small portion of the total regional production of wood-based products.

Fuelwood: Economic Demands and Environmental Issues

Fuelwood has long been an essential commodity. In 1880, it was estimated that New Englanders cut 4.1 million cords for residential fuelwood use alone.[28] In rural areas, woodpiles never completely vanished from farm and village dooryards. Hikers in remote areas of abandoned farms and villages find rusting cookstoves left in cellarholes and dumps. These relics symbolize a change in energy technology and way of life that many of us took to be permanent.

U.S. Forest Watershed Certificate Marks an Environmental Milestone

The beautiful 58,000 acre Quabbin Reservation is not just a natural wilderness. It is a working forest, managed as a watershed in order to guarantee pure drinking water for 2.5 million residents of Eastern Massachusetts.

While 3 to 5 million board feet are harvested annually from the area under active management, no specific revenue goals drive Quabbin silviculture. This has allowed careful, sustainable practices to prevail for almost four decades.

Indeed, for the past 30 years the Metropolitan District Commission's chief forester, Bruce Spencer, has coordinated a team of staff foresters and private contractors to manage the Quabbin Reservation with great care, balancing watershed protection and conservation of biological diversity with controlled timber harvesting. Goals which made certification to FSC's standards highly attractive.

And now, as a result of Forest Stewardship Council (FSC) certification, Quabbin Reservoir — at 24,000 acres the largest unfiltered surface drinking water source east of the Mississippi River — and its forest lands looks set to be used as a role model for evaluation of watershed management at a number of major public water supplies in the northeast. Already with the scheme in place, the forest's manager, the Metropolitan District Commission, is able to save more than $800 million in water treatment costs each year. Timber from Quabbin also earns the Metropolitan District Commission between $300,000 — $500,000 annually.

The Metropolitan District Commission is now hoping to tap into the green market with certified timber carrying the FSC logo. It already provides timber to around 40 forest product companies.

The assessment at Quabbin Reservoir was carried out by the SmartWood program. "Quabbin is an incredibly important resource for Massachusetts residents and visitors," says Alan Calfee from the National Wildlife Federation, "and during certification assessment we received valuable public outreach assistance from local conservation interests such as the New England Small Farm Institute."

The certificate was secured in June last year.

Source: FSC Notes, February/March 1998: Issue 7, p. 8–9.

The Northeast, particularly New England, is playing a leading role in wood energy use (Table 10.3). The tonnage of wood used each year in the region displaces roughly a billion gallons of fuel oil — largely imported — every year. Unless oil prices jump sharply we may never reach a level of residential fuelwood usage much higher than during the 1980's.

Until the mid-nineteenth century, wood and water were the premier energy sources. The aggregate economic cost of woodfuel, counting the la-

TABLE 10.3. Wood Energy Use in the Northeast, Early 1990s

States	Wood Used/Year, Green Tons, 1992		Biomass Fired Electric Capacity (megawatts)
	Industrial/ Commercial	Residential	
Maine	6,945,033	1,267,300	453.0
New Hampshire	2,154,471	848,700	130.0
Vermont	652,000	761,200	91.0
N. New England	9,751,504	2,877,200	674.0
Massachusetts	127,050	937,000	178.6
Connecticut	13,352	915,731	169.4
Rhode Island	0	127,000	13.0
S. New England	140,402	1,979,731	361.0
New York	1,118,000	4,444,000	90.0
Pennsylvania	477,320	3,600,000	75.0
New Jersey	65,000	1,518,000	134.0
Mid-Atlantic	1,660,320	9,562,000	299.0
Northeast Region	11,552,226	14,418,931	1,334.0

Source: National Wood Energy Assn., *State Biomass Statistical Directory, 1993,* and Coalition of Northeastern Governors, *Economic impact of wood energy in the Northeastern states,* vol. I, Washington, DC, 1994, processed.

bor time spent in its production, must have been enormous — a hidden item in the region's gross product. Even into the twentieth century, fuelwood was important in major industries. Fully one-third of the regional timber cut in 1920 was for fuelwood, and about one fourth of total drain in 1944 was for fuel. Estimates of varying quality have been made since, but they all reflect a steady level of use until the 1940s. Consumption then fell until the early 1970s. In the mid-1970s, oil prices soared. Thrifty Northeasterners hauled cookstoves and heating stoves out of barns or from the hardware store and restored chimneys to service. They installed copper tubing in wood furnaces to heat water and fostered a booming business in renting woodsplitters. Households began bidding hardwood pulpwood away from the papermills, and biomass electric plants were built.[29] Close to many cities, hardwood stumpage today sells for better prices as firewood than as pulpwood. As oil prices fell again after the mid-

1980s, fuelwood use declined. In the Northeastern recession of the early 1990s, utilities came under heavy pressure over high rates, and began buying out high-cost contracts with biomass power plants. Wood use for energy by the mid-1990s was definitely on the wane.

Residential use of wood for fuel has been most visible and is a standard topic of coffee break and cocktail party discussion today. But a sizable amount of industrial and commercial energy is being generated from wood. Churches, schools, and public buildings have been converted. Most of the large and medium size sawmills of the region use woodwaste for steam power. Surplus power is sold back to the area electric utilities.

Equipment designed for supplying chips to pulpmills can chip trees in the woods, including bark, branches, and leaves. When the whole tree is chipped, along with all small trees, the biomass removed per acre substantially exceeds the wood volume removed for traditional products. A large part of the nutrients in a stand are in the leaves and branches, which remain on the forest floor after traditional cutting. Conservationists and foresters are concerned that the large wood volumes required by such systems will not be harvested in a sound silvicultural manner, with minimal impact on soil, aesthetics, wildlife, and forest productivity. Several states attempt to ensure forester involvement in wholetree chipping timber sales.

As more wood-fired electric generating plants appeared, wholetree chipping increased. This raised concerns over possible nutrient depletion of the soils by nutrient removal. In cutting for traditional products, a residual stand is often left, which protects the site and retains nutrients. As branches and tops decay, they restore the nutrients to the ecological cycle of the site. Many Northeastern forest soils are shallow and infertile. Because of their high moisture content and acidity, some nutrients are not readily available to plants. Clearcutting accelerates organic matter decomposition and the loss of nutrients to volatilization and erosion. On some soils, repeated removal of all forest biomass — on short cutting cycles — would lead to reduced soil productivity. Such management practices would damage most other forest values as well. Sustaining forest ecosystem functions requires leaving some large log-sized material to decay naturally on the forest floor. As knowledge increases, it will be possible to design forest harvesting practices to avoid depleting soil nutrients.[30] Ensuring that such practices will be implemented will be far from simple, however.

In parts of the Northeast, urban woodwastes from yard and street trees and from demolition debris are becoming important sources of biomass fuel and other recycled products. Major efforts are being made to keep such wastes out of landfills, where tipping fees exceed $100 per ton in some suburban areas. In some areas, the avoided costs from disposal charges enables this material to be delivered to power plants at costs equal to or below the cost of chipped trees from the forest.

Spurred by the rising oil prices of the 1970s and 1980s, investigators have studied the feasibility of short-rotation biomass forests for energy. These are plantations of fast-growing genetically improved trees like poplar, sycamore, and cottonwood. They are carefully planted and tended, and mowed every three to five years, regenerating immediately from the rootstocks. High yields of biomass can be obtained in this way. Although regional energy costs are high, I do not see short-rotation biomass forests being planted here on a noticeable scale, until dramatic changes in energy costs occur. The promise of wood energy for this region can be amply fulfilled without biomass plantations. Despite current conditions, the outlook is for future wood use for energy to be significantly higher than at present.[31]

Tree Mortality, Forest Health, and the Global Commons

The Northeastern forest, when Europeans first arrived, was in a dynamic equilibrium. Wood growth roughly balanced mortality over the landscape as a whole. A patchwork of stands, many of which were extremely old, held high standing volumes and were gaining little or no net growth. Periodic fires, major storms, and insect outbreaks flattened large areas, restoring young vigorous stands. Spruce budworm outbreaks, every 40 to 80 years, are most dramatic. The 1912–1920 outbreak killed enough spruce and fir to keep Maine's current paper industry going for ten years.[32]

In the late nineteenth century, mortality was low in the rural forest outside of the wildlands. Because of heavy cutting, few trees lived long enough to suffer old age decline, insect and disease damage, or windthrow. In the early twentieth century the chestnut blight eliminated within 15 years the region's most important hardwood timber tree. Nationally, chestnut lumber output was 664 million feet in 1909, an amount that would make it one of the leading hardwoods today. The introduced white pine blister rust inflicted serious damage.

After its introduction in Massachusetts in 1869, the gypsy moth killed large volumes of oak. After slowly spreading across the region, the gypsy moth outbreak erupted in the early 1980s to cover almost 13 million acres before receding once more (Figure 10.5). By the early 1990s, the insect was widely established in the Northeast with outliers elsewhere (Figure 10.6).[33] The gypsy moth is an important pest of park and urban trees.

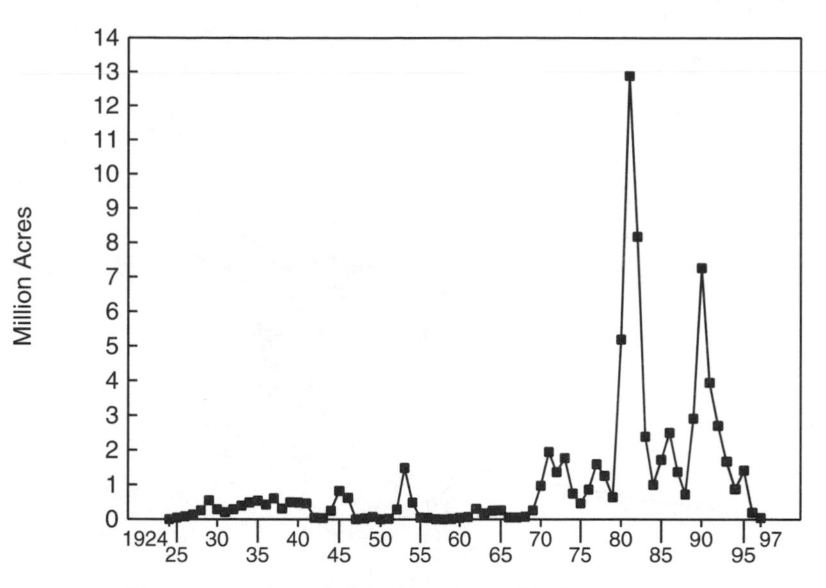

Figure 10.5 Acres Infested by Gypsy Moth, 1924–1997
Source: USDA-FS, Gypsy Moth, 1985; and USFS Website.

And Dutch elm disease has ravaged the region's streets and country roads — some of which are still littered with silvery elm skeletons. New Haven, Connecticut, the Elm City, now holds that title in name only. Dramatic losses in white birch were caused by the birch dieback, a syndrome of unknown cause, in the late 1940s. This epidemic swiftly killed mature birch over northern portions of the region. Later, a beech scale epidemic moved westward from New Brunswick and Maine, killing and stunting beech. Since beech, because of poor markets, was seen as a weed tree at the time, the damage was of little concern to foresters and landowners. Newly killed beech makes good firewood. But in many northerly areas, beech is an important mast producer, relied on by bear, deer, and other wildlife. Hemlock woolly adelgid is threatening hemlock in some areas. The role of insects in shaping the forest has often been ignored. Two scientists suggest that additional imports of harmful insects can be expected.[34]

Trees in the Northeast are subject to massive hurricane damage, an average of once per century on exposed sites. The famous 1938 hurricane blew down 2.5 billion board feet of New England timber, much of it pine. At the lumber production rates of that time, that volume represented perhaps a decade's cut of timber. Large fires in the past consumed much timber. They often started in budworm-killed or blown-down stands, or in logging slash. The damage from the great Ice Storm of January 1998 is still

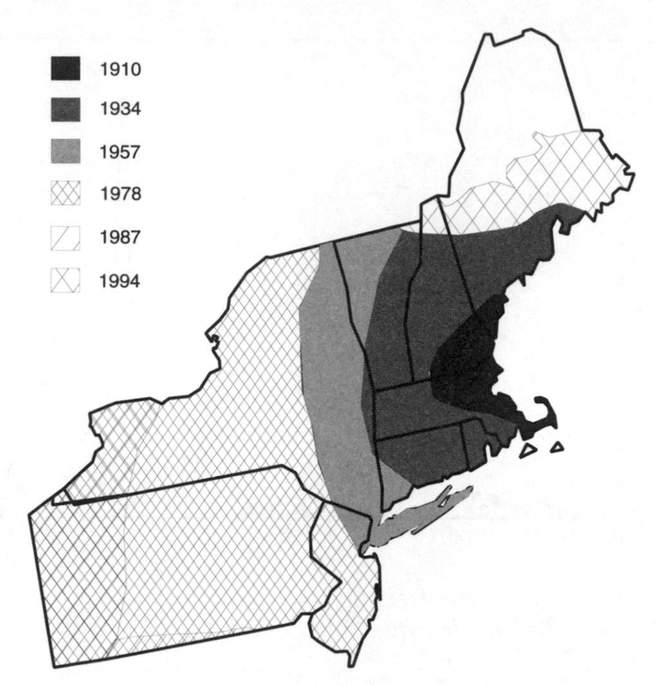

Figure 10.6 Spread of Gypsy Moth
Source: USDA-FS, Draft Impact Statement, April, 1995.

being tallied as this is being written. Intensive fire control programs have made fire risks low today despite the high hazard presented by extensive motorized recreational uses.[35]

The periodic forest surveys have made estimates of mortality losses. In 1986, the USDA-FS estimated that mortality was equivalent to 27 percent of total removals in the 13 Northeastern states. Mortality exceeded the total harvest of pulpwood. The impact of damaging agents on the forest remains huge, despite a growing level of cut and improved methods of forest protection. An encouraging trend is the increased emphasis on forest health as an integrative concept replacing the previous focus on individual insects, pathogens, and damaging agents.[36]

During the 1980s, some scientists began to argue that acids or other components of polluted rainfall were implicated in the declines of high-elevation Eastern forests. Some scientists believe that this fear was exaggerated, and the debate continues.[37]Scientists are debating the outlook for global climate change, with an apparent consensus emerging that global warming is likely to continue. If this outlook is correct, it would mean climate changes at a pace far exceeding any experienced in postglacial times.

Health of Spruce-Fir Forests: Adirondack Region of New York

In the late 1800s, concerns over the future of the wilderness resulted in the creation of the first forested area in the United States to practice conservation. The Adirondack Park established state-owned forest preserves where reforestation and watershed protection were practiced and the preserves were to be kept forever as Wild Forest land. The Park now contains over 2.5 million acres of state-owned preserves and in the 1970s became the first significant area to experience comprehensive land use controls on private lands. The area surveyed in New York is comprised primarily of the state lands in the Adirondack Park region. . . .

. . . The relatively older age of the spruce-fir forests in the Adirondack Mountains, compared with the rest of the survey area, contributes to the greater moderate to heavy mortality in the mixedwood and spruce-fir slope cover types. Competition among the trees decreases growth and vigor which leads to increased susceptibility to insects and pathogens. In some cases, large proportions of moderate and heavy mortality in mixedwood and spruce-fir slope is the result of an epidemic spruce beetle infestation — a significant mortality factor in larger trees. Dwarf mistletoe and root rots are also factors in these older stands.

About two-thirds of the spruce-fir slope occurred above 2,600 ft and a large proportion of the area exhibited heavy mortality. Below 2,600 ft, however, a greater proportion of light mortality occurred. Perhaps the reasons for this lower mortality is the better quality of the sites at lower elevations, which provides better growing conditions, and consequently more resistance to forest stressors.

Mixedwood was most prevalent below 2,600 ft where two-thirds of the type occurs. Another one-third was in the 2,600 to 3,600 ft zone. Very little mixedwood occurred above 3,600 ft. Over three-quarters of the mixedwood area had moderate or heavy mortality. As in the spruce-fir slope, the mixedwood had more areas, with moderate and heavy mortality as elevation increased (up to 3600 ft). . . .

Spruce-fir swamp generally was healthy, with about one-half of its acreage light mortality and one-third moderate mortality. Compared with the other states, New York contained the largest acreage of spruce-fir swamp and almost all of that is concentrated at elevations below 2,600 ft, due to the topography of the Adirondacks. The mountains are more dome-like with fewer steep slopes than the other states. Spruce-fir swamp was distributed across the Adirondacks, occurring in the low mountains and western hills in flats near streams, and in notches in the high peaks. . . .

. . . on Whiteface Mountain . . . nearly 92 percent of the total spruce-fir forest was in the moderate and heavy mortality classes (with many of the standing dead trees in the balsam fir type located within fir waves at the higher elevations). Numerous research projects are being conducted on Whiteface Mountain to determine the status of the forest resource and assess various forest stressors.

Source: USDA-FS, 1993. *Aerial assessment of red spruce & balsam fir condition.* NESPF, NA-TP-16-93. p. 27–29.

Standing dead trees, Whiteface Mountain, New York.
Courtesy of Yuen-Gi Yee, USDA-FS.

Potential impacts on Northeastern forests could be dramatic. This adds another important uncertainty to the future of the forest.

Summary

It is difficult to imagine the horrific condition of significant areas of Northeastern forest just a century ago or less. The photos of most stump-lands and burns simply fail to register on the mind. The restoration of extensive forests, wildlife populations, and scenic vistas is an extraordinary

fact. So is the fact that a forest that was at risk of even surviving in 1920 is today showing a positive balance in the regional timber budget.

The Northeast's timber budget has reflected the region's changing needs for wood, rapid natural regrowth after cutting or burning, and the retreat of farming. In this century, wood products from Canada, the South, and the West have been so cheap that Northeastern landowners have just let the woods grow. The next century will bring new concerns — controlling overcutting in local areas, harmonizing conflicting forest uses, and adjusting industry to a changing timber supply.[38] Active management — of high or low quality — will play an increasing role in shaping the forest.

Improving timber growth even more in the Northeastern forest would take many decades under the best conditions. The biological potential is there, and economic factors are improving. What will determine the timber budget in the future will be landowner choices and social decisions affecting forest practices. The improved markets for wood of the 1990s are a two-edged sword for the future forest. On the one hand, they enable conscientious landowners to practice better management. Yet, the strong markets provide temptations to cut quickly for a fast gain. Too many owners and loggers are taking the fast buck. Designing public policies to restrict exploitive cutting is a complex challenge.

The future of the Northeast's timber budget will be determined by a host of interactions and uncertainties. Our metaphors come into play as a shorthand listing for them. Future forest growth will be affected by changes in global temperatures and pollutant inputs from the atmosphere — the Global Commons. Future land availability will be determined by the outcome of Town Meeting policymaking. Future demands will be affected by the economic interdependencies within the region, and between the region and other supply regions. We have already seen how energy prices can change forest utilization. New technologies and products will emerge in the twenty-first century as global supplies of high-quality old growth timber continue to decline. The productive potential of the forest is an essential resource to be bequeathed to the future.

11 Wood Products and the Northeastern Economy

PAPERMAKING and converting, sawmilling, logging, and wood product fabrication employ roughly a quarter of a million Northeasterners. Of these, perhaps half depend directly on the region's forests. The output of these industries totals about 9 percent of the region's manufacturing production.[1] If federal projections for the 1986–2040 period come true, the region could substantially increase its timber harvest and its industrial capacity using wood. This chapter describes the changing economic importance of these industries and explores their future for the next half century.

Vanished Industries

Northeastern forests supplied raw materials for four industries that aided the settlement of the forested uplands, provided seasonal employment, and yielded export crops in areas poorly suited to farming. While these industries are now gone, they helped to shape the region's economy in many ways. These industries were ice, tanning, charcoal iron, and wooden shipbuilding.[2]

Thoreau wrote of the ice-cutters on Walden Pond. This industry was developed in Eastern Massachusetts by Frederic Tudor of Boston and N. J. Wyeth. In 1833, Tudor sent a shipload of ice to Calcutta. Thus began a worldwide trade. The Penobscot, Kennebec, and the Hudson were major sources once ponds near the seaports could no longer meet the demand. The river ice companies had an advantage in the large supplies of sawdust needed to store and ship the ice. Refrigeration and water pollution killed the industry, which largely vanished by World War I. The trade in sawdust to Boston was at times more profitable for Maine schooner owners than was lumber.

The tanning industry was one of the nation's largest in the mid-nineteenth century, providing an essential raw material needed by many

industries. The tanners felled hemlock for the bark, leaving the unwanted logs in the woods. It took 5 trees or more to yield a cord of bark; a tannery might use 6,000 cords in a season. In the Adirondacks, a tannery could cut out the bark within a ten-mile radius in ten years. In the late nineteenth century, one of the nation's largest concentrations of tanneries was in Pennsylvania, close to hemlock supplies and to markets. Some tanneries owned large areas of land to supply their needs. Changing hide sources, new technologies, and industry consolidation ended the industry's dependence on hemlock and on forests.

The charcoal business, largely associated with iron and brass industries, shaped forests across the region. An iron furnace might own 15,000 to 40,000 acres to supply its charcoal needs. The need for wood explains the persistence of large properties in New Jersey until the late nineteenth century. About 45 percent of all the iron in the U.S. was smelted by charcoal in the mid-1850s, and the total output smelted with charcoal rose until 1890. Before the Civil War, the inefficient furnaces consumed 150–250 bushels of charcoal per ton of iron produced. This fell below 100 by 1900. Probably millions of acres of sprout hardwood forests were shaped by the charcoal trade.

The shipbuilding industry used not only pine and spruce for masts and spars, but oak for planking, larch for knees, and imported woods for flooring and finish work. A good-sized schooner might require 200–300 white oaks, a quantity beyond what nearby forests could supply for long. The Northeast was the nation's financial, commercial, and trading center for generations, so it naturally contained the leading shipyards. Philadelphia, Boston, New York, and Maine dominated U.S. wooden shipbuilding. The clippers of Baltimore, Philadelphia, and Boston, and the schooners from the Maine yards showed the U.S. flag around the world, well into the age of steel and steam. Maine led the nation for years, its annual output exceeding 300 ships in 11 of the years between 1820 and 1890. They were, as Coffin called them, "ships with bones bred in Maine's forests"[3] though by the 1890s much of the planking and masts came from other regions.

As these industries waned, their draft on the region's forests declined, and trees could grow to larger sizes. At the same time, the completion of the expansion of farming in the Midwest and Plains after 1910 eliminated a lumber market of some 4 billion board feet annually, and the development of cardboard boxes eliminated the box market for pine boards.

Sawmilling and Logging

Sawmilling is one of the region's oldest industries, second only to fishing. South Berwick, Maine, is said to have been the home of the nation's first

Maine was a leader in shipbuilding throughout the nineteenth century. The industry had nearly vanished when this was taken in 1908 at Rockland.
Source: Maine Forest Commissioner's Report, 1912, p. 30.

sawmill, in the early 1630s. For generations, sawmills were important to rural communities. Towns offered millmen water rights, free timber, and local monopolies as inducements to build. Pit sawn lumber was produced for a time in the colonies, but watermills, often built with gristmills, dominated the region. Water-powered mills still were numerous as late as 1870. Of the 3,577 sawmills counted in New England in the census of that year, only 269 had steam power. These would have been the market-oriented mills on major rivers.[4]

The region has always had two distinct sawmill industries. One has been the large market oriented mills, which prepared large volumes for distant markets — the staves and headings, pine boards, and spruce lumber of previous centuries; or the pine clapboards, spruce 2 × 4's, and maple and cherry boards produced in the mills of the 1990s. The other sawmill industry, comprising most of the mills, operates tiny mills, often roofless, cutting several million feet or less per year, usually on a part-time basis. From the 1900s to the 1950s, portable mills were common. These small rigs were moved from woodlot to woodlot with the loggers, leaving piles of sawdust and slabs and edgings. Even today, these rotting piles of waste can still be found in the woods.

In 1849, before the Great Lakes and southern forests were exploited, the Northeast produced about half of the nation's lumber. At the time of the national peak of production in 1909, the region's contribution had fallen to about 10 percent. Today, the Northeast supplies only 4 percent

TABLE 11.1. Lumber Production, Northeastern States, 1996 (million board feet)

	Total	Softwood	Hardwood
Maine	1,069	939	130
New Hampshire	275	244	31
Vermont	228	124	104
Massachusetts	*	*	*
Connecticut	40	6	34
Rhode Island	11	3	8
New England	1,623	1,316	307
New York	534	91	443
Pennsylvania	1,033	37	996
New Jersey	8	1	7
Mid-Atlantic	1,575	129	1,446
Northeast Total	3,198	1,445	1,753
U.S. Total	44,699	34,025	10,674
Northeast as a % of total U.S.	7.2%	4.2%	16.4%

*Data withheld to avoid disclosing figures for individual companies.

Note: Most of the numbers below 200 have sampling errors above 15 percent. Because of the high sampling errors for Vermont and Rhode Island, they are listed together by the Census Bureau. On the widespread underestimates of hardwood production, see, W. G. Luppold and G. P. Dempsey, *New estimates of central and Eastern hardwood lumber production.* North. J. App. For. (13) (1989): 120–122.

Source: U.S. Dept. of Commerce, Bureau of the Census, Current Industrial Reports, *Lumber production and mill stocks,* 1996, MA-24T, July 1997.

of national softwood lumber output, and 16 percent of the hardwood (Table 11.1). This is less than proportionate to the region's forest area, and less than the region's 20 percent of the national population, which is a rough measure of its proportion of wood consumption. In addition, several hundred million feet per year are produced in Quebec and New Brunswick from New England and New York logs.

Because of the small size and part-time nature of many mills, census workers still have difficulty counting sawmills, so employment and production data must be handled with care, even today. Nevertheless, some broad generalizations are possible. Regional employment in sawmilling grew slowly to a peak in 1900–1910 of about 20,000 jobs, not counting logging. In 1900, in New York and Pennsylvania alone, there were more than 9,000 planing mills, many in urban areas. These mills employed more than 20,000 workers. Since that time, lumber production has fallen by nearly two-thirds, and major classes of products — boxboards, staves, lath — have vanished. Employment in sawmilling today is probably no greater, and may be less, than at the industry's historic 1909 peak.

Employment in logging is the most difficult to count of all sectors of the forest products business. Included in the lumber and wood industry, the census figures definitely understate logging employment. The region must employ at least 10,000 people in logging each year on a full- or part-time basis.[5]

Panel Products

During the 1980s, three plants were built in Maine to produce structural panels. These three plants now produce more than 500 million sq. ft. of oriented strand board (OSB) per year for construction, remodeling, and home craft uses. At times, modest volumes are exported. These panels are made by slicing logs into narrow "strands," which are oriented in a special press that cures the glue holding the panels together. These panels can use woods not in high demand for other purposes, such as aspen pulpwood and low-grade white pine.

Modern hardwood sawmill: Monadnock Forest Products, Jaffrey, New Hampshire. *Courtesy of Monadnock Forest Products.*

In addition, nonstructural panels have seen dramatic market growth. These panels are used in furniture, cabinets, and a wide variety of industrial markets. They can be produced in a variety of sizes. They can be laminated with veneer, melamine, or other surfaces to provide moisture resistant, colorful, and economical finishes. These panels can be made from sawmill residues, but in the Northeast they are made from roundwood. Plants at Deposit, New York, Mt. Jewett, Pennsylvania, and another near Clarion, Pennsylvania, comprise this segment of the industry at present. Allegheny Particleboard's Mt. Jewett mill is one of the largest of its kind in the world. These plants are using a raw material that is in substantial surplus, providing a local supply for the region's large furniture and cabinet industries. A new plant near Buffalo, New York plans to rely on urban wood waste.

Pulp and Paper

The region's pulp and paper industry is diverse. It includes mills relying solely on purchased pulp and mills that ship only pulp but do not make paper (Tables 11.2 and 11.3). The mills specialize in higher-grade papers

TABLE 11.2. Number of Papermills, 1992

	Pulpmills (SIC 2611)	Papermills (SIC 2621)	Paperboard Mills (SIC 2631)
Connecticut	—	4	5
Rhode Island	—	—	—
Massachusetts	—	26	4
Vermont	—	4	2
New Hampshire	—	10	3
Maine	1	13	—
New England	1	57	13
New York	—	32	13
Pennsylvania	—	16	12
New Jersey	—	6	8
Mid-Atlantic	0	54	33
NORTHEAST TOTAL	1	111	46
U.S. TOTAL	45	280	204

Source: U.S. Dept. of Commerce, Bureau of the Census, 1992 Census of Manufacturers, Pulp, paper and board mills, MC92-1-26A.

TABLE 11.3. Northeastern Paper Industry, 1996

	Paper Production (1,000 tons)	Paperboard Production (1,000 tons)
Maine	3,201	(D)
New Hampshire	370	(D)
Vermont	235	(D)
Massachusetts	341	379
Connecticut	inc. in VT	666
New England	4,148	1,329
New York	1,517	632
Pennsylvania	1,543	793
New Jersey	430	420
Mid-Atlantic	3,490	1,846
Northeastern Region	7,637	3,175
Percent of U.S.	18.0%	6.6%

D = Withheld to avoid disclosure.
Source: American Forest and Paper Association

and in converting. The region can no longer compete with the South and Eastern Canada in commodity items like newsprint, kraft bag paper, and paperboard used for cardboard boxes. Further, the large paper industry of the urban areas now relies almost exclusively on purchased pulp. Those mills make printing, writing, and art papers, which demand extremely high standards of manufacturing technology and quality control.[6]

The Northeast produces only 7 percent of the nation's pulpwood, but accounts for 22 percent of the nation's paper output. The region's paperboard output is nominal, although paperboard makes up half the output of the national pulp and paper industry. Paper mills bring in substantial tonnages of market pulp from other regions. The region also relies heavily on recycling, with more than half of the region's wastepaper usage (1992) in Pennsylvania, New York, and New Jersey (Table 11.4). The increased use of recycled fiber is creating new jobs as new mills are built, and is helping to solve solid waste disposal problems. Closing the loop in fiber is good business for the Northeast.

The paper industry emerged in colonial times, using rags, straw, and other raw materials. Papermakers were so highly valued that they were

Georgia-Pacific OSB plant, Woodland, Maine.
A plant like this employs about one hundred people.
Courtesy of Georgia-Pacific Corporation.

TABLE 11.4. Wastepaper Usage in Northeastern States (1,000 tons)

	1988	1994	Percent Change
Connecticut	370	520	40.6
Maine	164	1,063	547.0
Massachusetts	447	474	6.1
New Hampshire	121	*	
New Jersey	807	894	10.7
New York	595	964	62.0
Pennsylvania	864	1,099	27.2
Vermont	70	*	
Region Total	3,437	5,015	45.9

*Included in Maine.

Source: U.S. Bureau of Census, 1989 MA26A-(89)-1, p. 8; and American Forest & Paper Association. (RI not shown separately)

exempted from the draft during the Revolution. Early centers were in Eastern Pennsylvania and in Massachusetts along the Connecticut River. Papermills in the colonial period were small (employment as few as six persons) and numerous. Mills grew larger in the nineteenth century. The 1890 census counted 180 mills in New England (85 in Massachusetts), employing 12,304 persons, or 68 per mill. At this time, most of the region's paper was being made from nonwood materials. The development of woodpulp paper in the 1880s produced a spectacular boom in the region's paper industry, and the industry moved north to find the wood. In 1919, average employment in New England was 143 per mill, and Maine mills employed 300 people each on average.

The industry's maintenance of sustained employment over an entire century has been of critical importance in the region's economic health. It has sustained employment in milltowns buffeted by declining leather, textile, and food processing industries. It has helped support economic opportunity in rural areas, where it pays among the highest wages of any industry. As the Maine Commissioner of Industrial and Labor Statistics wrote in 1906, the paper industry "is constantly adding wealth to our state. It is adding prosperity to our old settled towns and building up new towns in the wilderness. It is giving employment to thousands of our

Recycled paper provides an increasing share of raw material for the region's paper mills. *Courtesy of USDA-FS, Forest Products Laboratory.*

young men who otherwise might be obliged to seek a livelihood in distant parts of the country."[7] In fact, from 1905 to 1991, the share of Maine manufacturing employment originating in the lumber and paper industries actually increased slightly, while their share of the state's value of manufacturing production rose significantly.

Importance to the Northeastern Economy

Total employment based on wood in the region exceeds 270,000 persons. The total output of the region's wood-based industry was almost $38 billion in 1992. Industries based on wood account for about 2 percent of total regional employment and personal income. They are a significant part of the region's manufacturing sector, especially in the three Northern New England states (Tables 11.5, 11.6, and 11.7). While wood-based employment is approximately equal in Massachusetts and Maine, the wood industries are 29 percent of all manufacturing employment in Maine, compared to 5 percent in Massachusetts. Considering their heavy urbanization, it is not surprising that the Mid-Atlantic states show relatively low percentages.

Great Northern Paper Company mill at Millinocket, Maine. *Courtesy of Gordon Manuel, Great Northern Paper Company.*

TABLE 11.5. Employment in Northeastern Lumber, Furniture, and Paper Industries, 1992 (Thousand Workers)

	SIC 24 Lumber & Wood	SIC 25 Furniture & Fixtures	SIC26 Paper & Allied	Total 24, 25, & 26	All Mfg.	% of all Mfg in SIC 24, 25, 26
Maine	10.7	1.1	15.8	27.6	91.1	30.3%
New Hampshire	3.9	0.9	4.8	9.6	93.5	10.3%
Vermont	3.2	2.1	2.0	7.3	44.5	16.4%
N. New England	17.8	4.1	22.6	44.5	229.1	19.4%
Massachusetts	4.1	4.8	19.7	28.6	480.3	6.0%
Connecticut	2.2	2.7	7.2	12.1	320.5	3.8%
Rhode Island	1.0	1.3	2.4	4.7	88.5	5.3%
S. New England	7.3	8.8	29.3	45.4	889.3	5.1%
New York	13.5	18.5	33.8	65.8	1,046.0	6.3%
New Jersey	3.6	7.1	21.5	32.2	573.8	5.6%
Pennsylvania	27.0	18.2	35.5	80.7	949.8	58.0%
Mid-Atlantic	44.1	43.8	90.8	178.7	2,569.6	7.0%
Northeast	69.2	56.7	142.7	268.6	3,688.0	7.3
U.S. total	655.8	471.1	626.3	1,753.2	16,948.9	10.3%
Northeast as % of U.S.	10.6%	12.0%	22.8%	15.3%	21.8%	—

Source: U.S. Department of Commerce, Bureau of the Census. 1992 Census of Manufactures, General Summary MC92-S-1, Table 1-2. Rhode Island estimated by author.

This contribution to employment and output has steadily increased over this century, making the lumber and paper industries a stable, slowly growing sector of the region's economy.

Dependence on wood products for employment varies within the region. It remains high in the old papermaking regions of central Massachusetts and in Northern New England, the Adirondacks, and several Pennsylvania counties. Because of their large service sectors, and diverse economies, the contribution of the forest-based industries to Gross State Product is extremely small in states outside of New England.

TABLE 11.6. Value of Manufacturing Shipments in the Northeast, SIC 24, 25, 26, 1992 (Million Dollars)

State	Lumber & Wood	Furniture & Fixtures	Paper & Allied	Total	Pct. of All Mfg in SIC 24, 25, 26	All Mfg.
Maine	1,474.7	94.6	3,480.0	5,049.3	43.5%	11,600
New Hampshire	376.2	91.5	818.0	1,285.7	11.4%	11,300
Vermont	300.2	134.3	442.0	876.5	13.7%	6,400
N. New England	2,151.1	320.4	4,740.0	7,211.5	24.6%	29,300
Massachusetts	458.1	455.7	2,935.8	3,849.6	5.9%	65,500
Connecticut	213.0	327.9	1,634.6	2,175.5	5.4%	40,400
Rhode Island	100.0	162.7	283.6	546.3	5.7%	9,600
S. New England	771.1	946.3	4,854.0	6,571.4	5.7%	115,500
New York	1,271.7	1,522.9	4,990.3	7,784.9	5.1%	153,300
New Jersey	370.8	761.7	3,355.2	4,487.7	5.2%	86,800
Pennsylvania	2,762.8	1,727.1	7,595.2	12,085.1	8.7%	139,200
Mid-Atlantic	4,405.3	4,011.7	15,940.7	24,357.7	6.4%	379,300
Northeast Region	7,327.5	5,278.4	25,534.7	38,140.6	7.3%	524,100
U.S. TOTALS	81,564.8	43,825.9	133,200.7	258,591.4	10.4%	3,004,723
NORTHEAST AS % OF U.S.	9.0%	12.0%	19.2%	14.7%	—	21.2%

Source: U.S. Department of Commerce, Bureau of the Census. 1992 Census of Manufactures, General Summary, MC92-S-1-Table 1-2. Rhode Island estimated by author.

Economic Development: Revival of an Old Concern

Fostering economic development was a policy concern for monarchs, governors, town officials, and farmers from 1620 to the early nineteenth century. Sawmills and gristmills were encouraged with monopolies and free timber and water rights. Public lands were granted for support of schools, churches, and other public purposes, and roads were built to promote

TABLE 11.7. Gross State Product by Industry and State, 1996 (Million dollars)

	Total GSP	Mfg.	Lumber & Wood	Paper & Allied	Total Forest-Based	Forest as % of Total	Forest as % of Mfg.
Maine	28,894	5,333	542	1,680	2,222	7.7%	41.7%
New Hampshire	34,108	7,557	233	298	531	1.6%	7.0%
Vermont	14,611	2,645	183	180	363	2.5%	13.7%
N. New Eng.	77,613	15,535	958	2,158	3,116	4.0%	20.1%
Massachusetts	208,591	32,265	213	1,162	1,375	0.7%	4.3%
Connecticut	124,046	20,712	140	929	1,069	0.9%	5.2%
Rhode Island	25,629	4,282	35	94	129	0.5%	3.0%
S. New Eng.	332,637	52,977	353	2,091	2,444	0.7%	4.6%
New York	613,287	72,154	566	2,071	2,637	0.4%	3.7%
Pennsylvania	328,540	68,074	1,632	2,785	4,417	1.3%	6.5%
New Jersey	276,377	37,985	190	1,206	1,396	0.5%	3.7%
Mid-Atlantic	1,218,204	178,213	2,388	6,062	8,450	0.7%	4.7%
Region Total	1,628,454	246,725	3,699	10,311	14,010	0.9%	5.7%

Source: U. S. Dept. of Commerce, Bureau of Econ. Analysis website.

settlement. Towns regulated the removal of roadside trees, trespass, export of wood, and measurement of wood.

An explanation of the reasons why an age of complete laissez faire followed this period would require a social and economic history of the nineteenth century. The climate of the age was one of public promotion of private economic objectives, of reducing public landownership, and minimal government supervision or concern with land use. An attitude that natural resources were endless was certainly in part responsible for these views.

By the 1950s, everyone realized that the region's decaying farm economy and collapsing milltown industries left behind serious poverty and economic loss.[8] New efforts to promote state and regional economic development were made, mostly using federal funds. A new emphasis on economic development seized state policy. Governors began to run for office on issues of jobs and favorable business climates. Strangely, however, the green backdrop concept of the forest and the poor regional competi-

Papermaking jobs are some of the highest-paying in the region's economy.
Courtesy of International Paper Company.

tive setting for forest products meant little public policy emphasis on wood as a source of products and jobs. Partly because of poor wages and working conditions in some branches of the wood-using industry, and due to its low level of technology and slow growth, the sector did not seem an attractive economic development prospect.

By the late 1970s, however, this had changed. Public officials were relearning what their nineteenth century predecessors knew about the utility of forests for products and appreciated the jobs that wood processing can bring. Many communities suffered when obsolete sawmills or papermills were closed. Even the boom of the 1980s did not put an end to this interest. An increased federal interest in rural areas helped support more emphasis on rural industries, including wood. State governors occasionally became involved, and a broad range of state and federal programs were developed.[9]

Today, then, a new awareness of the income-producing potential of the forest is emerging, bringing the region back to its colonial interest in the commodity values of the forest. The economic development focus of state governments merges with several important social concerns: the retention of economic opportunity in rural areas, and helping states adapt to economic change. Retaining some measure of local supplies of materials and energy is useful for its own sake as well as to reduce outflows of cash for basic materials, although few perceive it as a policy priority.

Norbord Particleboard mill, Deposit, New York, produces engineered panels for industrial users. *Courtesy of Fred Hathaway, Norbord Industries, Inc.*

Outlook for Output and Employment

In the mid 1990s, about a quarter of a million Northeastern residents were directly employed in the region's wood processing plants. Most of those in the urban paper and converting industry, plus a substantial fraction of those in secondary wood employment, do not depend on the region's own forests for raw material. Whether the other industries will use more local timber in the future is hard to guess. Also, the region depends heavily on pulp, lumber, panel products, and manufactured wood products imported from other regions. The Northeast has seen slow growth and investments in its forest-based industry in recent decades. More attractive investment opportunities were available in Canada, in the Pacific Northwest's large timber, and overseas. Markets elsewhere, especially in the Sunbelt, grew rapidly.

Several papermills closed in the 1970s because of high costs and their inability to justify the expense of modernizing and installing up-to-date pollution control equipment. In a few instances, companies were faced with the practical requirement to shut down old mills or rebuild them to meet modern environmental standards. Every few years, another old mill is closed in an industry downcycle. Yet, numerous major modernizations and expansions were accomplished in the 1970s through the 1990s. Not least of these were rebuilds of energy systems and new recycling lines.

Experts argue that the competitive balance for the forest products industry in the Northeast has improved and looks favorable for future growth, based in part on continued market growth and on higher production costs for other producing regions.[10] The region's outlook now appears even better, with the large declines in softwood lumber and pulp production that are occurring in the Northwest and in British Columbia. Also, lumber and paper output in Eastern Canada will peak soon; the South is reaching its production limits; and supplies in the Russia and the tropics are declining. These trends will increase the costs to Northeastern consumers of importing their wood products and will improve the financial prospects for regional wood processors. Furthermore, market forces will increase stumpage prices, thereby providing incentives both for better forest management and for faster high-grading. On the other hand, global restructuring of wood supplies in the wake of late 1990s economic weakness in Asia could significantly modify this outlook.

Recent federal projections for the 13 Northeastern states, which contain about 82 million acres of forest, illustrate the likely trends (Table 11.8). The percentage increases in output over the next 50 years are less than the national rate for pulp, but greater than the national rate for lumber and panels. The region's panel plants supply a needed market for aspen, low-grade hardwoods, and mill residues. They will increase regional self-sufficiency in an important raw material for manufacturing, construction, and shipping. An interesting development is the rising interest in timber bridges, which last longer, can be installed faster, and can be cheaper than steel and concrete for short spans.

Estimated 1990 regional timber consumption is roughly 2.2 billion cubic feet (Table 11.9). This is roughly a 50 percent increase since the mid-1940s. Using the Forest Service's output projections for the North, we would expect a doubling in regional wood cut. This large an increase is certainly possible based on demand. But for working purposes, a projection of a 22 percent increase seems a useful point of departure.[11] This recognizes that expanding output in the coming half century will more difficult than in the past, especially with the inventory peaking in Maine. The assumptions used may turn out to be mistaken, but they give us a simple way to assess the region's timber growing opportunities and constraints. Anyone preferring other assumptions can adjust these figures accordingly.

In recent decades, employment per unit of wood cut has fallen significantly. In the late 1970s, however, a national slowdown in labor productivity occurred.[12] It is impossible to predict the future trend of productivity and hence of jobs per unit of wood cut. We can say that, if productivity trends of 1950–1974 continue, there will be fewer jobs in

TABLE 11.8. Wood Products Output Levels, United States and Northeastern States,[a] 1990–2040 (projections based on future equilibrium price levels).

Item	1990[b]	2040	% Increase 1990–2040
U.S. Pulpwood Consumption (mill. cu. ft.)`	7,184	10,535	47%
Northeast Roundwood Consumption (mill. cu. ft.)	728	1,159	59%
Structural Panel Production (bill. sq. ft.)			
U.S.	24.1	39.8	65%
Northeast	0.6	5.7	850%
Northeast Fuelwood Used, All Sources (bill. cu. ft.)	.98	1.59	62%
Lumber, Hard and Soft (bill. bd. ft.)			
U.S. Consumption	56.7	74.0	31%
U.S. Production	48.0	63.3	32%
Northeast Production			
Softwood	1.5	2.0	33%
Hardwood	2.5	3.1	24%

[a]Northeastern states: New England plus New York, Pennsylvania, New Jersey, West Virginia, Maryland, and Delaware.

[b]Certain items are 1989 or 1991. See original source.

Note: These estimates were made before major supply changes of the early 1990s occurred.

Source: R. W. Haynes, D. M. Adams, and J. R. Mills, The 1993 RPA Assessment Update, USDA-FS, RMFRES, Gen. Tech. Rept., RM-GTR-259.

regional forest industries by the year 2030. If continued development of new products occurs and labor productivity growth slows, employment could rise somewhat. If the timber cut and output levels fall short of these projections, however, there will be fewer Northeasterners employed in wood products businesses in the year 2040 than today. Of course, the large number of forest products jobs that depend on imported pulp, lumber, and panels will be governed by interregional competition in those particular industries, which is extremely difficult to assess.

TABLE 11.9. Northeastern[a] Timber Production Outlook, Stylized Data

	Million Cubic Feet	
	1991	2040
Lumber and veneer	700	900
Pulpwood	522	700
Miscellaneous products	45	100
Fuelwood	940	1,000
Total	2,206	2,700

[a]Northeastern states: New England plus New York, Pennsylvania, New Jersey, West Virginia, Maryland, and Delaware.

Source: Author estimate based on USDA-FS data. For a more detailed analysis of the states of New England plus New York, see Irland and Whaley, in T. Lepisto, ed. 1993, in which we suggest that removals will rise by 50 percent. The Base Case USFS projection shows an increase of 75 percent for the North, before taking into account higher recycling.

Northeastern Wood and the World Economy: Log Exports

Economic interdependence is nothing new in the Northeast. In the eighteenth and nineteenth centuries, Northeastern lumber went to Midwestern markets down the Ohio, and to overseas destinations by schooners from Baltimore, Albany, and Bangor. Logs moved down the St. John to New Brunswick. In the twentieth century, lumber production fell. A few pulpmills exported pulp. The urbanizing Northeast met its newsprint needs from Quebec and its lumber needs from Oregon and British Columbia. Its second growth forests yielded little that other countries wanted. The region's trade balance in wood fell into deficit.

After 1985, however, the exchange rate on the dollar fell, and European, Asian, and Scandinavian wood buyers came shopping. In a short while, hardwood lumber exports were booming, and logs were moving to seaports for shipment to Japan, Korea, Belgium, and Germany. Regionwide, log prices rose, and local mills often had to struggle for raw material and to adjust to smaller and poorer logs, because the export market was taking the highest quality logs.[13] More and more sawmills learned to meet importers' demanding quality requirements, and to manage the financing, shipping, and documentation for exporting. A number of companies grew rapidly, based on exporting skills, helping restore a global business orientation that had been common in the 1880s lumber business.

The improved exchange rate for U.S. exporters was accompanied by dramatic shifts in management of tropical hardwood forests, on which Asian and European markets traditionally depended. Asian nations embarked on programs to increase prices of their log exports, and to force more domestic processing of the logs. Some dramatically cut their exports, and a few nations literally ran out of wood. Prices of tropical woods rose, and supplies became less dependable. Buyers became more interested in temperate hardwoods, and sales of North American oak, ash, maple, and cherry increased (figure 11.1). export markets improved not only for hardwood logs but for other products.

While the Northeast's ports are not particularly well-suited to exporting bulk commodities, offshore exports have increased, often through Canadian ports. They have naturally been concentrated in veneer logs and in high grades of lumber. This means that sawmills and furniture makers have seen their raw materials bid away by foreign buyers. At the same time, it has created broader, better paying markets for logs and creates the potential for moving downstream into producing wood parts and components that could generate more employment. Some mills are taking advantage of this demand for processed wood products now. To succeed in exporting "value-added items," however, it will be necessary to outcompete low wage, timber rich, tropical countries. But at present, wood prod-

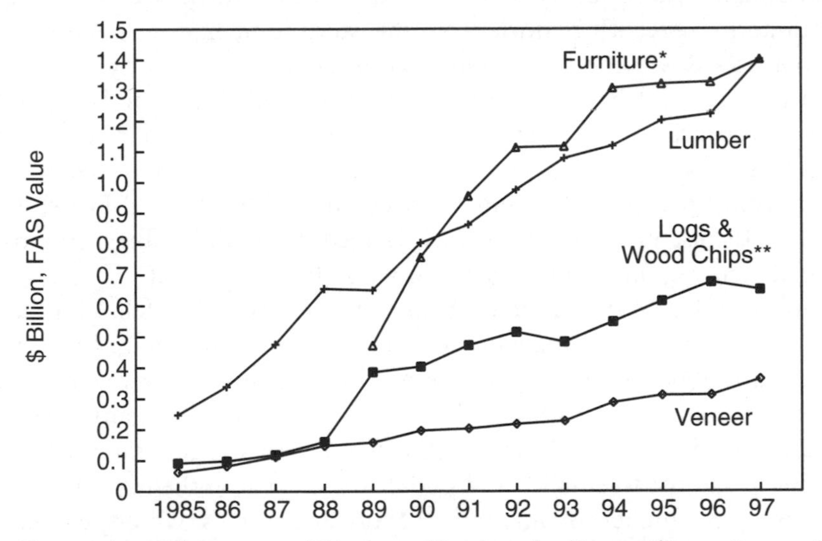

Figure 11.1 U.S. Exports of Hardwood Products by Commodity, 1985–1996
Source: Compiled by FAS from official statistics of the U.S. Department of Commerce.
*Household furniture.
**Chips data is incomplete due to a change in reporting styles in 1989, previously hardwood and softwood were combined as a single category.

ucts are moving overseas from the interior of North America, so it can be done.

Recycling: A New Resource Base

The region's overloaded dumps, improving paper technology, and rising public concern are sparking a strong rebound in paper recycling. In recent years, several major recycling projects have been announced in the region. At this writing, many are already in operation, while others are on hold due to soft markets. As consultant James McNutt points out, the "urban forests" of Manhattan "grow" as much fiber per acre as a healthy forest.[14] Collection and sorting systems are emerging to help close the region's paper cycle. Trucks that formerly returned to the paper mill towns empty now return with loads of wastepaper. So we have the spectacle of coal being hauled to Newcastle — trucks filled with wastepaper turning into millyards in places like Millinocket, Maine, surrounded by nothing but forest for miles. Increased recycling is expected and is largely responsible for a one-sixth reduction in projected 2040 U.S. pulpwood usage compared to 1990 projections.

Society is benefitting in many ways by closing the fiber cycle, for paper as well as other products. Instead of being landfilled or just incinerated, urban demolition and landclearing waste is being hauled to rural areas for use as fuel. As disposal costs rise and policies fostering source reduction, reuse, and recycling become more comprehensive, per capita consumption of paper, lumber, and other resource-based products may decline. Hopefully, the renewability and low energy intensity of forest products will receive due consideration when such policies are set.[15]

Can the Northeastern Forests Meet Growing Demands?

The current wood production in Northeastern forests is well below what could be grown if most of the forest were carefully managed. Net growth is reduced by high mortality, by cull increment (growing wood on valueless trees), by poor stocking of desirable trees, by overstocking or overmaturity, and by poor cutting practices. Current annual growth is just about half a cord per acre, but two-thirds of the region's forest should be capable of exceeding this significantly under sound management (Table 11.10).

Short-term, nonsustainable increases in timber harvests can come from three sources: the harvest of mortality caused by storms or pests, landclearing during periods of active development, and the use of surplus trees that are of low quality or represent excessive stocking for optimum value growth.

Exports of logs generated controversy across the Northeast as U.S. Forest Products exports rose in the 1990s. *Courtesy of Maine Department of Conservation.*

Owen Herrick of the USDA-FS has estimated that a high level of softwood timber consumption in the year 2000 could be met by placing less than 25 percent of the Northeast's forest land under intensive management.[16] The outlook for hardwoods was not as bright, but under intensive management, projected hardwood needs for year 2000 could be met with less than 75 percent of the land. The most recent Forest Service summary indicates that if fully stocked and well-managed, 22 percent of the forest land in the Northeast could grow more than one cord per acre each year.

The Forest Service analysts expect that net growth will fall significantly in the Northeast as a result of stand maturity and overstocking. Because of the large current growth/cut surpluses, however, a significant percent increase in wood products output can be produced while total regional timber inventories (in cubic feet) increase. These projections are based on detailed economic analyses by the USDA-FS (Figure 11.2). This outlook for production is higher than my own estimate given above. The Forest Service's inventory projections treat the entire North as a unit, and interpolating from those estimates to the Northeast would be hazardous. So I venture no projections of inventory here.

TABLE 11.10. Commercial Forest Land in 13 Northeastern States by Productivity Class, 1987

Potential Growth, Cords per Acre per Year	% of Acres
1.4 or more	4%
1 to 1.4	18%
0.58 to 1.0	44%
0.25 to 0.58	33%
Less than 0.25	1%

Source: Waddell, 1989, table 4.

Wholetree chips for energy raise new policy and environmental issues. *Courtesy of Maine Department of Conservation.*

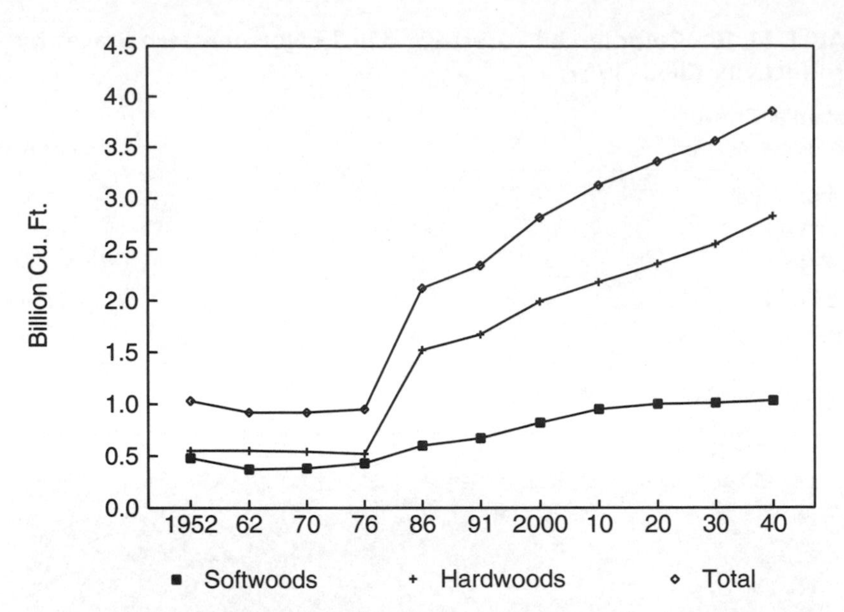

Figure 11.2 Historic and Projected Roundwood Harvests, Northeast, 1952–2040
Source: R. W. Haynes, D. M. Adams, and J. R. Mills, 1995. The 1993 RPA Timber Assessment Update. USDA-FS RMFRES, Gen. Tech. Rept. RM-GTR-259, Table 27.

There are several barriers to such a large increase in cut. First, the region's spruce-fir resource has been under heavy stress from spruce budworm, and the inventory in those species, at least in Maine and New Hampshire, declined in the following two decades, and may decline somewhat further. In the pine-hardwood regions, timber quality is generally poor. The larger diameter classes of pine and some hardwood species are being overcut in some areas. Second, outside the industrial forest, forest parcels are small and getting smaller. Loggers and mills will find expanding log supply and output costly and difficult. Finally, intensive management practices are not currently being implemented at a pace that could significantly boost overall timber growth on the region's forest lands in the next half century. The most intensively managed properties will undoubtedly increase their growth significantly, but I doubt that this will have a significant effect on regionwide growth. On the other hand, production declines in the West and in Canada, plus expanding export markets suggest that the region's competitive position is likely to improve.

Let us conduct a short arithmetic exercise. We can estimate the region's effective commercial forest land base at about 44 million acres. This figure assumes that one-half of the recreational and suburban forests and two-thirds of the rural forest are available for sustained timber produc-

tion. If the region harvested 1,800 million cubic feet (growing stock) from this resource in 1986, this was a rate of 41 cubic feet per acre, or about half a cord. This rate is roughly equal to the region's current average rate of wood growth. A rate of cut higher than this can be sustained for a decade or two more by using surplus growing stock, low-quality trees, and non-growing-stock timber. And how future growth will be affected by forest maturation, by atmospheric pollutants, by current cutting practices and management investments, and by climate change is uncertain. But these figures clearly suggest that a major increase in the cut by the year 2040 cannot be sustained over the long run at present levels of management, especially if the available landbase declines.

To produce a sustainable timber harvest at an increased level, there are four fundamental requirements:

1. The current growth rate would have to be significantly increased, on the average, over the entire commercial forest area. This would have to be done using exemplary forest practices with reasonable regard for landscape management considerations.
2. The acreage of forest land available for harvesting must be maintained to the maximum extent possible, although I would still urge an increase in the region's preserved wilderness.
3. Utilization of new species and of trees cut must be improved.
4. Efficiency of conversion into products and energy must be increased.

It is not, therefore, a question of biological potential. It is a question of landowner motivation to grow and harvest timber and to do it in ways that increase future growth while protecting environmental values. It is also a matter of size and quality. Average log sizes in several major species will peak or even decline, and average log quality is not likely to improve very much in the next few decades. Lastly, it is a matter of land use planning and of social tolerance of forest management. Ironically, as Strauss's example for part of Pennsylvania shows (Chapter 4), our failure to harvest surplus, low quality growing stock now will compromise future production potential.

All of these trends are now visible in their early stages. Serious efforts in technological improvement, industrial development, and improved forest practice can allow the Northeast to significantly boost its product yield, even from a declining forest land area. The region could then take better advantage of its recently improved competitive position in forest products manufacturing, and maintain or even increase the forest's contribution to regional employment and economic well-being. It could also aid in meeting its raw material and energy needs from its own resource base.

Impacts of Higher Harvest Level

Would a significant increase in the region's timber cut be an ecological disaster? Some fear it might. Personally, I do not think it needs to be. A higher timber cut will definitely change the forest, but not in ways that most travelers and recreationists would notice. A higher harvest level, if properly managed with due regard for landscape management principles could benefit some species of wildlife. Den trees and buffer areas can be preserved wherever needed. More intensive management need not produce longterm degradation of site potential. Horror-show scenarios can be imagined, but I believe they can be prevented. In sum, increased yield from the forests can be consistent with a continued yield of high value amenities, recreation opportunity, and wildlife habitat, but only if determined owners, loggers, foresters, and sound public policies make it so. Watchful scrutiny by public agencies will help it happen. Artful nudging by well crafted regulations and incentives may be required in some areas, such as ensuring that watercourses are well protected.

It remains to be seen whether the wood-using capacity needed to boost regional wood consumption will actually be built. That will depend on competitive forces, local zoning decisions, national and international markets, and the investment decisions of large corporations. Also, the investments in silviculture required may not be economical for some owners, and the funds and the determination to make them work may not materialize. In view of all the uncertainties, it will be essential for public agencies to more closely monitor forest inventories, wood use, and trends in forest management.

Equally, even a continuation of the present level of regional timber harvest could be seriously damaging in its impact on streams, wildlife, and the scenery, if cutting is conducted abusively and thoughtlessly. Whether future wood harvests degrade or improve the landscape will depend on thousands of decisions made daily by loggers, landowners, foresters, wood users, and local and state governments. As I argue in the next chapter, we cannot secure favorable performance simply by passing laws or subsidizing forest practices. It can be done only by developing the ecological conscience of all people concerned with managing the forest.

Further, the region's communities need the employment, income, and tax revenues that could be generated by a prosperous wood-based industry based on sustainable use of a renewable resource. Since the 1880s, the region has counted on outside sources to meet its wood and paper needs,

usually by liquidating virgin forests elsewhere. The time has come to turn to its own backyard to help meet these needs, even as aggressive efforts proceed to recycle lumber and paper, and to use forest products more efficiently.

The future of the region's wood using industries, then, interweaves important themes of this book. The Economic Interdependence of country and city, and of our economy on outside resource supplies, will continue. The resolution of many issues depends on Town Meeting forms of politics. Changes in the Global Commons could profoundly affect national wood supplies. The Bequest Motive requires vigilance to ensure that rising demands are met in ways that are consistent with sustainable yields of timber and all forest values. Alertness is required to what the Worm's Eye View and the Eagle's Eye View are telling us about how natural processes and cycles limit what we can do.

There is likely to be more wood in the region's forests in 50 years, unless demand prospects shown here are far exceeded, or Global Commons effects cause severe damage. But failure to improve current levels of forest practice, to foster investment in growing for quality, coupled with projected rising timber demands, will seriously degrade the forest for all of its values.

12 Forest Policy: Themes, Challenges, and New Directions

THE HISTORY OF forest policy in the Northeast presents a series of major themes, changing slowly as economic conditions and public attitudes changed over the decades. Although forests and wood products have always been important, there was no institutional focus for forestry programs until the 1890s, nearly three centuries after settlement. By this time, many of the forest's most valuable products had been liquidated, and 40 percent of the original forest had been cleared. I will outline the challenges of the future and my own general ideas on how to meet them.[1] This is not the place for a summary and critique of the forest and land use policies of nine states, or a detailed legislative program. Because of the limits of time and space, federal forest policies as they affect this region are not discussed; this is not to suggest that federal policies have been unimportant.[2] They are likely to increase in importance in the future.

Changing Forest Policy Themes

Changing themes in forest policy have been based on public attitudes about the role of government and on perceptions about the serious forest problems of the day (Table 12.1).

In the colonial period, forest products were critical to individual welfare, to trade, and to the promotion of economic activity. Colonists were accustomed to government regulation of the economy. For example, laws requiring fire control measures were common, although we know little of how they were enforced.[3] In this age, there was a veritable "compulsion to control."[4]

Through their policies on land distribution, the colonies tried, with varying degrees of success, to promote farming and settlement, and to prevent widespread absentee ownership, land speculation, and large consolidated ownerships. Lands were also granted in support of canals, sem-

TABLE 12.1. Outline History of Northeastern Forest Policy

Period	Themes
Colonial	Land disposal by governments to promote settlement Promotion of manufacturing Supervision of utilization and trade Laws against fire Minimal reservations for public use
Revolution to 1890	Complete land disposal to raise revenue and place land in private hands Low rates of forest taxation Occasional tax preferences for tree planting
1890–1917	Establishment of state forest agencies Establishment of forestry research and educational institutions Beginning of extensive state land acquisition programs (excluding Maine) Institution of pest control programs: blister rust, gypsy moth
1917–1950	Increased federal cooperation Tree planting, nurseries Continued land acquisition
1950–1970	Establishment of small-landowner programs Openspace programs Parks land acquisition Land and Water Conservation Fund and SCORP process Multiple use concept
1970–1980	Air and water quality Revitalization of traditional programs Concern with effects of land booms Emergence of Worm's Eye View
1980–Present	Biodiversity and Eagle's Eye View Concern over wildness Emerging debate over federal role Global Commons issues.

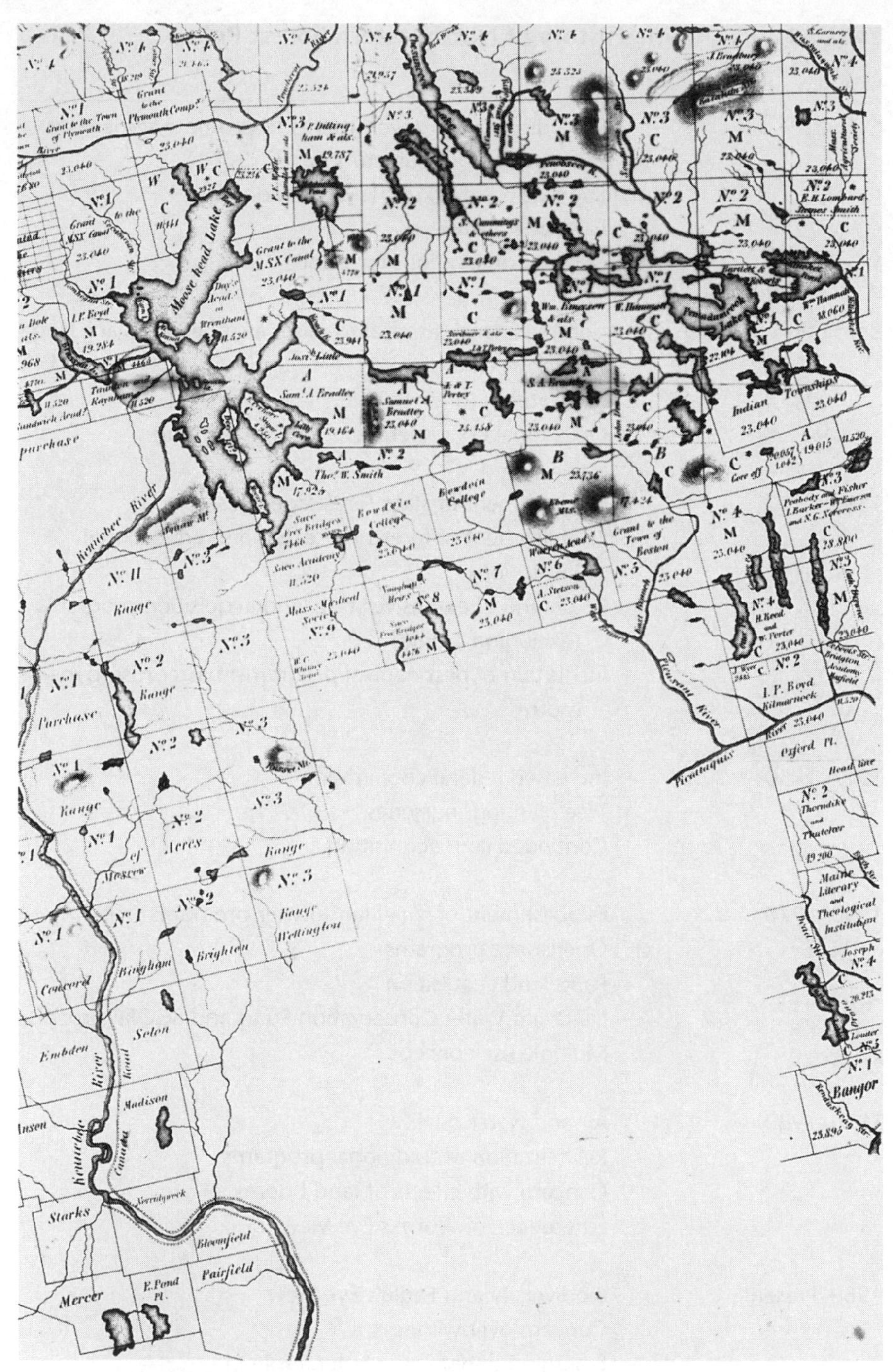

G. Coffin, *Map of Maine's Public Lands*, 1835. *Courtesy of Concord Free Library. This was made from Thoreau's personal copy.*

Earliest known forest type map of the U.S. by Joseph Henry, secretary of the Smithsonian Institution. Shows Maine and adjacent areas clothed with forest of "coniferous everygreens" and "deciduous trees."
Source: US Congress, House, Report of Commissoner of patents, 1858.

inaries, and colleges.[5] In New York and Pennsylvania, however, vast land grants tended to slow the pace of settlement until after the Revolution.

In legislating about the forest itself, many towns and colonies took a meticulous interest in enforcing the quality of products; controlling prices; controlling trade; promoting mills by grants of monopolies, timber and water rights; and ensuring close utilization of timber. In places, timberlands were reserved from sale to be used to support local government, schools, and the ministry.[6] Colonists regulated forest use as open range for livestock, and even in one case granted bounties for the removal of barberry, an alternate host for a wheat rust.

Efforts to conserve longterm timber supply were nonexistent, except where coastal towns and colonies controlled the export of various wood products. William Penn's requirement that planters retain land in woods is often cited, but there is little evidence that it was enforced. Even the famous Broad Arrow Acts only viewed pines as a stockpile of material for short-term use and made no effort to assure a longterm supply. Even these limited efforts, however, were a complete failure.

In the nineteenth century, the colonial concepts of control withered. The period may be characterized as one of indifference; public opinion and legislative inclinations favored private landownership and the unhindered use of resources. Public opinion supported the free use of private capital. But, as J. W. Hurst argues, "a crude laissez faire was never the guiding doctrine."[7] Legislatures and courts supported the speedy disposal of public lands. States subsidized railroads, canals and turnpikes. The law was supportive of industry objectives, in allocating water rights, adjudicating conflicts over navigation, and in a thousand other ways.

The period from the 1880s to the 1920s saw rapid change. Northeastern lumber production waned, as large trees were cut and cheap Western lumber began to reach Midwestern markets. Public and legislative interest in conservation rose, and state forest agencies, nurseries, forestry schools, and research stations were created. In this period, the decline of the region's farm economy was painfully apparent. Trees seemed a fine way to use abandoned pastures. A new consciousness of timber supply and the environmental values of forests emerged. Northeastern states were the first in the country to authorize land purchases for conservation, to establish game laws, and to organize forestry services. It was this period that gave the region the nucleus of its national and state forest and park systems. It also spawned the private organizations that have contributed so much to forestry and land conservation in New England.

The Great Depression of the 1930s saw a brief increase in rural populations, extensive concern with "land use adjustment," and a dramatic collapse in consumption of lumber and paper. The federal government led in conservation and economic recovery programs. The Civilian Conservation Corps built trails and parks, and planted trees. State and federal agencies expanded their land holdings for forests, parks, and wildlife management. The hurricane of 1938 stimulated a cooperative venture, the federally funded New England Timber Salvage Administration, to help salvage blown-down pine timber. World War II produced labor shortages in the northwoods and stimulated a new exodus of population from the remote hill farms and logging towns. Many of the soldiers who grew up in these areas never went back. What many of them wanted instead was a house in the suburbs.

The late 1940s, then, sparked the new development pattern that created, and now threatens to destroy, the suburban forest. Though the hard times of the 1930s were a clear memory in rural areas, forests continued to play a key economic role. But in suburbia, they came to be seen by many merely as potential houselots and green backdrops for suburbia.

There, the public and policy makers almost forgot about the products of the forest. To the extent that public opinion in general troubled itself at all with the forests, the concern was only casual. The forestry agencies were active during these years in setting up service forestry programs, modernizing fire control, and planting soil bank plantations.

The reinvigorated conservation movement of the 1960s brought a new emphasis on open space and on wildlife and air and water quality. The Land and Water Conservation Fund and the natural beauty programs of Lyndon Johnson's administration helped spark a new interest in land acquisition for recreation and conservation. Funds went to towns and cities for open space uses. The Cape Cod and Fire Island National Seashores, among many new conservation areas, date from this period.[8]

By the 1970s, the political interest in amenity values had reached a new peak.[9] A new interest in management of public lands arose. Federal and state agencies revived land management programs with new emphasis on land use planning and multiple use. The preservation wing of environmental groups asked that two-thirds of the White Mountain National Forest be converted to wilderness. The surging real estate market brought forest taxation again to the fore, and many states adopted new ways of taxing woodlands.

By the end of the 1970s, new attitudes toward the forest were emerging. More people took chainsaw, axe, and splitting maul in hand to heat

Forest mosaic, Bigelow Mountain, Maine. *Photo by Roger Lee Merchant.*

A Brief History of the New Jersey Pinelands and the Pinelands Comprehensive Management Plan

In the center of America's most populous region lies over a million acres of forests, farms, and scenic towns — the New Jersey Pinelands.

The Pinelands is a patchwork of pine and oak forests, tea-colored streams and rivers, spacious farms, crossroad hamlets, and small towns stretched across Southern New Jersey. In the country's early years it had been a place where fortunes were made from lumber, iron, and glass. But the early industries died out, and as the state's major roads bypassed the area, the "Pine Barrens" gradually became known as a remote part of New Jersey abounding in local legends like the "Jersey Devil.". . .

. . . So in 1978 Congress created the Pinelands National Reserve, the country's first. The Reserve was to be a place where governments at every level — from Washington down to local planning boards — could help shape the Pinelands' future in keeping with some basic guidelines. The state was to take the lead in evaluating the Pinelands' resources and planning how best to balance their protection with new development. As provided in the federal law, Governor Brendan T. Byrne established the Pinelands Commission by executive order on February 8, 1979 and gave it responsibility for these tasks. The Pinelands Commission consists of fifteen members. Seven are appointed by the Governor, and seven are appointed by each of the counties within the Pinelands. One member is appointed by the U.S. Secretary of the Interior.

The New Jersey Legislature, at Governor Byrne's request, supplemented the federal law by passing the Pinelands Protection Act in June 1979. The Act affirmed the temporary limitations on development which the Governor had put into effect while a plan to protect the Pinelands was being created. It also established a requirement that county and municipal master plans and land use ordinances be brought into conformance with the Comprehensive Management Plan which the Commission was developing.

The boundaries of the Pinelands National Reserve and the Pinelands Area as defined by the state legislation differ somewhat. The Reserve, totaling 1.1 million acres, includes land east of the Garden State Parkway and to the south bordering Delaware Bay which is omitted from the 937,000 acre state Pinelands Area. The two major jurisdictions together cover all or parts of 56 municipalities spread across seven counties — Atlantic, Burlington, Camden, Cape May, Cumberland, Gloucester, and Ocean.

The state law makes another important distinction between the remote interior of the Pines and the surrounding portions. Development is to be highly regulated in the 368,000 acre Preservation Area which encompasses the largest tracts of relatively unbroken forest and most of the economically vital berry industry. The larger surrounding area, known as the Protection Area, contains a mix of valuable environmental features, farmland, hamlets, subdivisions, and towns, making the Commission's task there more complex. . . .

- Wildlife — The Pinelands is home to 39 species of mammals, 299 types of birds, 59 reptile and amphibian species, and 91 different types of fish. Development must be carried out in a manner which does not disturb habitats for significant wildlife populations.

Of the approximately 500 fish and wildlife species in the Pinelands, 34 are in danger of becoming extinct in New Jersey. These threatened and endangered animals are listed by the New Jersey Department of Environmental Protection's Office of Endangered and Nongame Species. Included in this list of endangered species are the majestic bald eagle and the elusive Pine Barrens treefrog. Particular attention must be devoted to the design of projects so that habitats critical to these species are protected.

- Water Resources — The Pinelands is famous for its vast underground water supply. The Cohansey aquifer in the Pinelands contains over 17 trillion gallons of pure water, enough to cover the entire State of New Jersey with ten feet of water. This underground reservoir feeds most of the area's streams, supports its agricultural industry, maintains the ecological balance of our coastal estuaries, and provides drinking water for hundreds of thousands of people. The high water table and porous, sandy soil of the Pinelands make this aquifer particularly vulnerable to pollution. Pollutants move quickly through this sandy soil into the ground water.

To ensure that this water is not substantially degraded, new sewer plants and on-site septic systems must be designed so that discharges do not raise pollution levels on the property to more than two parts per million nitrate-nitrogen. Nitrate-nitrogen, a primary pollutant in the Pinelands, can cause serious health problems and environmental damage.

Source: Pinelands Commission, 1989. *A brief history of the New Jersey pinelands and the pinelands comprehensive management plan.* August. p. 1–2, 10.

their homes and found themselves interested in the productive potential of the forest. A revived competitive position of the forest industry brought investment and jobs to rural areas that badly needed economic balance but that were not close enough to the cities to benefit from growth in electronics, defense, growing services, employment, and rapidly growing employment in suburban retailing.

During the 1970s and 1980s, federal and state policies turned sharply toward increasing regulation of private forest practices, to protect water quality, aesthetic values, and wildlife habitat. Several states adopted formal forest practices acts, or tightened the eligibility requirements for use-value tax provisions. Federal programs under the Clean Water Act of 1972 and the Coastal Zone Management Act addressed water quality issues in every state. Regulations and Best Management Practices were adopted to ensure that loggers kept slash and sediment out of waterways. A proper treatment of this movement and its effects would require an entire volume.[10]

During the 1980s, scientists and journalists debated the likely effects of air pollution and global warming on the region's forests. By the early 1990s, however, a consensus seemed to be emerging that acid rain was not a major factor affecting forest health outside of limited alpine areas. But scientific support seemed to be swinging behind a predicted continuation of global warming. If this prediction is borne out, it could have profound effects on the composition and productivity of Northeastern forests. This change in climate would occur at a rate unprecedented in the planet's history, with ecological consequences that are difficult to predict.[11] Some experts counsel immediate preventive responses, while others urge accelerated research to be certain that the problems are real and that the cures are cost-effective. These debates will not soon be resolved, which leaves us with a significant range of uncertainties in predicting the outlook for the region's forests. The history of the debate over acid rain should make us careful on this and future issues.

Wetlands of various forms are intermingled with the Northeastern forest. Farming, roadbuilding, dredging, filling, and development have significantly reduced the area of wetlands. In New York, New Jersey, and Pennsylvania, more than half the original area of wetlands has been lost.[12] Significant areas of forest are at least seasonally wet or grow on soils that would be considered wet. Roadbuilding and harvesting can affect these as well as nearby swamps and marshes. During the early 1990s, a major debate arose over how to use the Federal Water Pollution Control Act to regulate activities affecting wetlands. Federal agencies proposed exceedingly broad definitions of wetlands, leading to a backlash. How federal law and regulations will address this issue remains unsettled, although many states regulate wetlands under their own laws. Strong support has developed for a concept of "no net loss" of wetlands. This recognizes that some use of wetlands is necessary, but that losses should always be offset by mitigation measures.[13]

A series of integrative paradigms are being applied to our thinking about forest policies. At the stand level, the concepts of nutrient cycling and stand dynamics became prominent — our Worm's Eye View. The landscape management model became widely advocated, emphasizing the importance of a mosaic of vegetation across a large area — our Eagle's Eye View. Finally, the concept of biodiversity as a management objective was widely discussed.

How have past policies actually affected events? Were they successful in terms of their original objectives? We hardly know. The kings saved few

Fire control has been one of the more successful forest policies, Buck Hill Tower, Rhode Island. *Courtesy of Rhode Island Division of Forest Environment.*

pines with their Broad Arrow Policy. Many of the soil bank plantations are forgotten and overgrown with brush. The forest tax programs are controversial, and it is difficult to tell whether they have promoted very much forest management or helped retain any very large areas in forest.

Beyond doubt, the most successful policy was the fundamental one of avoiding public landownership. The region today, particularly in New England, contains less public land than most other forested regions. Most of its state and federal forests, parks, and wildlife refuges were purchased or donated. Yet these lands offer the possibility for regional and national leadership in multiple use management under growing pressure for wood products, intense recreational use, biodiversity, and other social values.

Perhaps the next most visible improvement in Northeastern forest conditions has been in fire damage. In 1903, nearly a million acres burned regionwide; in the 1980s, only 30,000 acres a year. Clearly the end of massive old growth logging with its attendant slash and steam locomotives played a role. Damage due to fire has been reduced to its practical minimum — in fact, dangerously so, because there is no longer serious legislative and public concern with maintaining fire control funding. Only after rare major fires do the fire control units receive funds for equipment and maintenance backlogs. As the early foresters believed, effective fire control is an essential precondition for private investment in forest management. Also, with the sprawl of the suburban and recreational forests into remote areas, the values at risk have never been higher.

The programs helping small landowners have been a mixed success. Many acres of land have been thinned, planted, and cut successfully with the help of the state and county foresters. Probably these efforts help bring a better appreciation of forest and wildlife conservation to thousands of landowners each year. Yet the programs are criticized by some observers for an excessive timber orientation, lack of provision for longterm follow-through, and subsidy of management practices that make little economic sense. Most of the practices applied have been on such tiny areas that they cannot affect future timber production very much. Finally, many forest landowners, after thirty years of these programs, remain unaware of the services available.[14]

The really troubling fact about all these forestry programs, however, is that we cannot evaluate them fairly. We do not have the data. So the record is mixed, controversial, and the facts are not all in. But, we must ask, do these policies seem to fit the needs of the next few decades?

The answer, I believe, is that they don't.

Challenges of the Coming Half-Century

My review of the trends tells me that the forest of the future will face the following challenges:

1. Suburban sprawl, poorly controlled and lacking provision for forests as a positive landscape feature, will continue.
2. Public values about forests will change, with greater acceptance of logging by most people, but with escalating opposition by groups committed to a preservation philosophy. Public concern for water quality, forest health, wildness, and biodiversity will increase.
3. Timber prices and demands will rise, which will increase harvesting pressure throughout the region. During this period, it will become evident that the productive capacity of remaining forests is being overtaxed.

4. Recurring speculative booms in rural land are likely. Even without them, parcel size and length of ownership tenure will fall.
5. Mechanization will reduce the number of jobs per cord of wood processed.
6. The region's current competitive position as a prime location for making pulp into high-grade papers will slowly diminish.
7. The forest will slowly change its health and composition, as successional changes, cutting, and other disturbances proceed.
8. The resource of wildness that has persisted in the Industrial and Wild Forests will continue to leach away under pressure of rising mechanized recreational uses, a slow but steady increase in backcountry use, and the cumulative effects of forest management, harvesting, and land use changes.
9. Potential threats from the Global Commons will continue to generate scientific debate and public concern. An early resolution of the significance, impacts, and best responses to these issues is not likely.

These trends, if they materialize, will pose several practical problems. Current policies are not capable of addressing them.

Today, shoddy harvesting, high-grading, and inappropriate clearcutting are more serious threats to future timber productivity than are failure to plant, thin, or improve timber stands. The rapid turnover in ownership, land use conversion, and parcel fragmentation will continue to erode the future landbase for wood growing, wildlife habitat manipulation, and public dispersed recreation. Too many subsidized plantations are forgotten, to be overwhelmed by aspen or stripped for subdivisions. Too many are losing value growth to overstocking because timely thinnings are ignored by owners who never invested much in the stands anyway.

To overcome these difficulties, forest management assistance must develop in landowners a sense of their own responsibility for their forests. Today's assistance programs encourage the idea among landowners and the public that good forestry is government's responsibility, not theirs. Foresters and public leaders can help lead us out of this dead end, if they want to. Some of them are trying hard right now. But designing regulations and incentives to foster quality forestry is an awesome challenge. In North America to date, it has never been done.

Today's forest tax, land use planning, and land acquisition programs are unable to deal with the effects of suburbanization and future rural land subdivision booms. We cannot keep lands in forest, assume that they will be managed, and maintain them in a developed landscape simply with low taxes. More is required. The short time horizons of landowners will be an increasing problem. Today's programs, which do not effectively tie

Wildlife is an important resource of New England's forest. Efforts to maintain public access to private land for hunting and other wildlife-related activities will be a key policy issue.
Courtesy of Massachusetts Division of Fish and Wildlife.

tax breaks and subsidized management practices into future follow-through, simply will not meet the needs of the future.

In time, as demand begins to clearly outstrip sustainable timber yields, the issue of controlling the overall volume of timber cut will have to be faced. This is an exceedingly difficult problem for which off-the-shelf solutions are absent. This problem will occur with increasing frequency in this country, but virtually no effort is being made to develop and test policies for coping with it.

Finally, nowhere is there even a rough framework for addressing landscape-level forestry, habitat, and public access issues.

What Can Be Done?

The first need is to develop a shared professional recognition among resource managers, planners, and political leaders of the stakes, and to con-

vey the resulting ideas to the public. The vigorous public leadership displayed by foresters in the turn-of-the-century conservation movement and in the 1930s must be restored.

Fourteen ingredients of a forest policy for coming decades provide a useful starting point (Table 12.2). Spelling out all the details would far exceed the scope of this book.

1. Reform taxation and small-owner assistance programs to require management follow-through. Contracts, refund clauses, or other devices should be used to bind recipients of public funds to deliver the follow-through to final harvest and regeneration. No more forgotten soil bank plantations. This is the first step in the larger process of landowner education. With the high ownership turnover, this will not be easy. Educate landowners, through all available channels, to take responsibility for their land, without holding their hands out for government funds or tax breaks.

2. Re-educate professional foresters. Our dogmas of low taxes and handouts for forest landowners as entitlements demean our trade. We should believe that forestry is worth paying for, and that good forestry is responsible land stewardship — not a favor to the public that owners must be bribed to bestow. Also, more foresters need to respond to the landscape-level perspective that manages forests as an ecosystem. Foresters who cannot accept these ideas are one of our greatest impediments to progress.

3. Abandon subsidies that undermine responsibility, such as federal funds for fire control, and for spraying gypsy moths in suburbia and state parks.

TABLE 12.2. Elements of a Forest Policy for the Coming Fifty Years

1. Taxation and assistance programs: require follow-through.
2. Re-educate foresters.
3. Abandon or re-design subsidies that undermine responsibility.
4. Increase emphasis on job creation based on wood.
5. Promote public access to private forest land.
6. Acquire land and interests in land.
7. Upgrade research and technology transfer.
8. Improve forestry education.
9. Prepare policy responses to potential overcutting.
10. Support nonprofit sector.
11. Support training and other needs of loggers.
12. Implement a landscape-wide management perspective.
13. Promote growing wood for quality.
14. Protect and enhance the remaining resource of wildness.

4. Increase the emphasis on job creation based on wood. The economic development offices of each state should have effective forest industry marketing and development programs.

5. Promote a return to the old New England ethic that allows free, responsible, public recreational use of undeveloped land. Seek ways to persuade owners to take down the "No Trespassing" signs, building on programs already being developed by wildlife agencies and forest industry groups.[15] Recreational users need to be made more aware of responsible use of forest lands.

6. Based on carefully developed regional plans with clear goals, vigorously acquire land and less-than-fee interests such as easements and development rights.[16] Emphasize acquisition by towns, local districts, and the nonprofit sector. The objective should be a significant increase in state and local landownership and control by the year 2010. In Chapter 7, I advocate a 2.5 million acre increase in the Wild Forest. One-half of the total program should be devoted to the suburban areas and one-half to scenic and other significant natural areas and river corridors across the landscape, outside the industrial forest. Much of this land would be retained in one form or another of active management. In some cases, the lands would be added to the Wild Forest.

7. Upgrade forest research, assessment, and information transfer. Conduct detailed surveys of timber cut no less than every three years. Apply increased funds to evaluating economical means of boosting forest yields of timber, wildlife, and recreation with minimal environmental impact, and for preserving and enhancing biodiversity. Design and test more effective programs of landowner assistance and taxation. Improve and expand coverage of price reporting for standing timber. Vigorously expand research and policy analysis on Global Commons effects. Upgrade the transfer of emerging scientific knowledge to resource managers and landowners. Conduct penetrating and tough-minded reviews of existing policies and programs to develop improved approaches. Think deeply about how conditions, challenges, and needs will change in the future. A recent National Research Council Committee report made a strong case for strengthening and redirecting forestry research (Table 12.3).[17]

8. Improve education in forestry and related fields. We train many forestry, wildlife, and allied professionals in the Northeast, but too few of them are well prepared to meet the region's future needs. The turn-of-the-century foresters spent most of their time convincing landowners that forestry was worthwhile. The foresters of the future should be equipped to do the same. Today, all too often, forestry students are told that forestry on small lands is not worthwhile without government subsidy. Careful sil-

TABLE 12.3. Gordon Committee Recommendations

- More scientists are needed to conduct forestry research.
- Involve nontraditional researchers from the scientific community.
- Maximize the benefits from increased forestry research.
- Implement research by improving extension and technology transfer.
- Increase funding: $100 million USDA forestry competitive grants. USDA-FS budget up 10 percent per year for 5 years. Increase McIntire-Stennis to $109 million in 5 years.

Source: National Research Council, *Mandate for change,* 1990.

vicultural analysis does not receive adequate emphasis in forestry education, nor does the challenge of responding to the need for improved landscape management. Educators should correct this.

9. Start thinking now about how states will respond if developing market forces lead to overcutting. Maine's forests lost softwood volume between 1987 and 1995. This is not conclusive evidence of overcutting because there are many causes. But it is a dramatic alert. Will the states stand aside and watch? Will they appeal to theories that "the market will fix it"? What limits can be applied to prevent sustained episodes of overcutting?

10. Strongly support the region's nonprofit sector in its educational and land protection efforts.

11. Attend to the information, training, and other needs of logging contractors, who dispense much of the practical forestry advice in the region. Develop ways to recognize professionalism and good craftsmanship among loggers. In particular, roadbuilding practices and skidding, which cause much of the aesthetic and environmental damage in logging, are under the logger's control. Several states have established programs of logger certification.

12. Support well-designed programs of land use control and regulation that recognize and protect the landscape-wide benefits of forests and related landscapes. Values such as wildlife habitat, open space, informal recreation, groundwater protection, and wood products need protection. Progress in this field will be difficult in the region's Town Meeting political culture. But it is the only way to save the parts of the forest threatened by unwise subdivision and development.[18]

13. Vigorously promote the concept that the future of the region's forest lies in managing for quality logs, design educational and other programs to support this concept.

Research is a key government contribution to progress in forestry. USDA-FS scientists tally trees, Bartlett Experimental Forest, New Hampshire. *Courtesy of USDA-FS.*

14. An active, innovative, and sustained approach to retaining and enhancing the region's resource of wildness must be pursued, as outlined in chapter 8 above.

These programs will have to be pursued in an economic climate of staff and budget stringency. Federal and state programs have been cut, and the cuts are probably not finished. Despite improving state and federal finances in the late 1990s, there seems to be little hope in sight for improved funding for conservation purposes. To meet the needs outlined above, persons and groups concerned with forest conservation will have to work more effectively, finding help and allies wherever they can.

A Case for Laissez-Faire?

Some observers believe that the above lists are simply imagining a lot of problems. They say that when wood gets expensive enough, people will manage the forests. Don't worry about the suburban forest — that land is worth more for houses anyway. If folks can cut up rural woodlots, strip

them, and hustle them to distant buyers for $5,000 per acre, why not? Landscape values are just a matter of taste, in which government should not interfere.

Why not, indeed? A productive and attractive forest landscape is an essential part of the region's quality of life — for the scenery, fuelwood, jobs, and walking it provides. I cannot imagine that these values will be conserved by the unrestrained forces of private speculation, public agency indifference, and a landowner and professional attitude that places responsibility for conservation entirely on the government.

A Case for Expanded Subsidies?

The received wisdom of government forestry agencies today is that in the future we will want a lot more wood. This is probably true. According to this doctrine, to have that wood we must subsidize woodgrowers. Why not? We subsidize suburban homeowners, farmers, fishermen, and nearly everyone else.

Great programs have been planned to subsidize forestry on private lands, and administrators continue to hope for new infusions of funds. Their continued appeals for tax breaks and subsidies seem to keep this major client group happy. But the national mood in today's budget climate will not support such outlays, even if they are tied to retaining land in open space.

These programs, as used in the past, will be of little help. They undermine responsibility. They provide no built-in assurance that the land benefiting will ever produce a timber crop or be available for public use. They encourage the view that conservation is the government's job. And finally, programs like these are simply not going to get enough funding today to make any difference.

In the past, even the USDA-FS, currently a leading advocate for subsidies, did not believe in their effectiveness: "after careful study it was concluded that incentive payments do not form a sound major approach to forest conservation."[19] This statement was no doubt made to scotch the major competing alternative to the agency's then favored program of land acquisition and regulation, but I think time will prove it correct nonetheless. Aldo Leopold, decrying the bureaucratizing of conservation, characterized the attitude behind this trend: "whatever ails the land, the government will fix it."[20]

A Case for Government Regulation of Forest Practice?

In the past, American forestry leaders believed that private owners would not voluntarily manage forests properly. The profits were too limited, the risks too high; most forest landowners held land simply for liquidation and not for longterm management. Leaders of the USDA-FS publicly advocated government regulation of private forests until the late 1940s.[21]

I do not believe that government regulation of private forest practice can meet the array of challenges listed above. Tempting as the idea may be to some, it is administratively impossible to implement. Giving this task to already understaffed agencies would produce little but frustration. The tremendous adversary debate that would ensue would accomplish nothing. The tremendous, skillfully orchestrated landowner backlash against Maine's proposed "Forest Compact" illustrates this. In short, general regulations, controlling all aspects of private forest practice, would be a misallocation of resources. They would not succeed in building a sense of landowner responsibility, but would undermine it further. They would not address the largest problems, which are landowner indifference and the conversion of forest to speculative lots and other uses, and the potential risks posed by the Global Commons.

A good case exists, however, for regulating forest practice *in particular areas of environmental or scenic importance,* like roadsides, streamsides, lakeshores, wetlands, and sensitive wildlife habitats. This is being done now in a variety of ways, most thoroughly under the jurisdiction of the Maine Land Use Regulation Commission, and in the Adirondack Park. While these agencies have been criticized, they are in place, and alternatives are not obvious. This approach does not seek to control forest practices on all acres, but only seeks to protect environmental values in limited areas of true importance. Setting minimum standards for building, maintaining, and "putting to bed" skid trails and logging roads is an important role for regulatory and forest agencies. Of course, the use of chemical insecticides and herbicides should continue to be subject to strict public regulation.[22] Policy attention to the worst forest abuses is warranted.

The English monarchs and their governors were unable to impose the Broad Arrow policy on enterprising and ornery colonial woodcutters — and those pines were the king's own legal property. We would experience much more serious difficulty in imposing general forest practice regulation on the descendants of those unruly colonials. George Perkins Marsh, the nineteenth-century conservationist, had little faith in regulating forest practice:

> For prevention of the evils upon which I have so long dwelt, the American people must look to the diffusion of general intelligence on this subject, and to the enlightened self interest, for which they are remarkable, and not to the action of their local or general legislatures.[23]

Notwithstanding what is said above, preparations must be made for a public role in ensuring that future increases in wood demand do not overwhelm the forest's growth potential, aesthetic values, wildlife habitat, potential biodiversity, and its resource of wildness. This is a challenge for which American public policy provides no background or experience. Dealing with these challenges will undoubtedly require an intelligent mix of regulatory, landownership, and other programs.[24]

Further, environmental regulations in other spheres will be necessary if the scientific consensus concerning threats from the Global Commons continues to gain ground.

Landscape Management Areas: What Will the Eagles See?

The greatest gap between policy concerns and traditional policies concerns what the eagles see. The overall management of entire landscapes or drainage basins to meet wider habitat and biodiversity goals is one of the principal challenges of the coming century. Yet continued fragmentation of ownerships and escalating demands mean that in many areas, landscape-level goals for forest management will not be met. How to identify and meet landscape level goals in a region of mixed and often very small ownerships will be one of the region's largest challenges.

To meet this challenge in the Northeast's largely private forest landscape, I propose the creation of *Landscape Management Areas* (LMAs). These would be identified zones in which public support would be provided for groups of landowners that enter into voluntary agreements to carry out certain management practices. These would be significantly more protective of forest values than normal private management would be. The practices might include:

- recreation and trail development and management;
- holding stands to very long rotation ages to restore patches of mature forest;
- practices to improve wildlife habitats;
- more stringent practices protecting wetlands and waterways;
- design measures to maintain larger areas of interior habitat;
- design measures to minimize visual impacts;
- minimal use of short rotation even-aged practices and type conversions;
- other practices suited to the local situation.

These LMA's could be developed adjacent to major public ownerships and serve in effect as "buffers" for them. They should emphasize prominent areas important for wildlife, scenic, and recreational values. Hopefully, they would cover substantial areas, 50,000 to 100,000 acres or more. Smaller LMA's may be desirable on the fringes of the rural and suburban forests.

An assessment of the potential role of public lands in a portion of Massachusetts was conducted by Golodetz and Foster. They exhaustively studied land use history and conservation ownership in a region with an unusually high 37 percent of its land in protected status. Their review showed that threats and opportunities remained that could only be dealt with by using a regional perspective and involving multiple ownerships.[25]

What kinds of public support would be involved? This would have to be tailored to the need and to each state's capabilities. Plainly, funds are simply not available today. In time, however, public support might consist of targeted tax breaks, "right to practice forestry" protection, technical assistance, authorization to charge recreation user fees, and selective cost sharing, plus management and law enforcement aid. As part of the bargain, states might seek assurances regarding public access and conserva-

Patterns of cutting units across a landscape affect aesthetics and habitat values. This northern Maine clearcut has linear boundaries, no "beauty strip," and the green tree retention was probably unplanned. Linear haul roads may cause unnecessary interference with streams and ponds. Strip cuts are visible in the near background. *Courtesy of Maine Department of Conservation.*

tion easements as well as rights of first refusal for acquisition if parcels come onto the market.

The LMA would be a highly flexible concept, a model around which to organize ideas. Beginning from such an idea, states, landowners, and conservation groups could develop many effective projects.[26]

Conserving Wildness

The region's resource of wildness is what gives the more remote forests their powerful mystique and sense of place. The threats to wildness are many. Again, most of these threats lie outside the range of traditional forest policies. They must be addressed through land use controls, regulations, and acquisition, as well, perhaps, as through carefully designed taxes and subsidies. Yet policies must go far beyond this, into realms that public leaders are unwilling to enter. One of the biggest threats to wildness is overuse by the public, as Aldo Leopold noted years ago:

> Thus always does history, whether of marsh or market place, end in paradox. The ultimate value in these marshes is wildness, and the crane is wildness incarnate. But all conservation of wildness is self-defeating, for to cherish we must see and fondle, and when enough have seen and fondled, there is no wilderness left to cherish.[27]

There is no escaping the conclusion that in the long run, users of the Wild Forest will have to accept increasing restrictions on their activities, or the wildness will be severely diminished. On some major rafting streams, restrictions are already in effect. But many visitors to the backwoods are not really seeking wildness, but other forms of recreation. If we were to own the entire northwoods, yet high powered motorboats still race around Lobster Lake, what will remain of wildness?

What Role for the Federal Government?

Tension between state, local, and federal governments is a longtime theme in the Northeast. Efforts to formulate the challenge of the northwoods as a federal problem are illustrating this once again.[28] The attempt to federalize the problem in order to tap federal dollars faces formidable political obstacles. Ironically, this drive to bring the federal government directly into the region's forest conservation problem occurred just as Washington's ability and willingness to pay the tab has hit all-time lows. As a result, advocates of Northern Forest Lands protection have had to settle for unsatisfying declarations of concern, and laundry lists of options with no clearly defined follow-through.

It remains to be seen whether an expanded federal role in the region's forest problems, beyond its existing activities, will emerge. The region's Town Meeting tradition has led to ambiguous results in the past. The history of past regional efforts suggests that the odds are low. As this work is completed, the federal treasury has turned at least temporarily into surplus, but there seems little likelihood that this happy condition will lead to more federal spending on traditional forest and conservation issues.

A Program in Brief

Today, most states possess effective fire control organizations, some form of open space or use-value taxation for rural land, and a corps of trained foresters available to assist small landowners. They are upgrading management of their own public land. Efforts to secure public access are underway. Programs to protect watercourses are in place, and research continues. The upsurge in wood demand presents opportunities for better forest management and helps more consultants earn a living. Yet it also threatens to increase overcutting and high-grading. Also, dramatic land use and ownership change is likely.

The successes of the past cannot carry us through the next half-century. Today's programs and policies were not designed to meet the challenges we now face. They should not be scrapped, but intelligently overhauled. What I am urging is based on a complete reform of landowner attitudes toward their forests. In the future, the only guarantee of good forest stewardship will be the widespread acceptance of what forester Aldo Leopold called, "an ethical relation to the land."[29]

But we cannot rely exclusively on improving landowner attitudes and practices. The incentives to strip land have not gone away; they are likely to increase. The region needs strong leadership from the states to devise the right mix of regulation, voluntary cooperation, public ownership, research, technical assistance, subsidies, and tax policy to meet the diverse needs within the region. The right policy mix will vary between the region's five forests, and from state to state.

A key policy challenge is to design new approaches to preserving key forest values while at the same time keeping the land at work providing products, services, tax revenues, jobs, and recreation. Quiet leadership on this approach is being exercised around the region.[30] A few examples:

- The ambitious program to reduce nutrient content of New York City Water from the Catskill Watersheds.

- Maine's Bigelow Preserve, where a carefully designed balance of uses maintains forested conditions, and a setting for a key Appalachian Trail segment, in a self-funding program of state management.
- The Western Pennsylvania Conservancy's management program on its recently acquired 11,300 acre H. J. Crawford Reserve near President, Pennsylvania. In this case, the Conservancy sold the land to an industrial company while retaining a broad conservation easement.
- The Atlas Timberland property in Vermont (26,000 acres), purchased by The Nature Conservancy and the Vermont Land Trust. This land will continue to produce commercial timber.
- The acquisition by the State of Vermont of a 31,500 acre conservation easement on timberland managed for investment income by Hancock Timber Resources Group.

During 1998 and early 1999, conservation groups embarked on land purchases on an unprecedented scale:

- The conservation fund purchased 330,000 acres of Champion International Lands in New York, Vermont, and New Hampshire for a mix of conservation and timber management.
- The Maine Nature Conservancy purchased 185,000 acres of northern Maine timberlands from International paper.
- The New England Forestry Foundation announced a deal to acquire a conservation easement on the entire ownership of the Pingree Heirs in northern Maine, roughly 900,000 acres.

The forest challenges for the twenty-first century defy simple diagnosis and simple solutions. Certainly they are beyond the reach of any one level of government, program, or agency.

13 Northeastern Forests: Past and Future

THE FUTURE FORESTS of the Northeast will be shaped by the same external and internal forces that have shaped them in the past. Two shorthand scenarios suggest how much the forest is likely to change by 2040. Also, several abiding conditions will continue to influence the role of the forest in the region's cultural and economic life. Through this overview, the metaphors used throughout this book (Table 13.1) will come together.

External Forces

The external forces affecting the forest include, first and foremost, those of the market. Market forces were evident in the early export trade, shipping lumber, masts, and other forest products overseas and to other coastal towns. The cutting and land abandonment of the Robber Baron period left behind millions of acres of slash, burn, and devastation, and millions more of depleted forest. Later, cheaper midwestern farm products and the demand for labor in the industrializing cities triggered a massive decline in farming that returned some 20 million acres of cleared land to forest. Substitutes eliminated some markets for the region's wood. The markets for recreational lots and suburban homesites have again modified the forest, redirecting its principal uses, and reducing public access, over millions of acres. The "shadow conversion" effects of these land use changes affected even larger acreages. Other market influences have brought about a high level of corporate ownership and a limited degree of foreign ownership of land in some areas.

In the 1970s, the dramatic increase in oil prices brought an ancient wood use — energy — to the fore again for many residents, businesses, and policymakers. The national wood products market, with rising demand, escalating costs in other regions, and new technology, contributed

TABLE 13.1. Metaphors for Thinking About the Future of Northeastern Forests

Metaphor	Some of the Meanings for the Future
Global Commons	Shifts in composition and productivity are possible.
Worm's Eye View	Nutrient status/cycling may increase in importance.
Eagle's Eye View	Landscape management must be implemented.
Economic Interdependence	Imported raw materials will be more costly; opportunities to improve self-sufficiency will need to be taken; urban-rural interdependence will continue.
Town Meeting View	Public programs will need to regain lost ground; policy debates will be more contentious.
Cultural Heritage	Wildlife; biodiversity, and wildness will grow in value.
Bequest	Future biodiversity and forest health; future raw material and energy supplies; future access for recreation; economical land use pattern in an age of more expensive energy; preservation of options; sustainability; conservation of wildness; retaining public access; retaining materials production potential will all increase in importance.

to a resurgence of the region's wood-using industries that is likely to continue into the new century. While this was principally visible in the northern forests, throughout the region sawmills were modernized, fuelwood demand grew, and a boom in export log and lumber markets boosted the sawlog market.

Northeastern forests are home to many native forest pests, such as the spruce budworm, beech scale, and birch dieback. But the region's forests have been afflicted and re-worked by imported pests as well. The gypsy moth has killed susceptible hardwoods across Southern New England and the Mid-Atlantic states, and periodically makes itself a major suburban nuisance. The most serious losses, however, have been the chestnut, a dominant hardwood tree of the region's hardwood forests, highly valued for its timber and nuts, and the American elm. No one can be certain that

Symbols of pastoral life and a new industrial society in *The Lackawanna Valley.* George Inness's painting, 1855. Much of the farm landscape depicted in 19th century paintings is now forest. *Courtesy of National Gallery of Art, Washington. Gift of Mrs. Huttleston Rogers, © 1995 Board of Trustees.*

similar disasters will not befall other important trees in the future, despite our improved watchfulness.

Internal Forces

From within the region have emerged other forces affecting the forest. Colonial policies governing landownership and settlement left half of the forest dominated by 100-acre blocks, relics of past farming and land sales. The wildland holdings of the north are descended from land sales and grants to speculators, mostly made for public revenue rather than immediate settlement. The abandonment of settlement as a goal anticipated the later retreat of farming from the fringes of the wildlands. Semi-feudal ownership patterns in the Mid-Atlantic states left their marks as well. These patterns of ownership will see further fragmentation as generations of owners pass. They will constrain management and policy options.

From within the region's communities have emerged major waves of regional and national conservation leadership. The early conservation movement resulted in the establishment of forestry schools and state forestry agencies. The Pinchot estate in Milford, Pennsylvania, symbolizes

this period. To the same period, we owe the rich diversity of private nonprofit groups who have recreated a public forest estate in this century. They acted by pressuring governments into action, by acquiring land themselves, and by inspiring philanthropists to donate land.

Unique to New England is its system of town government with limited county government. This means that there are few government agencies capable of taking action on a regional basis on natural resource issues, such as park acquisition. It means that towns ignore needs of wider areas, while the state is excessively distant and distrusted by most citizens. In rural portions of New Jersey, New York, and Pennsylvania, despite stronger county government, local governments weigh heavily in many matters affecting land use and the forest. Virtually every state is divided along urban-rural lines on a range of environmental and land use issues. The Town Meeting metaphor summarizes the increasing difficulties the region will have in making decisions about both local and regional issues.

The region has produced important scientific knowledge of forests. Of a series of scientific research stations, the best known are perhaps the Harvard Forest (Petersham, Massachusetts) and the Hubbard Brook Experimental Forest in the White Mountains. At Hubbard Brook, scientists have worked out with meticulous care the inner ecological cycles about which Marsh wrote a century earlier. These concepts have contributed greatly to our understanding of the forest's relation to the Global Commons and to

---

Journey to the Outer Edge of Cape Cod

. . . It is not that our ancestors were any more dependent upon the land or sea than we are, but that they possessed a sense of direct, repetitive involvement, not only of remembered processes, but of the remembered texture that they brought to life's activities, which in turn gave them a strong feeling of identification with the land — a literal "sense of place."

It is true, of course, that it was in many ways a very narrow sense. On Cape Cod a life of subsistence farming or fishing demanded that the inhabitants see and know the landscape primarily in terms of extracting a living from it.

. . . Rarely did these country people, these rural New Englanders, experience an aesthetic apprehension of their surroundings unyoked to some task or necessity it signified, and yet because of this the landscape was palpable, three-dimensional to them in a way we have largely lost, however much we have refined our "appreciation" of its visual beauty and increased our "respect for the environment."

Source: Robert Finch, *Outlands: Journeys to the Outer Edge of Cape Cod,* Boston: David R. Godine, 1986, pp. 15–16.

---

the full development of the Worm's Eye View. These insights are being applied to implement the implications of the Eagle's Eye View of the forest, most prominently on federal and state forest lands.

The literary, visual arts, and scientific works originating in the region's forests have continued to inspire people across the nation who care about forests. These have helped not only to preserve but more widely develop the Cultural Heritage associated with the region's forests.

Economic interdependencies have been recognized a bit more fully, especially as experts and citizens have sifted over the meaning of the late 1980s and early 1990s economic bust that afflicted most of the region. In state capitals at least, the importance of rural economies and manufacturing for the cities and suburbs is somewhat better appreciated. Still to be widely understood, however, are the manifold economic and environmental implications of failure to take full advantage of the region's forests as a source of raw material.

The Northeast's Future Forests

A few speculative scenarios may help suggest what the region's forests might be like in fifty years. The year 2040 will be more than four centuries after Europeans settled at Massachusetts Bay. Despite its high level of urbanization, in fifty years the Northeastern landscape will still be dominated by its forest. The principal difference between optimistic and pessimistic scenarios is not in total forest *area*. Rather, it will be the kinds of *values* these forests will be able to serve (Table 13.2).

The optimistic scenario projects no revival of subdivision of rural forests, a modest erosion of the industrial forest, and a doubling of the Wild Forest. It assumes that a more land-conserving pattern of suburban development will prevail.

The pessimistic scenario envisions a significant reduction of the industrial forest by conversion to recreational subdivisions, recreational reserves for small groups, and Wild Forest units. The suburban forest dwindles in the face of inefficient, land-consuming, unplanned sprawl. There is only modest expansion of the Wild Forest. The parcel size distribution of the rural forest flip-flops, to one-third of the acreage above 100 acres in size, from the present two-thirds. Public access to remaining forest land plummets.

These outcomes will mostly be determined by land markets and the values placed on land by its owners. They are thus beyond the reach of traditional forest policy tools like favorable taxation, financial aid to landowners, or information programs. They can only be affected by planning and zoning, by public and nonprofit landownership, and by developing more widely in landowners an ethical relationship to the land.

TABLE 13.2. Two Speculative Scenarios for Northeastern Forests: 2040 (million acres)

		2040	
	1997	Optimistic	Pessimistic
Industrial	14	12	8
Recreational	7	7	11
Suburban	10	8	6
Wild	5	8	6
Rural	34	30	30
Total	70	65	61
Developed	10	14	18
Percent avail. for wood production	80%	65%	50%

Source: Author estimates.

Sound tax policies for rural land can help but are insufficient by themselves. At the margin, various public regulations and other policies might improve matters a bit. The makeup of the region's forests in 2040 will be determined by literally millions of tiny decisions made by individual landowners, town and county governments, and state agencies.

The Fate of the Forest Matters

What difference does the fate of the forest make? We can briefly compare one aspect, the commercial forest land base available for longterm timber production. The 70 million acres shown for 1997 are now only partially available. Perhaps one-half of the suburban and recreational forests might be used for occasional commercial timber crops, and two-thirds of the rural forest (the amount exceeding 100 acres in size). This would place the current wood production land base at about 44 million acres. Smaller lots will offer more limited opportunities for intensive forestry, recreational use, and active management for other purposes.

Under the pessimistic 2040 scenario, expanding speculative subdivisions, plus highly space-consumptive patterns of suburban sprawl would reduce total forest acreage to 61 million acres. Of this, perhaps only 30–35 million acres might be available for serious timber management, recreational uses, and active wildlife habitat management. The important point is that while the area of forest land may change little, the potential uses and values of the forest may change a great deal.

Sprawling, uncontrolled suburban growth and future rural land subdividing booms will destroy forest values and community amenities on an unprecedented scale. The addition of 6 million people to the Northeast's population in the next 25 years could cause as much environmental and visual damage as was caused by the entire previous population growth of this century. The expected loss of open space, forest production, and wildlife habitat should motivate vigorous action by those concerned with the natural values of farm and forest.

The centuries old tradition of being able to take a walk in the woods almost anywhere was not only a benefit to walkers and birdwatchers, it was an element of civility in a rural society. It is now at risk. This would be the ultimate privatization of the landscape, achieved by millions of individual Posted signs, not by the machinations of big corporations.

In the coming half-century, significant changes will occur in the forest's composition as a result of natural ecological succession and timber harvesting. In the industrial forest, stands on average will be younger than today, especially for the intensively used softwoods. Much of the spruce-fir acreage will be managed on a thirty- to sixty-year rotation for pulp and small sawlogs. Managed stands will be increasingly evident. By the year 2040, the extensive hardwood resource of that region will be displaying the effects of increased management. For some, the aesthetic appeal of this "damp and intricate wilderness" will decline noticeably.

Despite increased management intensity in local areas, the total volume and average age of much of the Northeastern forest will increase significantly. The birch-aspen stands of the north will be replaced by more tolerant softwoods and maple and beech, unless heavy cutting and soil disturbance or fires regenerate them. Whether silviculturists will find ways to maintain or increase representation of black cherry and other valuable hardwoods remains to be seen. Early effects of climate change and atmospheric deposition, if they indeed occur as many scientists now expect, will begin to become evident. Old-field brushlands will return to hardwood and mixed forests. Many oak stands will become oak-pine as pine enters the overstory. Many pine stands, especially on old fields, will be converted to hardwoods and fir, as their understories are released by cutting or windthrow.

The increased areas and volumes expected are comforting, but the increased forest management, and the spread of development will change wildlife habitat values. Habitats will be ever more fragmented. In some areas, maturing forest cover and shifts in stand types will reduce habitat for deer, grouse, and other species benefitting from early successional condi-

Threats to the forest's future are not as visible as this example from early this century. High-grading, ownership fragmentation, shadow conversion due to sprawl, and atmospheric deposition, all defy clear illustration and simple solutions. Unknown Pennsylvania location. *Courtesy of Susan Stout, USDA-FS.*

tions or from abundant understory vegetation. Further decline of farming, if it occurs, would reduce populations of birds and other species that favor farm-forest margins.

Forecasting the level and types of recreation uses in fifty years is a chancy business at best. Yet most would agree that recreational pressures on Northeastern forests are likely to increase. In many areas, ponds and streams are crowded and overfished. Noisy motorboats or RV's are seen in remote settings, and campsites and trails are already overused. Unmanaged increases in recreational use in the region's more remote forests pose a significant threat to the wilderness experience. There is a significant urban-rural conflict here. Remote streams and ponds where urbanites seek a wilderness experience are also the backyard hunting and fishing grounds for the local residents, who by and large pursue their outdoor experiences in quite different ways.

As they respond to increased wood demand, the region's landowners, led by the large industrial concerns, will have to raise the quality and sophistication of their management practices. Careful management of the hardwood forests will become possible for the first time. We can probably expect an improvement in the general level of forest practice more striking than in any previous period in the region's history. While this may be true on an acre by acre basis, there is no way to foresee how well the region will deal with the landscape-level issues raised by the Eagle's Eye View of the forest.

But many landowners will not respond to better wood markets by upgrading forest practice and giving more attention to multiple forest values. Instead, they will take the opportunity for the fast dollar, and leave behind high-graded and rutted slopes. When the next land boom comes, these exploiters will be in the clover again. The challenge for the forestry community and for policy is to find ways to shift the balance steadily from exploitation toward husbandry. Unfortunately, as the discussion in this book argues, there is no silver bullet.

The Northeast's forest future cannot be defined in terms of a single narrow perception of forest values, or in terms of simple either/or conflicts. The reality is that both the employment and raw material production values of the forest, and its recreational and cultural values, especially of its remaining wildness — are going to increase in importance. As a result, an important role for active forest management should be accepted as an abiding condition important to the region's future welfare (Table 13.3).

Abiding Conditions

Most of this book has concentrated on change, its sources, and its effects on the forest. But besides constant change, several abiding conditions will influence the forest. These conditions will constrain policies attempting to conserve and improve the forest.

The basic conditions are ecological. These are expressed in the region's forest geography, in its abundance of early and mid-successional

TABLE 13.3. Why Forest Management Deserves a Place in Northeast Forest Landscapes in the Twenty-first Century

- Rural communities need jobs, tax revenues.
- The region needs local raw material supplies.
- Renewability and environmental benefits of wood are significant compared to other raw materials.
- Management can help shape desired future habitat conditions.
- Replacing fossil fuels with wood-based energy benefits the environment and the economy.
- Forest management may assist in carbon sequestration.
- Landowners have rights to use their land.

forests, and the resilience of those forests. The ecological potential for sustained production of wood, wildlife, recreation, water, and other forest services remains high, and may even increase. To what extent the cumulative effects of acid precipitation, climate change, or other factors will alter this basic ecological productivity remains to be seen.

The other abiding conditions are social. They begin with the historic landownership pattern and the system of town government. These conditions include widespread private ownership of forest land and diverse centers of policy decisionmaking. They assure that change will be slow, but they offer some useful resistance to the rapid diffusion of bad ideas. The town system of government allows flexible innovation by local groups but hinders consideration of regional concerns within states.

An abiding condition of great importance is the basic attitude of citizens about the forest. Individual attitudes vary, of course, from indifference, to toleration of gross exploitation, to strong preservationism. Especially in suburbia, tension between advocates of green backdrop policies and proponents of commercial interests will continue. But in rural areas, many residents are suspicious of public landownership and support the commodity uses of the forest. Few favor widespread use of forests for wilderness. The region's basic belief in the right of private landowners to do as they please conflicts with another widely held value, that recreational use of private land is a public right. While this value has long been forgotten by suburbanites regarding their own land, they vigorously advocate it for the owners of the industrial and rural forests, which they speak of as "our forests."

Northeastern citizens are willing to recognize legislatively that farms and forests should be taxed differently from house lots. But the economic pressures of the future mean that political conflict over forest taxation will continue. Wrangling over farm and forest taxation will be a permanent item on the agenda of legislators and tax assessors, as well as a perennial concern of the owners themselves.

The corollary of extensive private ownership is limited public ownership. Even if the most ambitious goals for expanded public ownership are met, the region would still only have a small public forest estate. This seems to be what most people want and is no problem in itself.

Another abiding condition will be an escalating world demand for wood products. This growth may stimulate continued industrial development based on wood. Unless the secondary processing of wood expands markedly, however, employment in wood-based industry could fall considerably. This would constrain community prosperity and individual economic opportunity across the region. The world's ability to respond to growing demand will be constrained as traditional old growth supply regions in Western North America, the tropics, and the former U.S.S.R. are unable to sustain past production levels. Not only will tighter world supplies raise the Northeast's costs for imported raw materials, but it will improve the economic climate for growing wood here. In addition, higher prices will prompt more emphasis on reducing the throughput of paper and solid wood through more re-use and recycling. These improvements do not diminish the region's need to improve its self-sufficiency in wood-based products, however.

Further, the region's dependence on imported energy is a longterm structural condition that will not change. The cost of imported oil is likely to rise again at some time in the future. This will put many idle woodstoves back to work. When this occurs, it will remind us of the importance of retaining options for wood production in the forest. A boost in oil prices could increase regional fuelwood usage by a million cords or more.

This is a mixed picture, a complex one. On the one hand, major forest values will be lost to exploitative cutting, suburban and recreational sprawl, land posting, and unproductive land speculation. On the other, expanded wood markets, greater knowledge, and better landowner attitudes will permit major improvements in the region's standards of forest practice. As citizens begin to see what has been lost, they will appreciate more what has been saved. The Allagash, National Forest wilderness, and the local land trust sanctuaries will be ever more highly prized as part of the region's natural heritage and its Bequest to the future. Never before has the forest promised to mean so much to so many people. But those meanings are neither simple nor one-dimensional, as our list of metaphors above suggests. Political divisions are of historic vintage in this region. Many towns owe their origin to divisions within a community over where to locate a church . . . often solved by the aggrieved parties seceding to form their own congregation and local government. There is little likelihood that a clear consensus will be any nearer on the major issues reviewed in this book in 2040 than it is today.

For foresters, legislators, landowners, and citizens, the coming decades promise abundant challenge. We must convince owners of their responsibility to upgrade management of the rural and industrial forests. The acreage of Wild Forest should be increased, perhaps by 50 percent. Vigorous steps are needed to save what remains of the suburban forest. The recreational forest needs to be better managed and its expansion restrained. The concerns of landscape management must be addressed, and the region must make its own maximum contributions to reducing ozone, sulfur, and greenhouse gas emissions instead of just pointing fingers at Ohio. Access to private forests for fishing, birding, hunting, and simply enjoying the woods must be ensured.

The ability of government to deal with these problems is not encouraging. The tighter budgets of the 1990s eroded the effectiveness and the reach of many existing programs. Federal funding cuts have hindered progress in land acquisition. Despite improving state and federal fiscal balances as this is written, the improvement could be temporary. Conservation and forestry could slide still lower on the priority lists of legislators and governors, who will face massive and baffling fiscal and policy problems.

The region's active and vigorous nonprofit sector will have to continue to lead. The conservation commissions, land trusts, and statewide groups will have to strive even harder to raise funds and snatch tidbits of open land from the path of bulldozers. They will have to provide most of the leadership in convincing their neighbors of the value of forest conser-

Toe of the Boot, Moosehead Lake, Maine, 1991.
Courtesy of U.S. Geological Survey.

vation, rational development, multiple use forestry, public access, and sensible preservation of outstanding areas.

In my view, the days are over when we can say that timber is more important than recreation, that pickerel are more important than payrolls. *The fact is that all of the values of the forest — as a storehouse of biodiversity, a source of jobs, taxes, and raw materials, as a recreational resource — are growing increasingly important to our society. This means that there are no easy choices.*

Moving policies forward will require a mobilized sense of urgency. While the issues may be different, the urgency expressed by James S. Whipple, Commissioner of Forest, Fish, and Game of New York at the 1909 Governors Conference is relevant:

> Is there any doubt about the fact that a country without wood and water is like a house without a roof — valueless and uninhabitable? Let me illustrate: Suppose that today by some great force in Nature every tree and shrub were swept away from the face of New York State, what would the condition be tomorrow? Would not chaos reign? The streams would be uncovered, the waters would dry up; God's reservoir under the trees on the forest floor would be destroyed; the farms would soon become unproductive, and in a year would shrink in value and in products at least half. The commercial supremacy of the State would be much injured, if not destroyed; and all on account of the loss of our forests . . .
>
> . . . Forests can be restored. That is the problem of the hour. You say the coal is going and do not tell us how to replenish it. We all say the forests are going, and I tell you how to save them, how to restore them. Save them by setting aside a forest reserve; restore them by planting trees. A tree crop can be raised as well as an oat crop. It takes longer; that is all. It is profitable business, once understood. Shall we, the supposed most intelligent people in the whole word, stand with folded hands and do nothing of importance to save and restore our forests?[1]

These remarks were delivered at the height of the nation's lumber output, when liquidation cutting was the order of the day. The paper industry in the Northeast was growing rapidly. Some of the notions about forest influences of the time may seem quaint or naive today. But if we look around, where are the political leaders prepared to speak with such urgency on the challenges of the region's forest future?

The policies of the past are not equal to these challenges. New leaders and new policies will be needed. In the region's best traditions, they will emerge. States in this region led the nation in responding to the forest crises of the turn of the century, with programs of public acquisition, fire protection, tree planting, scientific research, education, and landowner assistance. Our scientific understanding is not static. New metaphors for thinking about the forest will emerge, as they have done in the past.

And through all of this change and tension, one reality will abide: the Northeast will remain, as it was in 1620, a region of forests.

The forest of the future consists of the younger trees in this Western Maine mixed stand. If sustained by prudent landowners, sound silviculture, and sensible public policies, these young trees will be supporting forest values in the mid-twenty first century. *Photo by author.*

References

Notes on Chapter 1

1. Valuable background on modern forest resource conditions and issues can be found in D. W. McCleery, *American forests: a history of resiliency and recovery,* USDA-FS in coop. with For. Hist. Soc., Durham, NC, FS-450; K. D. Frederick and R. A. Sedjo, eds., *America's renewable resources: historical trends and current challenges* (Washington, D.C.: Resources for the Future, 1991), esp. chs. 1, 3, 6, and 7; and H. Kimmins, *Balancing act: environmental issues in forestry* (Vancouver: Univ. of British Columbia Press, 1992). Regionally, see D. R. Foster, *Thoreau's Country: Journey Through a Transformed Landscape* (Cambridge: Harvard Univ. Press, 1999). Also useful are J. G. Mitchell, "Yankee forest for sale by the Btu?," in *Dispatches from the deep woods* (Lincoln: Univ. of Nebraska Press, 1991); and "Yankee forest: for sale by the acre," in the same volume; R. D. Nyland, "Exploitation and greed in eastern hardwood forests," *J. For.* 90(1): 33–37, 1992; L. C. Irland, "Challenges for the North Woods," *Maine Policy Rev.* 1(1): 71–82; C. G. Klyza and S. C. Trombulak, *The future of the northern forest* (Hanover: Middlebury College Press, by Univ. Press of New England, 1994), contains useful topical essays. D. Dobbs and R. Ober, *The northern forest* (White River Junction: Chelsea Green Press, 1995), recounts vivid personal interviews with loggers, residents, and land managers in the northern forest.

2. The story of this forest as it was discovered by the settlers is a rich and fascinating one but cannot be recounted here. An excellent window into the subject up to about 1800 is William Cronon's acclaimed book, *Changes in the land: Indians, colonists, and the ecology of New England* (New York: Hill and Wang, 1983). An extremely detailed study of north central Massachusetts compares vegetation in the colonial period, in 1830, and at present, with an extensive listing of citations: D. R. Foster, G. Motzkin, and B. Slater, "Land-use history as long-term broad-scale disturbance: regional forest dynamics in central New England," *Ecosystems I:* 96–119 (1996); a readable and learned review is W. H. MacLeish, *The day before America: changing the nature of a continent* (Boston: Houghton Mifflin Company, 1994). See also, J. R. Stilgoe, *The common landscape of America 1580 to 1845* (New Haven: Yale Univ. Press, 1984). A valuable overview is provided in T. Horton and W. M. Eichbaum, 1991, *Turning the tide — saving the Chesapeake Bay* (Washington,

D.C.: Island Press), esp. pp. 130–141. Other useful sources are T. G. Siccama, "Presettlement and present forest vegetation in northern Vermont with special reference to Chittenden County," Amer. *Midland Naturalist* 85(1) (1971): 153–172; G. M. Day, "The Indian as an ecological factor in the New England forest," *Ecology* 32(2) (1973): 329–346; and C. G. Lorimer, "Presettlement forest of northeastern Maine," Ecology 58(2) (1977): 139–148. Maine's coastal forests are discussed in P. W. Conkling, *Islands in time: a natural and human history of the coastal islands of Maine* (Camden: Down East Books, 1981); and in P. W. Conkling, *From Cape Cod to the Bay of Fundy: an environmental atlas of the Gulf of Maine* (Cambridge: MIT Press, 1995). The postglacial history of New England forests is reviewed by T. E. Bradstreet and R. B. Davis, "Mid-postglacial environments in New England with emphasis on Maine," *Arctic Anthropology* 12(2) (1975): 7–22, with full references. For a current view on a continental perspective, see E. C. Pielou, *After the Ice Age, the return of life to glaciated North America* (Chicago: Univ. of Chicago Press, 1991); and D. Foster, "Land-use history and four hundred years of vegetation changes in New England," in B. L. Turner, et al., eds., *Global land use change . . . A perspective from the Columbian encounter* (Madrid: Superiorde Investigaciones Cientifica's, 1995); P. K. Schoonmaker and D. R. Foster, "Some implications of paleoecology for contemporary ecology," *Botanical Rev.* 57(3): 204–245 (1991).

3. S. E. Morison, *Oxford history of the American people* (New York: Oxford Univ. Press, 1965), pp. 141, 176.

4. A useful overview can be found in F. J. Deneke and G. W. Grey, *Urban forestry* (2d ed., Melbourne, FL: Krieger Publ. 6. 1992 reprint); R. H. Platt, R. A. Rawntree, and P. C. Muick, eds., *The ecological city: preserving and restoring urban biodiversity* (Amherst: Univ. of Mass. Press, 1994); and P. H. Gobster, ed., *Managing urban and high-use recreation settings,* in USDA-FS NCFES, Gen. Tech. Rept. NC-163 (1993).

5. D. W. Gross, H. Carter, C. Geisler, and G. R. Goff, "The northeastern forest as a new commons," in T. Lepisto, ed., *Sustaining ecosystems, economies, and a way of life in the northern forest* (Washington, D.C.: The Wilderness Soc., 1993).

6. An accessible introduction by leading experts is D. Botkin and L. Talbot, "Biological diversity and forests," in N. P. Sharma, *Managing the world's forests: looking for balance between conservation and development* (Dubuque: Kendall/Hunt, 1992), ch. 4; also J. C. Ryan, "Conserving biological diversity," in J. R. Brown, et al., *State of the world, 1992* (New York: W. W. Norton & Co., 1992), ch. 2. For a more complete treatment, see E. O. Wilson, ed., *Biodiversity* (Washington, D.C.: Nat. Acad. Press, 1988). Works oriented toward land management and planning include M. L. Hunter, *Wildlife, forests, and forestry: principles of managing forests for biological diversity* (Englewood Cliffs, NJ: Prentice-Hall, 1990). Society of American Foresters, *Task force on sustaining long-term forest health and productivity*

(Washington, D.C.: SAF, 1993); and D. J. Mladenoff and J. Pastor, "Sustainable forest ecosystems in the northern hardwood and conifer region: concepts and management," in G. H. Aplet, J. T. Olson, N. Johnson, and V. A. Sample, eds., *Defining sustainable forestry* (Washington, D.C.: Island Press, 1993). A well-illustrated overview is C. L. Shafer, *Nature reserves: Island theory and conservation practice* (Washington, D.C.: Smithsonian Press, 1990). Bill Drury's volume of essays, edited posthumously by J. G. T. Anderson, will interest many readers: W. H. Drury, Jr., *Chance and change: ecology for conservationists* (Berkeley: Univ. of California Press, 1998).

7. See, among the vast literature, G. M. Woodwell, "The role of forests in climate change," in Sharma op. cit. note 6 above, ch. 5; and; chs. 19, 23, 24, 25, and 26 in B. L. Turner, II, et al., eds., *The earth as transformed by human action* (New York: Cambridge Univ. Press, 1990); R. C. J. Somerville, *The forgiving air: understanding environmental change* (Berkeley: Univ. of California Press, 1996), is current and very well written; J. Houghton, *Global warming: the complete briefing* (New York: Cambridge Univ. Press, 1997, 2d ed.) is well-illustrated and authoritative. On policy implications, see U.S. Congress, Office of Technology Assessment, "Preparing for an uncertain climate: summary," Washington, OTA-0-563, Sept. 1993; J. Aber, "Forest ecosystem vulnerability," in *A regional response to global climate change: New England and eastern Canada* (Orono: Univ. Maine Water Resources Program, 1993), pp. 80–92. Also, see papers by Birdsey, Fernandez, Irland, and Wolfe in *New England Regional Climate Change Impacts Workshop, Summary Report* (Durham: Univ. of New Hampshire Inst. Stud. Earth, Oceans, and Space, Sept. 3–5, 1997); J. T. Overpeck, P. J. Bartlein, and T. Webb, III, "Potential magnitude of future vegetation change in eastern North America: comparisons with the past," *Science,* 254: 692–695, 1 Nov. 1991; and L. F. Pitelka and the Plant Migration Workshop Group, "Plant migration and climate change," Amer. Sci. 85: 464–473 (1997).

8. Useful overviews of forest nutrient cycling can be found in papers by Hornbeck, Smith, Martin, and Johnson in C. T. Smith, Jr., E. W. Martin, and L. M. Tritton, eds., Proc., *1986 symposium on the productivity of northern forests following biomass harvesting, USDA-FS NEFES NE-GTR-115, 1987; C. T. Smith, Jr., Literature review and approaches to studying the impacts of forest harvesting and residue management practices on forest nutrient cycles* (Orono: Univ. of Maine Agr. Exp. Sta. Misc. Rept. 305, 1985); A. Carlisle and L. Chatarpaul, *Intensive forestry: some economic and environmental concerns,* Agr. Canada and Canadian For. Serv. Inf. Rept. D1-X-43, Petawawa Nat. For. Inst., 1984; S. M. Maliondo, *Possible effects of intensive harvesting on continuous productivity of forest lands,* For. Canada Inf. Rept. M-X-171, Maritimes For. Res. Centre, 1988; M. K. Mahendrappa, C. M. Simpson, and G. D. Van Raalte, *Proc., Conference on impacts of intensive harvesting,* Fredericton, NB, 1990, For. Canada, Maritimes Region, 1991. More technical is R. H. Waring and W. H. Schlesinger, *Forest ecosystems, concepts, and management* (Orlando: Academic Press, 1985). A useful recent review is J. Dutch,

"Intensive harvesting of forests: a review of the nutritional aspects and sustainability implications," in M. K. Mahendrappa, C. M. Simpson, and C. T. Smith, 1994, *Proc. of the IEA/BEA workshop,* Fredericton, NB, May 16–22, 1993, Canad. For. Serv. Inf. Rept. M-X-191E.

9. See, e.g., A. E. Luloff and M. Nord, "The forgotten places of New England," in T. A. Lyson and W. W. Falk, *Forgotten places: uneven development in Rural America* (Lawrence: Univ. of Kansas Press, 1993), pp. 125–167. On how globalizing wood markets are tying the world together, see papers by Mather, Sutton, Michaelis, and Marshall, in *Globalization of wood: supply, processes, products, and markets* (Madison: For. Prods. Res. Soc. Proc. No. 7319, 1993).

10. The various essays of Thoreau are well-known, *The Maine Woods* less so. In the visual arts, Donald Keyes, curator, *The White Mountains, place and perceptions* (Hanover, NH: Univ. Press of New England, 1980), is a beautiful exhibit catalogue displaying depictions of the White Mountains in the visual arts. A. Skolnick, *Paintings of Maine* (New York: Clarkson Potter Publishers, 1991), illustrates contrasting visions of the forest by important national and local artists. G. B. Barnhiss, *Wild impressions: the Adirondacks on paper* (Boston: Adirondack Museum and David R. Godine, 1995), assembles black and white and color prints from the Adirondack Museum's collection. P. G. Terrie, *Forever wild: environmental aesthetics and the Adirondack Forest Preserve* (Philadelphia: Temple Univ. Press, 1985), studies the development of what he calls a "wilderness aesthetic" in the Park. A short, illustrated essay on the *Hudson River School is found in ch. 4 of Frances F. Dunwell, The Hudson River Highlands* (New York: Columbia Univ. Press, 1991). An excellent discussion of Thomas Cole's career and his inspiration in the Catskills is found in chs. 51 and 52 of Alf Evers, *The Catskills: from wilderness to Woodstock* (Garden City: Doubleday and Co., 1972).

11. C. S. Colgan and L. C. Irland, "The sustainability dilemma: observations from Maine history," in R. E. Barringer, ed., *Toward a sustainable Maine* (Portland: Univ. of So. Maine, Muskie Inst., 1993), pp. 53–116; Rept. of the "Brundtland Commission," The World Commission on Env. and Devel., *Our common future* (New York: Oxford Univ. Press, 1987).

Notes on Chapter 2

1. Readers interested in pursuing northeastern forests as naturalists will enjoy S. Connor, *New England natives: a celebration of people and trees* (Cambridge: Harvard Univ. Press, 1994), which is beautifully illustrated; N. Jorgensen, *Sierra Club naturalist's guide to southern New England* (San Francisco: Sierra Club Books, 1978); J. Burk and M. Holland, *Stone walls and sugar maples: an ecology for northeasterners* (Boston: Appalachian Mountain Club, 1979); Ann and Myron Sutton, *Eastern forests* (New York: Knopf, 1985), a well-illustrated natural history guide with color photos and includes animals, birds, and plants; and T. Wessels, *Reading the forested land-*

scape: a natural history of New England (Woodstock: Countryman Press, 1997). The classic is B. F. Thomson, *The changing face of New England* (Boston: Houghton Mifflin, 1977), a readable overview. A technical treatment of the ecology of the northern hardwood forest that helped develop the modern holistic view of forest dynamics is G. E. Likens and H. F. Bormann, *Pattern and process in a forested ecosystem* (New York: Springer Verlag, 1979). For a geographic perspective on the suburban forest, see J.-P. Gottman, *Megalopolis: the urbanized northeastern seaboard of the United States* (Cambridge: MIT Press, 1964), esp. chs. 4–8. A brief overview is found in J. W. Barrett, *Regional silviculture* (New York: John Wiley, 1980, 2d ed.). Anyone interested in this subject has probably read and reread Henry David Thoreau's, Walden, but may not have encountered his *Excursions* (New York: Corinth Books, 1962), esp. the essays entitled "The natural history of Massachusetts" and "The succession of forest trees." More technical treatments are J. L. Vankat, *Natural vegetation of North America* (New York: John Wiley, 1979); and R. Daubenmire, *Plant geography* (New York: Academic Press, 1978). The all-time classic is E. L. Braun, *Deciduous forest of eastern North America* (Philadelphia: Blakiston, 1950).

2. W. M. Denevan, "The pristine myth: the landscape of the Americas in 1492," Ann. Amer. Assoc. of Geogr. 82(3): 369–385 (1992); see also S. T. Pyne's classic, *Fire in America: a cultural history of wildland and rural fire* (Princeton Univ. Press, 1982), esp. pp. 45–64, "Fire history of the northeast," and pp. 71–83, "Our Grandfather fire: fire and the American Indian."

3. The forests of the northeast are so complex and so extensively disturbed that vegetation zones are difficult to define. Not only that, but classifications used for similar vegetation in Canada are different. Changing perceptions of forest values lead to very different ways of depicting forests over time. For examples from Maine, see L. C. Irland, "Maine's forest vegetation regions: selected maps 1858–1993," *Northeast Nat.* 4(4): 241–260 (1997). A current regional vegetation map is *Ecological units of the Eastern United States: First approximation,* edited and integrated by James A. Keys, Jr. and Constance A. Carpenter, USDA-FS, 1995, 1:3.500,000 map sheet.

4. An accessible overview is P. Marchand, *North Woods: an inside look at the nature of forests in the northeast* (Boston: Appalachian Mountain Club, 1987), which gives strong coverage to the mountain forests. Other valuable sources include L. C. Bliss, "Alpine plant communities of the Presidential Range, NH," *Ecology* 44(11) (1963): 678–691; D. E. May and R. B. Davis, *Alpine tundra vegetation on Maine mountains* (Augusta: State Planning Office, Critical Areas Program, Planning Rept. No. 36, Jan. 1978); and T. G. Siccama, "Vegetation, soil, and climate on the Green Mountains of Vermont," *Ecol. Monogr.* 44(3) (1974): 325–349. A useful discussion and map are given in C. V. Cogbill and D. S. White, "Latitude-elevation relationship for spruce-fir forest and treeline along the Appalachia Mountain Chain," *Vegetation* 94: 153–175 (1991). Turn-of-the-century forest conditions of the White

Mountains are summarized in A. K. Chittenden, *Forest conditions of northern New Hampshire,* USDA Bur. For. Bull. 55, 1905; those of the Adirondacks in R. S. Hosmer and E. S. Bruce, *Forest working plan for Township 40 . . . New York State Forest Preserve,* USDA Bur. For. Bull. 30, 1901; and H. S. Graves, *Practical forestry in the Adirondacks,* USDA Bur. For. Bull. 26, 1899; R. K. McGregor, "Changing technology and forest conservation in the Upper Delaware Valley, 1790–1880," *J. For. Hist.* 32(2): 69–81, 1988.

5. The literature is immense, scientific views keep changing, and many points are in contention. A good start may be had by consulting the following: J. H. B. Garner, T. Pagano, and E. B. Cowling, *An evaluation of the role of ozone, acid deposition, and other airborne pollutants in the forest of eastern North America,* USDA-FS SEFES Gen. Tech. Rept. SE-59, Dec. 1989; National Acid Precipitation Assessment Program Office of the Director, *Assessment highlights* (Washington, D.C., Sept. 1990); P. A. Addison and P. J. Rennie, *The Canadian Forestry Service air pollution program and bibliography,* For. Canada Info. Rept. DPC-X-26, 1988; *Effects of atmospheric pollutants on the spruce-fir forests of the eastern United States and the Federal Republic of Germany,* Proc. of the U.S./FRG Research Symposium, USDA-FS NEFES Gen. Tech. Rept. NE-120, Oct. 19–23, 1987, Burlington, VT; and R. D. Noble, J. L. Martin, and K. F. Jensen, eds., *Air pollution effects on vegetation including forest ecosystems,* Proc. of the second U.S./U.S.S.R. symposium, Sept. 1989. More recent are National Science and Technology Council Committee on Environment and Natural Resources, "National acid precipitation assessment program biennial report to Congress: an integrated assessment," May 1998. International Joint Commision, "United States–Canada air quality agreement: 1998 progress report," n.d.

6. J. M. Skelly and J. L. Innes, "Waldsterben in the forests of central Europe and eastern North America: fantasy or reality?" *Plant Disease* 78(11): 1021–1034 (1994). Another leading forest scientist concurs that the acid rain-forest decline linkage has been exaggerated, and argues that effects of ozone pollution on forests are more important: W. H. Smith, "Air pollution and forest damage," *Chemical and Engineering News,* Nov. 11, 1991, pp. 30–43. An exploration of subtle pathways of influence by acid precipitation in Pennsylvania, with numerous references is J. Lyon and W. E. Sharpe, "Effects of acidic deposition on forests, in S. K. Majumdar, E. W. Miller, and F. J. Brenner, eds., *Forests: a global perspective* (Easton: Pennsylvania Academy of Science, 1996, pp. 242–266). See also C. E. Little, *The dying of the trees: the pandemic in America's forests,* (New York, NY: Penguin Books USA, Inc., 1995).

7. An excellent overview on this lowland forest can be found in E. C. Pielou, *The world of northern evergreens* (Ithaca: Cornell, Comstock Press, 1988); and J. Parker Huber, *The wildest country: a guide to Thoreau's Maine* (Boston: Appalachian Mountain Club, 1981). More technical descriptions of these forests are found in the Forest Survey reports, cited in note 22, ch. 10;

R. M. Frank and J. C. Bjorkbom, *A silvicultural guide for spruce-fir in the northeast,* USDA-FS NEFES Gen. Tech. Rept. NE-6, 1973; P. S. White and C. V. Cogbill, "Spruce-fir forests of eastern North America," in C. Eager and M. B. Adams, eds., *Ecology and decline of red spruce in the eastern United States* (New York: Springer Verlag, 1994). L. C. Irland, et al., *The spruce budworm outbreak in Maine in the 1970s — Assessment and directions for the future* (Orono: Univ. of Maine Agr. Exp. Sta. Bull. 819, 1988), esp. chs. I-III; W. B. Leak, D. S. Solomon, and S. M. Filip, *A silvicultural guide for northern hardwoods in the northeast* (revised), USDA-FS NEFES Res. Pap. NE-603, 1987; D. A. Marquis, D. S. Solomon, and J. C. Bjorkbom, *A silvicultural guide for paper birch in the northeast,* USDA-FS NEFES Res. Pap. NE-130, 1969; R. W. Nash and E. J. Duda, *Studies on extensive dying, regeneration, and management of birch* (Augusta: Maine Forest Service Bull. No. 15, 1951); K. F. Lancaster and W. B. Leak, *A silvicultural guide for white pine in the northeast,* USDA-FS NEFES Gen. Tech. Rept. NE-41, 1978; and P. W. Conkling, *Old growth white pine stands in Maine* (Augusta: State Planning Office, Critical Areas Program, Planning Rept. No. 61, Aug. 1978); D. T. Funk, comp., *Eastern white pine: today and tomorrow,* Symposium proc., USDA-FS NEFES Gen. Tech. Rept. WO-51; W. B. Leak, J. B. Cullen, and T. S. Frieswyk, *Dynamics of eastern white pine in New England,* USDA-FS NEFES Res. Pap. NE-699; and R. S. Seymour, "The red spruce-balsam fir forest of Maine: evolution of silvicultural practice in response to stand development patterns and disturbances," in M. J. Kelty, B. C. Larson, and C. D. Oliver, eds., *The ecology and silviculture of mixed-species forests* (Dordrecht: Kluwer Academic Publ., 1992), pp. 217–244.

8. Accessible overviews on these forests include J. McPhee, *The Pine Barrens* (New York: Noonday Press, 1968); A. N. Strahler, *A geologist's view of Cape Cod* (Garden City, New York: Natural Hist. Press for the American Museum of Natural Hist., 1966); and P. W. Dunwiddie, "Forest and heath: the shaping of the vegetation on Nantucket Island," *J. For. Hist.* 33(3): 126–133; R. T. Brooks, D. B. Kittredge, and C. L. Alerich, *Forest resources of southern New England,* USDA-FS NEFES Res. Bull. NE-127, provides a detailed statistical view. In proportion to their area, these forests and their adjacent beaches and bays have attracted more than their share of literary attention, for example in works by Henry Beston, John Hay, David Finch, Thoreau, and many others.

9. The Pine Barrens have spawned a large literature: J. W. Harshberger, *The vegetation of the New Jersey Pine Barrens: an ecological investigation* (New York: Dover, reprint of 1916 ed., 1970, w. pocket map); H. J. Lutz, *Ecological relations of the pitch pine barrens of southern New Jersey* (New Haven: Yale School of For. Bull. 38, 1934); R. T. T. Forman, ed., Pine Barrens: ecosystems and landscapes (New York: Academic Press, 1979); J. Berger and J. W. Sinton, *Water, earth, and fire: land use and environmental planning in the Pine Barrens* (Baltimore: Johns Hopkins Univ. Press, 1985); and B. R.

Collins and E. W. B. Russell, eds., *Protecting the New Jersey Pinelands: a new direction in land-use management* (New Brunswick: Rutgers Univ. Press, 1988), provides a useful ecological overview and a detailed description and evaluation of the Commission's planning and regulatory efforts. Fire data in previous paragraph from New Jersey Dept. of Env. Prot., *Protecting New Jersey forests from fire* (Trenton: Div. of Parks and Forestry).

10. A large literature exists on the forests of this region. Summarizing its changes during the Colonial period and status as of about 1800 is W. Cronon, *Changes in the land: Indians, colonists, and the ecology of New England* (New York: Hill and Wang, 1983), see esp. ch. 7; see also D. M. Smith, "Changes in eastern forests since 1600 and possible effects"; and G. Stephens, "Forests of Connecticut," both in J. F. Anderson and H. K. Kaya, eds., *Perspectives in forest entomology* (New York: Academic Press, 1976); and G. G. Whitney and W. C. Davis, "From primitive woods to cultivated woodlots: Thoreau and the forest history of Concord, Massachusetts," *J. For. Hist.* 30(2): 70–81, which includes excellent illustrations. A classic on old field successions in southern New England is H. J. Lutz, *Trends and silvicultural significance of upland forest succession in central New England* (New Haven: Yale School of For. Bull. No. 22, 1928). G. E. Stephens and P. E. Waggoner, *A half century of natural transitions in mixed hardwood forests* (New Haven: Conn. Agr. Exp. Sta. Bull. 783, 1980); G. Motzkin, et al., "Controlling site to evaluate history: vegetation patterns of a New England sand plain, *Ecol. Monogr.* 66(3): 345–365 (1996) (The Montague Sand Plain in Massachusetts); D. R. Foster, "Land-use history (1730–1990) and vegetation dynamics in central New England, U.S.A.," *J. Ecology* 80: 753–772, 1990; and D. R. Foster, G. Motzkin, and B. Slater, "Land-use history as long-term broad-scale disturbance: regional forest dynamics in central New England," *Ecosystems I:* 96–119 (1998), provide authoritative and detailed analyses.

11. An excellent overview is S. Q. Stranahan, *Susquehanna* (Baltimore, MD: Johns Hopkins University Press, 1993). T. Horton and W. M. Eichbaum's *Turning the tide: saving the Chesapeake Bay* (Washington, D.C.: Island Press, 1991), contains useful sections on the forests of the Susquehannah; D. A. Marquis, *The Allegheny hardwood forest of Pennsylvania,* USDA-FS NEFES Gen. Tech. Rep. NE-15, 1975. D. S. Powell and T. J. Considine, Jr., An analysis of Pennsylvania's forest resources, USDA-FS NEFES Res. Bull. NE-69, 1982; and T. J. Considine, Jr., *An analysis of New York's timber resources,* USDA-FS NEFES Res. Bull. NE-80, 1984. Somewhat more technical is G. G. Whitney, "History and status of the hemlock-hardwood forest of the Allegheny Plateau," *J. Ecology* (78): 443–458, 1990, which reconstructs the presettlement forests of what is now the Allegheny National Forest and compares them to present conditions; M. D. Abrams and C. M. Ruffner, "Physiographic analysis of witness-tree distribution (1766–1798) and present forest cover through North Central Pennsylvania," *Can.* J. For. Res. 25: 659–668 (1995).

12. Surprisingly, there is no available data source listing the area of planted stands in this region.

13. Adirondack Park Agency, *Citizen's guide to Adirondack wetlands* (Ray Brook, NY: APA, 1988), is very well-illustrated and supplies an excellent primer on wetlands generally. An excellent series of articles on forested wetlands is in the May 1993 *J. For.*; for national background, see T. E. Dahl, C. E. Johnson, and W. E. Frayer, "Wetlands status and trends," USDI Fish and Wildlife Service, 1991; and Anon., *Protecting America's wetlands: an action agenda: the final report of the National Wetlands Forum* (Washington, D.C.: The Conservation Foundation, 1988); see also F. C. Golet, et al., *Ecology of red maple swamps in the glaciated northeast: a community profile,* USDI Fish & Wildlife Service, Biol. Rept. 12, 1993.

Notes on Chapter 3

1. Moses Greenleaf, *Survey of the state of Maine* (Augusta, Maine State Museum, 1970), reprint of 1829 edition, p. 110ff; Rept. of the Forest Commissioner, 1902, Augusta, Maine. For a long-term historical perspective on the region, see B. McKibben, "An explosion of green," *Atlantic Monthly,* April, 1995; and H. M. Raup, "The view from John Sanderson's farm: a perspective for the use of land," *J. For. Hist.* 10: 2–11 (1966).

2. C. L. Alerich and D. A. Drake, *Forest statistics for New York — 1980 and 1993,* USDA NEFES Res. Bull. NE-132 (1995), table 148. Carbon statistics in succeeding paragraphs are from: N. D. Cost, et al., *The forest biomass resource of the U.S.,* USDA-FS Gen. Tech. Rept. WO-57, 1990; and R. A. Birdsey, *Carbon storage and accumulation in U.S. forest ecosystems,* USDA-FS Gen. Tech. Rept. WO-59, 1992.

3. R. T. Brooks, T. S. Frieswyk, and A. Ritter, *Forest wildlife habitat statistics for Maine 1982,* USDA-FS NEFES Res. Bull. NE-RB-96, 1986.

4. An authoritative discussion is found in chs. 4–6 of C. D. Oliver and B. C. Larson, *Forest stand dynamics* (New York: McGraw-Hill, 1990). A vivid example of the method of analysis, though applied in another region, is D. J. Mladenoff, et al., "Comparing spatial pattern in unaltered old-growth and disturbed forest landscapes," *Ecological Applications* 3(2): 294–306 (1992).

5. G. G. Whitney, "History and status of the hemlock-hardwood forests of the Allegheny Plateau," *J. Ecology,* 78:443–458 (1990). Whitney has assembled his work into a major volume, *From coastal wilderness to fruited plain: a history of environmental change in temperate North America from 1500 to the present* (New York: Cambridge Univ. Press, 1994). Examples from other parts of the region include J. R. Runkle, "Gap dynamics of old-growth eastern forests: management implications," *Nat. Areas J.* 11(1): 19–25 (1991); A. C. Cline and S. H. Spurr, *The virgin upland forest of central New England* (Petersham: Harvard Forest Bull. No. 21, 1942); and R. B. Gordon, *The primeval forest types of southwest New York* (Albany: New

York State Museum Bull. 321, 1940), which studies lands that later became Allegheny State Park. A valuable overview essay is D. G. Sprugel, "Disturbance, equilibrium, and environmental variability: what is 'natural' vegetation in a changing environment?" *Biological Cons.*, 58: 1–18 (1991).

6. A. F. Hough, "Some diameter distributions in forest stands of northwest Pennsylvania," *J. For.* 30(8): 933–943, 1932.

7. C. G. Lorimer, "Presettlement forest and natural disturbance cycle of northeastern Maine," *Ecology* 58: 139–148, 1977. The spruce stand shown in the photo is discussed in L. C. Irland, "A virgin red spruce and northern hardwood stand; Maine, 1902: it's forest management implications," *Maine Naturalist* 1(4): 181–192 (1993).

8. T. G. Siccama, "Presettlement and present forest vegetation in northern Vermont with special reference to Chittenden County," *Amer. Midl. Nat.* 85(1): 153–172 (1971).

9. D. Leopold, C. Reschke, and D. S. Smith, "Old-growth forests of the Adirondacks," *Nat. Areas J.* 8(3): 166–189 (1988); the Big Blow is discussed in Keller, "Adirondack wilderness," pp. 225–230; the 1995 storm data from *Insight*, Sept. 1995 newsletter of the Empire State Forest Products Assoc.; and L. C. Irland, "Ice Storm 1998 and the forests of the Northeast," *J. For.* 96(9): 32–40, 1998.

10. C. S. Sargent, *The forests of the United States* (Washington, D.C.: Rept. of the 10th Census, 1884), pp. 494–510; and USDA-FS, *Timber depletion, lumber prices, lumber exports, and concentration of timber ownership, Rept. on Sen. Res. 311* (Washington, D.C.: GPO, 1920).

11. Cronon, p. 121.

12. D. A. Gansner, et al., "Cutting disturbance on New England timberlands," *North. J. Appl. For.* 7(3) Sept. 1990.

13. Whitney, p. 455 (note 5 supra).

14. D. R. Foster, "Land-use history (1730–1990) and vegetation dynamics in central New England, U.S.A.," *J. Ecology* 80: 753–772 (1992), p. 768. For a detailed reconstruction of the 1938 hurricane's effects, see D. R. Foster and E. R. Boose, "Patterns of damage resulting from catastrophic wind in central New England, U.S.A.," *J. Ecology,* 80: 79–98 (1992), where the authors show that the hurricane's effects included a multitude of gaps and blowdown patches, few of which were large. Also, D. R. Foster, et al., "Forest response to disturbance and anthropogenic stress," *Bio Science* 47(7): 437–445.

15. J. J. Dowhan and R. J. Craig, *Rare and endangered species of Connecticut and their habitats* (Hartford: State Geol. and Nat. Hist. Survey, Rept. of Investigations, No. 6, 1976), p. 8. For Pennsylvania, see S. G. Thorne, et al., *A heritage for the 21st century: conserving Pennsylvania's native biological diversity* (Harrisburg: Penn. Fish and Boat Comm., 1995). Also, C. E. Williams, "Alien plant invasions and forest ecosystem integrity: a review," in

S. K. Majumdar, E. W. Miller, and F. J. Brenner, eds., *Forests: a global perspective* (Easton: Pennsylvania Academy of Science, 1996, pp. 169–185); Whitney, *From coastal wilderness to fruited plain,* provides a table showing estimates of introduced vs. extirpated species for several northeastern states at table 13.1. Also, E. W. B. Russell, *People and the land through time* (New Haven: Yale Univ. Press, 1997, ch. 6).

16. E. H. Bucher, "Causes of extinction of the passenger pigeon," *Current Ornithology,* 9: 1–36 (New York: Plenum Press, 1992).

17. An excellent overview of wildlife history and policy is P. Matthiessen, *Wildlife in America* (New York: Penguin Books, 1995, 3d ed.); also W. R. Mangun, ed., *American fish and wildlife policy* (Boulder: Westview Press, 1992); C. Foss, "Wildlife in a changing landscape," in R. Ober, ed., At what cost? *Shaping the land we call New Hampshire* (Concord: Soc. for the Prot. of N.H. Forests, 1992), pp. 15–22.

18. R. P. Curran, "Biological resources and diversity of the Adirondack Park," in Vol. I, *Technical Repts., The Adirondack Park in the 21st Century* (Albany, 1990), pp. 414–461.

19. An initial entry into a vast literature can be had through M. L. Hunter, *Jr., Wildlife, forests, and forestry: principles of managing forests for biological diversity* (Englewood Cliffs, NJ: Prentice-Hall, 1990); J. C. Finley and M. C. Brittingham, *Timber management and its effects on wildlife,* Proc., 1989 Penn. State Forest Res. Issues Conference (Univ. Park: Penn. State School of Forest Resources, 1989); J. A. Bissonette, ed., *Is good forestry good wildlife management?,* Conference proc. (Orono: Univ. of Maine Agr. Exp. Sta. Misc. Pub. 689, 1986); R. M. DeGraaf, et al., *New England wildlife: management of forested habitats,* USDA-FS NEFES Gen. Tech. Rept. NE-144, 1992; R. M. DeGraaf and D. D. Rudis, *New England wildlife: habitat, natural history, and distribution,* USDA-FS NEFES Gen. Tech. Rept. NE-108, 1986. J. E. Rodiek and E. G. Bolen, *Wildlife management in managed landscapes* (Washington, D.C.: Island Press, 1991), offers a number of useful case studies. For a less welcome result of the deer population recovery, see A. G. Barbour and D. Fish, "The biological and social phenomenon of Lyme disease," *Science,* 260: 1610–1616, 11 June 1993.

20. Just a few of many examples include: G. W. Gullion, "Aspen management for ruffed grouse," in T. W. Hoekstra and J. Capp, *Integrating forest management for wildlife and fish,* USDA-FS NCFES Gen. Tech. Rept. NC-122, 1988; G. F. Sepik, R. B. Owen, Jr., and M. W. Coulter, *A landowner's guide to woodcock management in the northeast* (Orono: Univ. of Maine Agr. Exp. Sta. Misc. Rept. 253, 1981). For pine marten, see J. A. Bissonnette, R. J. Frederickson, and B. J. Tucker, "American marten: a case for landscape-level management," in Rodiek & Bolen, note 19 supra; C. A. Elliott, *Forester's guide to managing wildlife habitats in Maine* (Orono: Maine Coop. Extension and Wildlife Soc., 1988). The USDA-FS project at Warren, Pennsylvania has circulated an informal unpublished compilation on forest

management and biodiversity in that area; it is undated, approximately 100 pages; abstracts are published in *Proc. Northeast Weed Sci. Soc.* Vol. 46, pp. 134–137.

21. C. W. Severinghous and C. P. Brown, "History of the white-tailed deer in New York," *NY Fish and Game J.* 3(2): 129–166 (1956). See also Don Stanton, *A history of the white-tailed deer in Maine* (Augusta: Dept. of Inland Fisheries and Game, Game Div. Bull. No. 8, 1963).

22. See papers by Bowyer, Shea and McKenna, Reay, and Marston in Bissonette, ed., note 19 supra, and by Hirth, in Finley and Brittingham, eds, 1989, note 19 supra. On cutting and deer habitat, see A. H. Boer, "Transience of deer wintering areas," *Can. J. For. Res.,* 22: 1421–1423 (1992); see also D. B. Kittredge and P. M. S. Ashton, "Impact of deer browsing on regeneration in mixed stands in southern New England," *North. J. Appl. For.* 12(3): 115–120 (1995); and J. Redding, "History of deer population trends and forest cutting on the Allegheny National Forest," in K. W. Gottschalk and S. L. C. Fosbroke, eds., *10th Central Hardwood Forest Conference,* USDA-FS NEFES Gen. Tech. Rept. NE-197 (1995).

23. See T. Wigley and M. A. Melchiors, "Wildlife habitat and communities in streamside zones: literature review for the eastern U.S.," in *Riparian ecosystems in the humid U.S.* (Washington, D.C.: Nat. Assoc. Cons. Dists., 1994), pp. 100–121, and other papers in same volume; USDA-FS, State and Private Forestry, Northeastern Area, *Riparian forest buffers: function and design,* Radnor, PA, NA-PR-07-91, which contains good illustrations and current references; and R. B. Newton, *Forested wetlands of the northeast* (Amherst: Univ. Mass, Env. Inst. Publ. No. 88-1). Also useful are V. Chase, L. Deming, and F. Latawiec, *Buffers for wetlands and surface waters: a guidebook for New Hampshire municipalities* (Concord: Audubon Soc. of New Hampshire, 1997); and S. C. Peterson and K. Kimball, *Citizen's guide to conserving riparian forests* (Gorham, NH: Appal. Mtn. Club, 1995).

24. R. A. Askins, "Hostile landscapes and the decline of migratory songbirds," *Science* 267: 1956–1957 (31 Mar. 1995). J. Terborgh, "Why American songbirds are vanishing," *Scientific Amer.,* May 1992, pp. 98–104; and is treated more fully in J. Terborgh, *Where have all the birds gone?* (Princeton: Princeton Univ. Press, 1990); R. A. Askins, J. F. Lynch, and R. Greenberg, "Population declines in migratory birds in eastern North America," *Current Ornithology* 7: 1–57 (New York: Plenum Press, 1990); B. G. Peterjohn and J. R. Saver, "Population trends of breeding birds from the Northern American Breeding Bird Survey," *Wild. Soc. Bull.* 22: 155–164 (1994); H. S. Crawford and R. W. Titterington, "Effects of silvicultural practices on bird communities in upland spruce-fir stands," in *Managing of northcentral and northeastern forests for nongame birds,* Workshop proc., USDA-FS NCFES Gen. Tech Rept. NC-51, pp. 110–119. D. DeCalesta, "Effect of white-tailed deer on songbirds within managed forests in Pennsylvania," *J. Wild. Manage.* 58(4): 711–718 (1994); D. H. Kuhnke, ed., *Birds in the Boreal forest* (Ed-

monton: North. For. Res. Centre, For. Canada, 1993); Scott Robinson, "The case of the missing songbirds," unpub. paper, 1997, Illinois Natural History Survey; R. M. DeGraaf and J. A. Rappole, *Neotropical migratory birds: natural history, distribution, and population change* (Ithaca: Comstock, Cornell Univ. Press, 1996).

25. C. J. E. Welsh and W. M. Healy, "Effects of even-aged timber management on bird species diversity and composition in northern hardwoods of New Hampshire," *Wildl. Soc. Bull.* 21: 143–154 (1993). Reporting on work done in Maine, M. Vander Haegen and J. M. Hagan, "2 papers on migratory land birds in an industrial forest landscape," in R. D. Briggs and W. B. Krohn, eds., *Nurturing the northeastern forest,* Maine Agr. and For. Exp. Sta. Misc. Rept. 382, 1993, pp. 195–216.

26. Maine Dept. of Inland Fisheries and *Wildlife, Wildlife Division Research and Management Report* (Augusta, ME, 1992, p. 45).

27. Overviews of national status include L. L. Langner and C. H. Flather, *Biological diversity: status and trends in the U.S.,* USDA-FS RMFRES Gen. Tech. Rept. GTR-RM-244 (1994); and C. H. Flather, L. A. Joyce, and C. A. Bloomgarden, *Species endangerment patterns in the U.S.,* USDA-FS RMFRES Gen. Tech. Rept. GTR RM-241 (1994), which contains excellent maps of endangerment patterns by major taxa and by county.

28. Anon., "The wildlands project," *Wild Earth,* Special Issue, 1993, Canton, NY; C. C. Mann and M. L. Plummer, "The high cost of biodiversity," *Science,* 260: 1868–1871, 25 June 1993, is a brief commentary on the concepts in the previous citation; see also Society of American Foresters, *Task force report on biological diversity in forest ecosystems* (Bethesda, MD, 1991). USDA-FS, "Conserving our heritage: America's biodiversity," Washington, D.C. in coop. with The Nature Conservancy, 1990; National Association of State Foresters Report, *The conservation of biological diversity on state and private lands,* Washington, D.C., 1993; and C. Petersen, *What is biodiversity?* (Amherst: Univ. of Mass. Dept. of For., Current Issues in Forestry II(2), 1991).

29. The literature of ecosystem management and sustainable forestry is growing at a rate beyond any one's ability to keep up. Further, state groups are at work on guidelines for sustainable forest practice or "ecosystem-friendly" forestry. A major textbook, attempting to codify ecosystem management for western forest is K. A. Kohn and J. F. Franklin, eds., *Creating a forest for the 20th century: the science of ecosystem management* (Washington, D.C.: Island Press, 1997); D. L. Bottom, G. H. Geeres, and M. H. Brookes, eds., *Sustainability issues for resource managers,* USDA PNW Res. Sta. Gen. Tech. Rept. PNW-GTR-370 (1996). Roger Sedjo "deconstructs" the concept effectively in "Toward an operational definition of ecosystem management," in J. E. Thompson, comp., *Analysis in support of ecosystem management* (Washington, DC: USDA-FS, Ecosystem Manage. Analysis Center,

April 1995). B. Freedman, S. Woodley, and J. Loo, "Forestry practices and biodiversity, with particular reference to the Maritime Provinces of E. Canada," *Environ. Rev.* 2: 33–77 (1992) is highly relevant to northerly portions of our region. R. H. Harlow, R. L. Downing, and D. H. VanLear, *Responses of wildlife to clearcutting and associated treatments in the Eastern United States* (Clemson: Clemson Univ. Dept. of For. Res. Tech. Pap. No. 19, 1997), includes material for the south but indicates the increasing complexity of the issues. P. DeMaynardier and M. L. Hunter, Jr., "The relationship between forest management and amphibian ecology: a review of the North American literature," *Environ. Rev.* 3: 230–261 (1995). A. J. Meier, S. P. Bratton, and D. C. Duffy, "Possible ecological mechanisms for loss of vernal-herb diversity in logged eastern deciduous forests," *Ecological Appl.* 5(4): 935–946 (1995). Focused on this region is R. S. Seymour and M. L. Hunter, Jr., "Practicing ecological forestry," Ch. 2, in M. L. Hunter, Jr. (ed.), *Managing biodiversity in forest ecosystems* (Cambridge: Cambridge University Press, in press); emerging topics, see J. M. Hagan and S. L. Grove, "Coarse woody debris," *J. For.* 97(1):6–11, 1999 and Y. Bergeron and B. Harvey, "Basing silviculture on natural ecosystem dynamics: an approach applied to the southern boreal mixedwood forest of Quebec," *For. Ecol. & Mgmt.* 92:235–242, 1997.

30. An essay offering a broad view is N. Johnson and D. Ditz, "Challenges to sustainability in the U.S. forest sector," in R. Dower, et al., *Frontiers of sustainability* (Washington, D.C.: Island Press, 1997, pp. 191–280). See also, *Land and Resource Management Plans* for the three National Forests and the *Annual Monitoring and Evaluation Repts.;* I. R. Noble and R. Dirzo, "Forests as human-dominated ecosystems," Science 277: 522–525 (25 July 1997); J. Franklin, *Forest stewardship in an environmental age,* SUNY CESF Faculty of Forestry Misc. Pub. 27, 1992; various papers in T. Lepisto, ed., *Sustaining ecosystems, economics, and a way of life in the northern forest* (Washington, D.C.: The Wilderness Soc., 1993); and G. L. Baskerville, "Designer habitats: edge, carrying capacity, diversity, and more," *Canadian Forest Industries,* Jan./Feb. 1992, pp. 29–32. R. M. DeGraaf and W. M. Healy, eds., *Is forest fragmentation a management issue in the northeast?,* USDA-FS NEFES Gen. Tech. Rept. NE-140, 1988. A landscape management view of future habitat needs, including farm and forest lands, is given in D. F. McKenzie and T. Z. Riley, *How much is enough? A regional wildlife habitat assessment for the 1995 Farm Bill* (Washington, D.C.: Wildlife Management Inst., 1995). A massive national "Gap Analysis" is being led by U.S. Dept. of the Interior and state agencies. There is no single regional overview; contact state university wildlife research units for details. On the method, see J. M. Scott, et al., "Gap analysis: a geographic approach to protection of biological diversity," *Wildlife Monogr.* No. 123 (Jan. 1993).

31. See, e.g., L. Alexander, "Nonindustrial private forest landowner relations to wildlife in New England," unpub. Ph.D. diss., Yale Univ., May 1986; also summarized in Alexander's paper in Bissonette, ed., 1986, note 19

supra; R. W. Kurtz and L. C. Irland, "Integrated timber/wildlife management through education of NIPF owners," *Trans. 53rd North Amer. Wildl. and Nat. Res. Conference* (Washington, D.C.: Wildl. Mgt. Inst., 1988), pp. 94–101.

32. M. L. Hunter, Jr., *Wildlife, forests, and forestry,* op. cit., note 19 above, esp. ch. 5; R. S. Seymour and M. L. Hunter, Jr., *New forestry in eastern spruce-fir forests: principles and applications to Maine* (Orono: Maine Agr. Exp. Sta. Misc. Pub. 716, 1992); and S. G. Boyce, *Landscape forestry* (New York: John Wiley, 1995). An especially detailed assessment of forest practice effects over a landscape over a long span of time is J. M. Hagan and R. B. Boone, *Harvest rate, harvest configuration, and forest fragmentation: a simulation of the 1989 Maine Forest Practices Act* (Manomet, MA: Manomet Center for Conservation Sciences, 1997). I. R. Noble and R. Dirzo give a broad view in "Forests as human-dominated ecosystems," *Science* 277 (25 July 1997): 522–525.

33. T. R. Cox, R. S. Maxwell, P. D. Thomas, and J. J. Malone, *This well-wooded land* (Lincoln: Univ. of Nebraska Press, 1985), the phrase is the title of ch. 9.

Notes on Chapter 4

1. USDI, NPS, 1982–83, *Nationwide recreation survey* (Washington, D.C.: GPO, 1986), p. 22.

2. General sources on the management of Maine's industrial forest include Austin Cary, Maine forests, their preservation, taxation, and value, *in Maine Bureau of Industrial and Labor Statistics, Twentieth Annual Rept.,* Augusta, 1906, pp. 181–193; D. C. Smith, *A history of lumbering in Maine, 1861–1960* (Orono: Univ. of Maine Press, 1972); and R. S. Wood, *A history of lumbering in Maine, 1820–1861* (Orono: Univ. of Maine, 1935). These deal almost exclusively with the industrial forest. A highly critical report is W. C. Osborn, *The paper plantation* (New York: Grossman, 1974); one-sided as it is, the book is informative and describes, if in a one-sided manner, the political power of Maine's leading landowners in the 1950s and 1960s. More recent is L. C. *Irland, Land, timber, and recreation in Maine's Northwoods,* (Orono: Maine Agr. and For. Exp. Sta. Misc. Rept. 730, 1996).

3. The story of log driving is found in A. G. Hempstead, *The Penobscot boom* (privately printed, 1975); E. D. Ives, "The Argyle boom," Northeast Folklore 17, 1976; and Smith, *Lumbering in Maine,* chs. 3 and 4, which cite full references. An excellent view of how Maine lumbermen carried these methods to the midwest is in *Michigan log marks,* WPA and Mich. Agr. Exp. Sta., Memoir Bull. No. 4, 1941. R. E. Pike, *Tall trees: tough men* (New York: W. W. Norton, 1984), is filled with vivid anecdotes and period photographs. On Maine's last log drive, see J. Nacthwey, "The last log drive, what price progress?" *New England Business* (Aug. 1976); and S. D. McBride, "America's last log drive," *Christian Science Monitor,* Oct. 29, 1976. Pennsylvania

log driving generated an immense literature. An accessible and detailed personal memoir is R. D. Tonkin, *My partner, the river: the white pine story on the Susquehanna* (Pittsburgh: Univ. of Pittsburgh Press, 1958). Conflicts between the rafters and drivers are reviewed in T. R. Cox, "Transition in the woods: log drivers, raftsmen, and the emergence of modern lumbering in Pennsylvania," *Penn. Mag. of Hist. and Biogr.* CIV(3): 345–365 (1980). On the Adirondacks, see Defebaugh's, *History of the lumber industry; and J. E. Keller, Adirondack wilderness: a story of man and nature* (Syracuse: Syracuse Univ. Press. 1980), ch. 8. Log driving in Pennsylvania and New York is also described in T. R. Cox, R. S. Maxwell, P. D. Thomas, and J. J. Malone, *This well-wooded land* (Lincoln: Univ. of Nebraska Press, 1985), pp. 93, 119, 157, and elsewhere. Cox, et al., note that log driving seems to have developed first along the Hudson (p. 157).

4. J. E. McLeod, *The Northern . . . the way I remember* (Millinocket: Great Northern Paper Co., n.d., ca. 1980). For earlier detail on Maine, see note 2, ch. 3, in *Wildlands and Woodlots.*

5. Data cited on the Allegheny unit are from C. L. Alerich, *Forest statistics for Pennsylvania — 1978 and 1989,* USDA-FS NEFES Res. Bull. NE-126, 1993; and D. S. Powell and T. J. Considine, Jr., *An analysis of Pennsylvania's forest resources,* USDA-FS NEFES Res. Bull. NE-69, 1982.

6. G. G. Whitney, "History and status of the hemlock hardwood forests of the Allegheny Plateau," *J. Ecology,* 78: 453 (1990).

7. Sargent, pp. 507–509, excerpts.

8. USDA-FS, *A national plan for American forestry,* 73d Cong. 1st Sess. Doc. No. 12, Mar. 13, 1933, pp. 966–967.

9. Whitney, p. 453.

10. W. H. McWilliams, et al., *Characteristics of declining forest stands on the Allegheny National Forest, USDA-FS, NEFES, Res. Note NE-360 (1996).*

11. C. H. Strauss and R. G. Lord, "Timber availability in North Central Pennsylvania — the next hundred years," *North J. Appl. For.* 6: 133–137 (1989).

12. Pennsylvania Bureau of Forestry, *Penn's woods sustaining our forests,* Harrisburg: Dept. of Conservation and Natural Resources, n.d.; and Citizens Advisory Council, *These woods are ours: a report on Pennsylvania's state forest system,* Harrisburg. See also L. A. DeCoster, *The legacy of Penn's woods: a history of the Pennsylvania Bureau of Forestry* (Harrisburg: Pennsylvania Historical and Museum Commission, 1995).

13. Frances Belcher, *Logging railroads of the White Mountains* (Boston: Appalachian Mountain Club, 1980).

14. Austin Cary, in *Report of the Forest Commissioner,* 1896, Augusta, p. 179.

15. C. S. Sargent, *The forests of the United States* (Washington, D. C.: Reps of the 10th Census, 1884) pp. 494-510.

16. Environmentalists, fisheries experts, and citizens have been highly critical of large clearcuts and incidents of poor care for erosion control. This is a complex field involving state and federal laws. For a regional review of the issues and state management programs, see L. C. Irland and J. E. Connors, "State nonpoint source programs affecting forestry: the 12 northeastern states," *North. J. Appl. For.* 11(1): 5–11 (1994); see also D. R. Satterlund and P. W. Adams, *Wildland watershed management* (New York: John Wiley, 1992, 2d ed.), esp. chs. 10 and 13. In Maine, a major Atlantic Salmon Program has been adopted to improve the level of protection for seven coastal streams sustaining natural salmon populations.

17. For a vigorously argued critique of industrial forestry, see A. M. Lansky, *Beyond the beauty strip* (Gardiner, ME: Tilbury House, 1992). For readable on-the-ground views, see D. Dobbs and R. Ober, *The northern forest* (White River Junction: Chelsea Green Publ. Co., 1995). For samples of a vast literature, see Anon., *Clearcutting in Pennsylvania* (College Station: Penn. State Univ., School of For. Res., 1975), which while not current has a broad overview on water, aesthetics, wildlife, and soil in addition to timber considerations. R. S. Seymour, "The northeastern region," in J. W. Barrett, ed., *Regional silviculture of the U.S.* (New York: John Wiley, 1994, 3d ed.), pp. 31–79. R. S. Seymour, "The red spruce-balsam fir forest of Maine: evaluation of silvicultural practice in response to stand development patterns and disturbances," in M. J. Kelty, B. C. Larson, and C. D. Oliver, eds., *Ecology and silviculture of mixed-species forests* (Dordrecht: Kluwer Acad. Publ., 1992), pp. 217–244; J. Pastor and D. J. Mladenoff, "The southern boreal-northern hardwood forest border," in H. H. Shugart, R. Leemans, and G. B. Bonan, *Systems analysis of the global boreal forest* (New York: Cambridge Univ. Press, 1992); C. W. Martin, C. T. Smith, and L. M. Tritton, eds., *New perspectives on the silvicultural management of northern hardwoods, Proc. of a symposium,* USDA-FS NEFES Gen. Tech. Rept. NE-124, 1988; The Irland Group, *Clearcutting as a management practice in Maine forests,* Report to the Maine Dept. of Cons., Augusta, 1988; F. Rubin and D. Justice, *New Hampshire clearcut inventory* (Durham: Univ. of N.H. Complex System Res. Ctr., 1995); R. S. Seymour and M. L. Hunter, Jr., *New forestry in eastern spruce-fir forests: principles and applications to Maine* (Orono: Maine Agr. Exp. Sta. Misc. Publ. 716, 1992); Papers in C. T. Smith, C. W. Martin, and L. M. Tritton, *Proc., 1986 symposium on the productivity of northern forests following biomass harvesting,* USDA-FS NEFES Gen. Tech. Rept. GTR-115, 1986; J. E. Johnson, et al., *Environmental impacts of harvesting wood for energy* (Chicago: Great Lakes Regional Biomass Energy Program, 1987); Anon., *Clearcutting in the Adirondack Park* (Ray Brook, NY: Adirondack Park Agency, 1981); R. F. Powers, et al., "Sustained site productivity in North American forests: problems and prospects," in S. P. Gessel, et al., eds.,

Sustained productivity of forest soils, Proc., 7th North American Forest Soils Conference, USDA-FS SOFES Proc. Reprint, 1988, which has abundant citations; and W. B. Leak, D. S. Solomon, and P. S. DeBald, *Silvicultural guide for northern hardwood types (revised),* USDA-FS NEFES Res. Pap. NE-603 (1987).

18. For a case study, see L. M. Tritton and P. E. Sendak, "Ecological aspects of forest management planning: a northern hardwood case study," *North J. Appl. For.* 12(3): 121–126 (1995), which also cites other references on long-term ill effects of mishandled partial cuttings.

19. See M. S. Greenwood, R. S. Seymour, and M. W. Blumenstock, *Productivity of Maine's forest underestimated — more intensive approaches are needed* (Orono: Univ. of Maine Agr. Exp. Sta. Misc. Rept. 328, 1988); R. S. Seymour, "Plantations or natural regeneration: options and trade-offs for high-yield silviculture," in R. D. Briggs and W. B. Krohn, eds., *Nurturing the northern forest* (Orono: Univ. of Maine Agr. and For. Exp. Sta. Misc. Rept. 382, 1993), pp. 16–32. H. Kimmins, *Balancing act: environmental issues in forestry* (Vancouver: Univ. of British Columbia Press, 1992), ch. 8, which deals with western issues but contains a balanced review of this issue. Much more technical is J. D. Walstad and P. J. Kuch, *Forest vegetation management for conifer production* (New York: John Wiley, 1987), esp. chs. 1, 4, and 14; see also K. V. Miller, et al., *Herbicides and wildlife habitat, an annotated bibliography,* USDA-FS So. Region Tech. Publ. R8-TP 13, Jan. 1990, which lists and cross-indexes almost 300 references. Office of Legal and Policy Analysis, *Final report of the Commission to study the use of herbicides* (Augusta: Maine Legislature, Dec. 1, 1990).

20. For a start on Maine's two-year referendum campaign, see Maine Forest Service, *The compact for Maine's forests: key elements,* Information package dated Sept. 12, 1996; Maine Forest Service and State Planning Office, *Economic impact of the citizen's initiative to promote forest rehabilitation and eliminate clearcutting,* Augusta: Maine Dept. of Conservation, 1996; Maine Council on Sustainable Forest, *Sustaining Maine's forests: criteria, goals, and benchmarks,* Augusta: Maine Dept. of Conservation, May, 1996; Maine Report, Vol. 103, *Maine 1907* (Opinion of the Justices on forest practice regulation); 117th Maine Legislature, *Resolution, proposing a competing measure . . . to implement the Compact for Maine's Forests,* HP.1390-LD#1892, Sept. 7, 1996; Mitch Lansky, "Forestry referendum commentaries: three perspectives on where to go from here," *Maine Policy Rev.* 5(3): 81–94 (1996); Lloyd C. Irland, "Forest policy is hard," *Maine Policy Rev.* 5(3): 85–89 (1996); Kevin Hancock, "Common sense over politics is the answer," *Maine Policy Rev.* 5(3): 89–90 (1996); J. M. Hagan III, "Clearcutting in Maine: would somebody please ask the right question?" *Maine Policy Rev.* 5(2): 7–20 (1996); and J. Carter, "Ballot measures as a political spike: referenda and reform in the Great North Woods," *Wild Earth,* Spring 1998.

21. See R. G. Healy and J. G. Rosenberg, *Land use and the states* (2d ed., Baltimore: Johns Hopkins Univ. Press, 1979), ch. 3; and relevant sections of the *Northern Forest Lands Study.*

22. T. M. Beckley, "Pluralism by default: community power in a paper mill town," *Forest Science* 42(1): 35–45 (1996), profiles one mill town from a sociological perspective.

23. The tax law is at 36 MRSA 571–584, amended by ch. 308, Sec. 1-18, PL 1973. No thorough, current discussion of this tax is readily available; but see D. B. Field, "Converting to a productivity tax: the Maine experience," in H. L. Haney, Jr. and W. C. Siegel, eds., *Proc. of Forest Tax Symposium II* (Blacksburg: Virginia Polytech. and State Univ., 1982), FWS-4-87, pp. 90–110. Excellent reviews of general issues are J. K. Wood, *Timber taxation in the states* (Lexington, KY: Council of State Governments, 1978); R. Boyd, *Forest taxation: current issues and future research*, USDA-FS SOFES Res. Note SE-338, 1986; W. D. Klemperer, "Taxation of forest products and forest resources," in P. V. Ellefson, ed., *Forest resource economics and policy research: strategic directions for the future* (Boulder: Westview Press, 1989), ch. 18; and J. Stier, *Forest taxation: a bibliography of literature for the period 1986–1988* (Madison: Univ. of Wisconsin School of Nat. Res. Staff Paper Series #38, 1989). Also, Resource Systems Group, "Forest property taxation programs," Report to the North. For. Lands Council, in Technical Appendix, North. For. Lands Council, Feb. 1994 (USDA-FS-SPF, Durham, NH).

24. Resource Systems Group, cited note 23 above.

25. A valuable review from the woodsworker's perspective is in J. Falk, *The organization of pulpwood harvesting in Maine* (New Haven: Yale School of For. and Env. Studies, Working Paper No. 4, 1977). General background on woodsworker problems is found in L. C. Irland, ed., *Manpower — forest industry's key resource* (New Haven: Yale School of For. and Env. Studies Bull. No. 86, 1975); and several papers in Proc., *A forest-based economy: Blaine House Conference on Forestry* (Augusta: Maine Dept. of Cons., 1986). For a historical perspective, see Cox, et al., pp. 170–175.

26. The story is well told in A. Wilkins, *Ten million acres of timber* (Woolwich, ME: TBW Books, 1978), ch. 10.

27. See C. J. Sanders, et al., eds., *Recent advance in spruce budworms research* (Ottawa: Canadian Forestry Service, Cat. No. FO18-5/1984, 1985), for a large compendium of technical materials; and L. C. Irland, et al., *The spruce budworm outbreak in Maine in the 1970s — assessment and directions for the future* (Orono: Univ. of Maine Agr. Exp. Sta. Bull. 819, 1988). The classic policy analysis of the budworm problem is *Report of the task force for evaluation of budworm control alternatives* (the "Baskerville Report"), Fredericton, N.B., Dept. of Natural Resources, 1976.

28. C. S. Holling, "The role of forest insects in structuring the forest landscape," ch. 6 in H. H. Shugart, R. Leemans, and G. B. Bonan, *Systems*

analysis of the global boreal forest (New York: Cambridge Univ. Press, 1992), is technical but very illuminating.

29. Osborn, *The paper plantation,* note 2 above.

30. See Land Use Regulation Commission, *Comprehensive land use plan* (Augusta: Maine Dept. of Cons., 1983).

31. A good sense of the current issues in the region's industrial forest can be obtained by perusing the *Northern forest lands study,* USDA-FS, Rutland, VT, 1990; the *Final report of the Northern Forest Lands Council* (Concord, 1994); T. Lepisto, ed., *Sustaining ecosystems, economics, and a way of life in the northern forest* (Washington, D.C.: The Wilderness Soc., 1993); and Dobbs and Ober, cited above, note 17.

Notes on Chapter 5

1. The Connecticut case study in this chapter is based on research supported by funds provided by the Northeastern Forest Experiment Station of the USDA-FS through the Pinchot Institute of Environmental Forestry Studies, Grant No. 23-820, by the Yale School of For. and Env. Studies, and by the Univ. of Maine Cooperative Forestry Research Unit. An abridged version of the report appeared in *Connecticut Woodlands* (Summer, 1980).

2. Commutersheds are used to define suburban areas in B. Pushkarev, "The Atlantic urban seaboard," *Regional Plan News No. 90,* Sept. 1969, pp. 14, 15. A detailed analysis of suburbanization is found in J.-P. Gottman, *Megalopolis: the urbanized northeastern seaboard of the United States* (Cambridge: MIT Press, 1964), esp. chs. 4 and 5, and p. 223. On population trends 1980–1990, see D. A. McGranahan and J. Salsgiver, "Recent population change in adjacent nonmetro counties," *Rural Devel. Perspectives* 8(3): 2–7. A recent ERS publication identified "commuting counties" adjacent to identified metropolitan areas: P. J. Cook and K. L. Mizer, *The revised ERS county typology: an overview* (USDA ERS Rural Devel. Res. Rept. No. 89, 1994). A valuable overview, including an update on the extensive "cost of sprawl" literature is in Office of Technology Assessment, *Technological reshaping of metropolitan America* (Washington, D.C.: Congress of the U.S., 1995).

3. Useful social histories of suburbia are J. R. Stilgoe, *Borderland: origins of the American suburb, 1820–1939* (New Haven: Yale Univ. Press, 1980); and K. T. Jackson, *Crabgrass frontier: the suburbanization of the United States* (New York: Oxford Univ. Press, 1985). For a challenging critique of the politics of suburban land use, see A. Downs, *New visions for metropolitan America* (Washington, D.C.: Brookings Inst., 1994).

4. Gottman, *Megalopolis,* pp. 181, 223; Pushkarev, Atlantic urban seaboard, p. 14.

5. R. E. Coughlin, E. Cohn, and E. J. Keagy, *Development in the Philadelphia-Trenton region, 1970–1980: a pilot study of land use change*

(Philadelphia: Dept. of City and Regional Planning, Univ. of Penn. Res. Rept. Series No. 13, 1989); and J. A. Michaels, et al., *New York-New Jersey highlands regional study,* USDA-FS, State and Private For., Durham, NH, n.d. (ca. 1992); see also R. W. Dill and R. G. Otte, *Urbanization of land in the northeastern United States,* USDA ERS Publ. 485, 1971; L. Susskind and C. Perry, "The dynamics of growth policy formulation in implementation: a Massachusetts case study," *Law and Contemporary Problems* 43(2) (1979): 147–148; M. Vesterby and D. H. Brooks, 1988, "Land-use change in fast-growth areas, 1950–1980," in R. E. Heimlich, ed., *Land use transaction in urbanizing areas: research and information needs,* The Farm Foundation in cooperation with the USDA ERS, Washington, D.C., 1989, p. 99. For other analyses of land consumption, see Pushkarev, *Atlantic urban seaboard;* and North Atlantic Regional Water Resources Study Coordinating Committee, *North Atlantic regional water resources study,* "Appendix G, land use and management" (New York: U.S. Army Corps of Engineers, 1972), esp. p. 6ff. Data on New York region cited at p. 7 of J. Michaels, *New York-New Jersey highlands regional study,* USDA-FS State & Pvt. For., Durham, NH, n.d. (ca. 1992); T. Horton and W. M. Eichbaum, *Turning the tide: saving the Chesapeake Bay* (Washington, D.C.: Island Press, 1991), pp. 200–218; data cited are from p. 206. For suburbanization in New Hampshire, see W. A. Befort, A. E. Luloff, and M. Morrone, "Rural land use and demographic change in a rapidly urbanizing environment," *Landscape and Urban Planning* 16: 345–356 (1988).

6. See note 4 *supra.*

7. See note 4 *supra.*

8. A vast amount of research has analyzed visual values of northeast landscapes. See H. Dominie, "Maine's changing landscape," in R. E. Barringer, ed., *Changing Maine* (Portland: Univ. of Southern Maine, 1990), pp. 89–106. A vividly illustrated history is R. Ober, ed., *At what cost? Shaping the land we call New Hampshire* (Concord, NH: Soc. for the Prot. of N.H. Forests, 1992).

9. See, e.g., K. E. Medleg, S. T. A. Pickett, and M. J. McDonnell, "Forest-landscape structure along an urban to rural gradient," *Professional Geographer* 47(2): 159–168 (1995), reporting on a study stretching from New York City to northwest Connecticut; also C. S. Robbins, "Effect of forest fragmentation on bird populations," in R. M. DeGraaf and K. E. Evans, *Management of northcentral and northeastern forests for nongame birds,* USDA-FS NCFES Gen. Tech. Rept. NE-51, 1979, pp. 198–213; M. A. Gratzer, *Land use inventory for open space planning in eastern Connecticut,* Storrs Agr. Exp. Sta., Univ. of Conn., Res. Rept. No. 38, July 1972, n.p.; and M. L. Hunter, Jr., *Wildlife, forests, & forestry* (Engelwood Cliffs: Prentice-Hall, 1990), ch. 8.

10. Connecticut Dept. of Finance and Control, Office of State Planning, *Proposed: a plan of conservation and development for Connecticut, Hartford, 1973.* For more detailed data, see R. M. Field and Assoc., *Land use:*

planning report for the Long Island Sound Study (Boston: New England River Basins Commission, 1975). An early and valuable series of reports is by N. C. Whetten, et al., *Studies of suburbanization in Connecticut,* issued in three parts by the Storrs Agr. Exp. Sta. Bull. 212 (Oct. 1936), 226 (May 1938), and 230 (Feb. 1939).

11. *Connecticut wildlife,* 15(1)/Jan./Feb. 1995, special issue.

12. Unpub. USDA-FS data tables, Radnor, PA. The previous survey was N. Kingsley, *The forest landowners of southern New England,* USDA-FS NEFES Res. Bull. NE-41, 1976.

13. USDA ERS, *Potential new sites for outdoor recreation in the northeast,* Outdoor Recreation Resources Review Commission Study Rept. No. 8 (Washington, D.C.: GPO, 1962).

14. Kingsley, op. cit., note 12 above.

15. A. F. Hawes and R. C. Hawley, *Forest survey of Litchfield and New Haven counties,* Connecticut, Conn. Agr. Exp. Sta. Bull. 162 (Forestry Publ. 5), 1909.

16. R. T. Brooks, D. B. Kittredge, and C. L. Alerich, *Forest resources of southern New England,* USDA-FS NEFES Res. Bull. NE-127, 1993; N. Kingsley, *The forest resources of southern New England,* USDA-FS NEFES Res. Bull. NE-36, 1974. The 1952–84 data are from R. L. Nevel, Jr. and E. H. Wharton, *The timber industries of southern New England — A periodic assessment of timber output,* USDA-FS NEFES Res. Bull. NE-101, 1988. On forest landowner attitudes, see T. Holmes and J. Diamond, *An analysis of non-industrial private woodland owners' attitudes towards timber harvesting and forest land use in Windham County,* Connecticut, 1979, Storrs Agr. Exp. Sta., Univ. of Conn., Res. Rept. 63, 1980.

17. P. M. Raup, "Urban threats to rural lands," AIP J. 41(6) (1975): 371–378, nicely summarizes the problem; see also A. A. Schmid, *Converting land from rural to urban uses* (Baltimore: John Hopkins Univ. Press, 1968); Maine State Planning Office, *Land use and cumulative impacts of development: a study summary* (Augusta, ME: Exec. Dept., 1987). A broad overview of land use and policies is R. H. Platt, *Land use and society: geography law, and public policy* (Washington, D.C.: Island Press, 1993).

18. A short summary, though dated, is useful: R. C. Sherman, N. C. Shropshire, P. S. Wilson, and A. C. Worrell, *Open-land policy in Connecticut* (New Haven: Yale School of For. and Env. Studies Bull. No. 87, 1974). A current review like this for each state would be very useful.

19. J. E. Connors, C. Murdoch, and D. Field, *Guide to forestry regulations in Maine* (Orono: Univ. of Maine, Margaret Chase Smith Center, 1992), Tech. Paper 07-01-92, 82 pp. + app.; see also F. W. Cubbage, "Public regulation of private forestry: pro-active policy responses," *J. For.,* 89, (12), 31–35 1991; and F. W. Cubbage and W. C. Siegel, "State and local regulation of pri-

vate forestry in the east," *North. J. App. For.* 5 (1988), pp. 103–108. A recent survey is C. E. Martus, H. L. Haney, Jr., and W. C. Siegel, "Local regulatory ordinances: trends in the eastern U.S.," *J. For.* 93(6): 27–31 (1995); other articles in the same issue provide a broad perspective on forest practice regulation. The number of local ordinances in Pennsylvania estimated in Anon., *Best management practices for Pennsylvania forests* (Univ. Park, Penn. State Coll. of Agr. Sci., 1996, p. 38); and Empire State Forest Products Association, *Issue briefing paper: local regulation of timber harvesting in New York State* (Albany, 1997).

20. J. J. Lindsay, A. H. Gilbert, and T. W. Birch, *Factors affecting the availability of wood energy from nonindustrial forest lands in the northeast,* USDA-FS NEFES Res. Bull. NE-122, 1992; T. W. Birch, "Landownership and harvesting trends in eastern forests," in Proc. *20th Annual Hardwood Symposium* (Hardwood Research Council, Memphis, TN, 1992), pp. 143–157, which documents a strong rise in intentions to harvest between the 1970s and the 1980s.

21. R. M. Stapleton, "Deep woods and clean waters: what price Sterling forest?" *Land and People* Fall 1996, pp. 2–8 (magazine is published by Trust for Public Lands, SF, CA). The State of Connecticut has set itself a goal to protect 10% of its land area as open space: Governor's Blue Ribbon Task Force on Open Space. *Open space in Connecticut: a legacy for life* (Hartford: Conn. DEP, 1998).

Notes on Chapter 6

1. E. N. Torbert, "Evolution of land utilization in Lebanon, NH," *Geographic Rev.* 25 (1935): 209–230. Studies of farm and forest land use change in the northeast are beyond counting; note 1 in *Wildlands and Woodlots* has a sample for New England. Town histories can be helpful, but they are usually weak in land use and economic matters. Current trends in rural communities are reviewed in D. L. Brown, D. Field, and J. J. Zuiches, eds., *The demography of rural life* (Univ. Park: Northeast Center for Rural Devel. Publ. No. 84, May, 1993), esp. essay by Calvin Beale; also, L. L. Swanson and D. L. Brown, eds., *Population change and the future of rural America,* USDA ERS Staff Rept. AGES 9324, Nov. 1993. A current overview is John Fraser Hart, *The rural landscape* (Baltimore: Johns Hopkins, 1998).

2. J. T. Lemon discusses varying spatial patterns of settlement in his "Spatial order: households in local communities and regions," in J. P. Greene and J. R. Pole, eds., *Colonial British America* (Baltimore: Johns Hopkins, 1984), pp. 86–122, with abundant references. J. F. Hart, *The land that feeds us* (New York: W. W. Norton, 1991), is readable and eloquent, for its broad overview and vivid portraits of farming in Pennsylvania (ch. 1), New York, p. 193ff, Massachusetts, p. 199ff, and the New York metro area, p. 205ff. Standard histories are J. D. Black, *The rural economy of New England*

(Cambridge: Harvard Univ. Press, 1950); and also, H. S. Russell, *A long deep furrow* (Hanover: Univ. Press of New England, 1976). On New York, an informal but useful work is U. P. Hedrick, *A history of agriculture in the State of New York* (Albany, for the NY State Agr. Soc., 1931); P. O. Wacker and P. G. E. Clemens, *Land use in early New Jersey: a historical geography* (Newark: New Jersey Hist. Soc., 1995). From a wider perspective, see C. H. Danhof, *Change in agriculture: the northern states, 1820–1870* (Cambridge: Harvard Univ. Press, 1969); and J. T. Schlebeker, *Whereby we thrive: a history of American farming, 1607–1972* (Ames: Iowa State Univ. Press, 1975), are useful; also, Stilgoe, *Common landscape,* pp. 198–208. For a useful overview of northeastern farming in the 1970s, see L. P. Schertz, "The northeast," in L. P. Schertz, et al., *Another revolution in U.S. farming?,* USDA ESCS Agr. Econ. Rept. No. 441, Dec. 1979, and other papers in this volume; Anon., *Agriculture 2000:* a look at the future (Columbus, OH: Battelle Press for the Production Credit Assoc., 1983), esp. pp. 21–26; Anon., "Toward 2005 — northeast agriculture-food-forestry issues and opportunities," 2 vols. (Washington, D.C.: Joint Council on Food and Agr., USDA, June 1987), which provides a detailed analysis. On farmland preservation techniques, see G. Carfogno and J. R. Robinson, 1986, *Farmland preservation directory northeastern United States* (New York: Natural Res. Devel. Council, 1986); and F. Schnidmon, M. Smiley, and E. G. Woodbury, *Retention of land for agriculture: policy, practice, and potential in New England* (Cambridge: Lincoln Inst. for Land Policy, 1990). A critical review of policies is in A. C. Nelson, "Economic critique of U.S. farmland preservation policies," J. *Rural Studies* 6(2): 119–142, 1990.

3. T. A. Majchrowicz and J. Salsgiver, *Changes in farm-related employment,* 1975–89, USDA ERS Rural Devel. Res. Rept. No. 85, 1993.

4. R. A. Gross, *The minutemen and their world* (New York: Hill and Wang, 1976), p. 74.

5. Ibid., p. 177; see also C. S. Grant, *Democracy in the Connecticut frontier town of Kent* (New York: W. W. Norton, 1972), p. 100.

6. H. D. Thoreau, *Walden,* illus. ed. (Princeton: Princeton Univ. Press, 1973), p. 263.

7. H. F. Wilson, *The hill country of northern New England* (New York: Columbia Univ. Press, 1936). For more recent historical analysis, see M. M. Bell, "Did New England go downhill?" *Geogr. Rev.* 79: 450–466 (1989); and R. A. Easterlin, "Population change and farm settlement in the northern U.S.," *J. Econ. Hist.* 36(1): 45–75 (1976).

8. From 1830 to 1985 the North Quabbin region of Massachusetts lost all but a few scraps of its farmland. A. D. Golodetz and D. R. Foster, "History and importance of land use and protection in the North Quabbin region of Massachusetts," *Conservation Biology* 11(1): 231 (1997).

9. H. D. Thoreau, "The succession of forest trees," in *Excursions* (New York: Corinth Books, 1962).

10. Maine Bureau of Industrial and Labor Statistics, *Fourth annual report,* Augusta, 1891, pp. 130–131. The reader of this document will be rewarded by an eloquent speech by William Freeman extolling the virtues and promise of Maine agriculture. The reader is moved to hope that Mr. Freeman never knew that he spoke when the state's agriculture stood at the threshold of its epic decline in the twentieth century.

11. The data are from the Forest Service's *Copeland report,* p. 156ff, and include Delaware.

12. J. F. Hart, "Loss and abandonment of cleared farmland in the eastern U.S.," *Annual Amer. Assoc. of Geogr.* 58(3): 417–440 (1968); this article has excellent maps.

13. Maine State Planning Office, *The Maine coast: a statistical source* (Augusta: Maine Coastal Program, 1978), p. 77; J. Benson and P. B. Frederic, *A study of farmland conversion in nineteen Maine communities* (Augusta: Maine State Planning Office, Aug. 1982), p. 6. For a longer perspective on Maine, see L. C. Irland, *Maine's forest area, 1600 or 1995* (Orono: Univ. of Maine Agr. and For. Exp. Sta. Misc. Publ. 736, 1998).

14. Chs. 7, 14, and 22 in D. M. Ellis, et al., *A history of New York State* (Ithaca: Cornell Univ. Press, 1967), tell the story of farming in the state.

15. E. L. Gasteiger and D. O. Boster, *Pennsylvania agricultural statistics 1866–1950,* Harrisburg 1954; and U.S. Bureau of Census, *Statistical abstract of U.S.,* 1993.

16. G. A. Schnell and M. Monmonier, "Spatial-temporal trends in agricultural woodland in Pennsylvania, 1950–1990," in S. K. Majumdar, E. W. Miller, and F. J. Brenner, eds., *Forests — a global perspective* (Easton, PA: Pennsylvania Academy of Sciences, 1996, pp. 359–376), contains useful graphics.

17. For details on habitat fragmentation, see notes in ch. 3.

18. Data on trends in posting over time are sparse. In New York, posting increased 63 percent from 1963 to 1972 (T. C. Brown and D. Q. Thompson, "Changes in posting and landowner attitudes in New York State, 1963–73," *NY Fish and Game J.* 23[2] [1976]: 101–137). The second round of state by state landowner studies by the USDA-FS's FIA unit is remedying this. Pennsylvania data are from T. W. Birch and C. H. Stelter, "Trends in owner attitudes," *Penn. State Forestry Issues Conference,* 1993; see also D. F. Dennis, "Effect of trends in forest and ownership characteristics on recreational uses of private forests," in G. A. Vanderstoep, ed., *Proc., 1991 northeastern recreation research symposium,* USDA-FS NEFES Gen. Tech. Rept. NE-160, 1991, pp. 116–117; and H. K. Cordell, D. B. K. English, and S. A.

Randall, *Effects of subdivision and access restrictions on private land recreation opportunities,* USDA-FS RMFRES Gen. Tech. Rept. RM-231 (1993).

19. These problems have been thoroughly reviewed in W. L. Church, *Private lands and public recreation,* National Assoc. of Cons. District, 1979; R. T. Quarterman, *Incentives to the use of land for outdoor recreation: insulation from Tort liability, tax relief,* Office of Special Projects, Univ. of Georgia School of Law, 1975; and G. L. Strodtz and C. W. Dane, "Trespassers, guests and recreationists on industrial forest land," *J. For.* 66(12) (1968): 898–901.

20. Useful, if dated, papers on this subject include O. W. Herrick, "Profile of logging in the northeast," *N. Logger* 24(4 and 5) (1975): 14, 15, 20, 21; W. J. Gabriel, et al., *Machines and systems suitable for logging small woodlots in the northeast,* State Univ. of New York, Applied Forestry Res. Inst. Rept. No. 26, Apr. 1975; and J. Falk, *Harvesting systems for silvicultural control of spruce budworm,* Univ. of Maine at Orono, Life Sci. and Agr. Exp. Sta. Misc. Rept. 221, Dec. 1979, which provides extensive references.

21. Resource Systems Group, "Forest property taxation programs: report to the North. For. Lands Council," in *Technical appendix,* North. For. Lands Council, Feb. 1994 (USDA-FS, SPF, Durham, NH).

22. P. E. Sendak and D. F. Dennis, *Vermont's use value appraisal property tax program: a forest inventory and analysis,* USDA-FS NEFES Res. Pap. NE-627, 1989; P. E. Sendak and N. K. Huyler, *Timber management and use-value assessment,* USDA-FS NEFES Res. Pap. NE-691 (1994); W. Holmes and T. Howard, *Effects of federal and state taxes on forest investments in the Northern Forest Lands Study area* (Durham, NH: Univ. N.H. Inst. for Pol. and Soc. Sci. Res., Rural Revitaliz. Rept. #1, 1989); Anon., *Report of the Forest Lands Taxation Commission,* 2d Regular Sess., 113th Legislature, Augusta: Maine Off. of Fiscal and Program Review, which includes several useful appendices; and J. Malme, *Preferential property tax treatment of land* (Lincoln, MA: Lincoln Inst. for Land Policy, 1993), working paper; see also H. O. Canham, "Property taxes and the economics of timberland management in the Northern Forest Lands region," in *Technical appendix,* North. For. Lands Council, Feb., 1994, (USDA-FS, SPF, Durham, NH).

23. The standard references on forestry co-ops are G. F. Dempsey and C. B. Markeson, *Guidelines for establishing forestry cooperatives,* USDA-FS NEFES Res. Paper NE-133, 1969; and G. F. Dempsey, *Forest cooperatives: a bibliography,* USDA-FS NEFES Res. Paper NE-82, 1967; see also D. M. Simon and O. J. Scoville, *Forestry cooperatives: organization and performance,* USDA, Agr. Coop. Serv. AES Res. Rept. No. 25, 1982; and D. A. Gansner and O. Herrick, *Cooperative forestry assistance in the northeast,* USDA-FS NEFES Res. Paper NE-464, 1980.

24. The problems of the small forest owner have received extensive professional attention, so the reading list is long. For literature of the 1960s and

1970s, see note 21 in *Wildlands and Woodlots.* A useful overview is M. J. Baughmann, ed., *Proc.: first national conference on forest stewardship* (St. Paul: Univ. Minn. Ext. Service, Aug. 1994). Useful recent items include: T. W. Birch and N. A. Pywell, *Communicating with nonindustrial private forests owners: getting programs on target,* USDA-FS NEFES Res. Pap. NE-RP-593, 1986; L. B. Snyder and S. H. Broderick, "Communicating with woodland owners: lessons from Connecticut," *J. For.* 90(3): 33–37, 1992. J. P. Royer and C. D. Risbrudt, eds., *Nonindustrial Private Forests: a review of economic and policy studies* (Durham: Duke Univ., School of For. and Env. Studies, 1983); P. V. Ellefson, M. D. Bellinger, and B. J. Lewis, *Nonindustrial private forests: an agenda for economic and policy research* (St. Paul: Univ. of Minn. Agr. Exp. Sta. Bull. 592 1990); W. F. Hyde, R. G. Boyd, and B. L. Daniels, "Impacts of public interventions: an examination of the forestry sector," *J. Pol. Anal. & Manage.* 7(1): 40–61, 1987; F. W. Cubbage and R. W. Haynes, *Evaluation of the effectiveness of market responses to timber scarcity problems,* USDA-FS Mktg. Res. Rept. No. 1149, 1988; B. N. Rosen and H. Carryl, "Price reporting for nonindustrial private owners," *J. For.* 91(1): 34–41, Jan. 1993; L. C. Irland, *Organizational roles in forestry extension in the U.S., in Extension activities for owners of small woodlands, Joint FAO/ECE Comm. on Forest Working Techniques and Training* (Fredericton, NB: Canadian For. Serv., 1988); L. C. Irland, ed., *Improving the effectiveness of public programs for private forest owners* (Orono: Univ. of Maine, Coll. of For. Res. Tech. Note 91, 1984). R. Neil Sampson and L. A. DeCoster, *Public programs for private forestry* (Washington, D.C.: American Forests, 1997). For a useful review on nearby Canadian activities, see D. S. Curtis, *Woodlot organizations in eastern Canada: historic development, legislation, structure, financing, and services,* Canadian For. Serv. Inf. Rept. M-X-162, 1987.

Notes on Chapter 7

1. R. W. Judd, "Reshaping Maine's landscape — rural culture, tourism, and conservation, 1890–1929," *J. For. Hist.* 32(4): 180–190 (1988); also his expanded threatment in R. W. Judd, *Common lands, common people: the origins of conservation in Northern New England* (Cambridge, Mass.: Harvard University Press, 1997). T. D. Seymour Bassett, "Documenting recreation and tourism in New England," *Amer. Archivist* (50): 550–569 (Fall 1987); and D. Brown, *Inventing New England — regional tourism in the nineteenth century* (Washington, D.C.: Smithsonian Press, 1998). See also H. K. Cordell, *Outdoor recreation in American life: a national assessment of demand and supply trends,* (Champaign, IL: Sagamore Publishing, 1999).

2. USDI, NPS, 1982–83, *Nation-wide recreation survey* (Washington, D.C., USDI, Apr., 1986). See also U.S. Dept. of Commerce, "1996 national survey of fishing, hunting, and wildlife-association recreation," Nov. 1997.

3. Extrapolated from estimates for the three northern New England states plus New York in The Irland Group, *Economic impact of forest*

resources in northern New England and New York, Rept. to the Northeastern Forest Alliance, Saranac Lake, NY, 1992. State level summaries of an updated version have been released and are available from State Forester's offices; see also K. J. Boyle, et al., *Economic values and economic impacts associated with consumptive and nonconsumptive uses of Maine's fish and wildlife resources* (Orono: Univ. of Maine, Agr. Exp. Sta. Staff Paper ARE 410 (Mar. 1990), Dept. Agr. and Res. Econ.). See also his expanded threatment in R. W. Judd, *Common lands, common people: the origins of conservation in Northern New England* (Cambridge, Mass.: Harvard University Press, 1997).

4. C. E. Thomas, "The Cape Cod National Seashore: a case study of federal administrative control over traditionally local land use decisions," *Boston Coll. Law Rev.* 12(1): 225–272 (1985).

5. For cases in other areas, see "Land boom in the Poconos," ch. 6 in Council on Env. Quality, *The Delaware River Basin* (Washington, D.C.: GPO, 1975); R. G. Healy and J. L. Short, *The market for rural land: trends, issues, policies* (Washington, D.C.: The Conservation Foundation, 1981), pp. 172–181 (on Plainfield, NH); and USDA-FS, *Northern forest lands study,* Rutland, VT, 1990, App. D (on sample counties from New York to Maine in the late 1980s).

6. See H. F. Wilson, *The hill country of northern New England* (New York: Columbia Univ. Press, 1936); J. D. Black, *The rural economy of New England* (Cambridge: Harvard Univ. Press, 1950).

7. B. Huffman, *The Vermont farm, and a land reform program* (Montpelier: State Planning Office, 1973), vol. 2, app. D. These land ownership trends are not equivalent to land use changes. While total farmland ownership has declined, agricultural land uses have shown varying trends. For the state as a whole, harvested cropland fell from 858,000 acres in 1949 to 511,000 acres in 1969, and 326,000 acres in 1996. From 1949 and 1969, land used for pasture actually increased in Vermont: from 219,000 acres to 279,000.

8. N. P. Kingsley, *The forest resources of Vermont,* USDA-FS NEFES Res. Bull NE-46, 1977; R. O. Sinclair and S. B. Meyer, *Nonresident ownership of property in Vermont,* Univ. of Vermont Agr. Exp. Sta. Bull. 670, 1972; F. H. Armstrong, *Valuation of Vermont forests,* 1968–1974, Univ. of Vermont, Dept. of Forestry, 1975. Armstrong estimated that 31 percent of the potentially marketable Vermont forest land was sold in the years 1968–74. Considering parcels ten acres or larger, the average parcel sold fell from 80 acres in 1968 to 45 acres in 1974; see also F. H. Armstrong and R. D. Briggs, *Valuation of Vermont forest 1968–1977,* Univ. of Vermont, Dept. of Forestry, 1978 (draft).

9. B. Payne and L. C. Irland, unpublished data on file at USDA-FS NEFES Amherst, MA.

10. U.S. Council on Env. Quality, *Subdividing rural America, Executive summary* (Washington, D.C.: GPO, 1976).

11. C. R. Goeldner and S. Standley, "Skiing trends," in W. F. LaPage, ed., Proc., *1980 national symposium on outdoor recreation trends,* USDA-FS NEFES Gen. Tech. Rept. NE-57, vol. 1, p. 105ff; see also USDA-FS, *Growth potential of the skier market in the national forests,* Washington, D.C., Res. Paper WO-36, 1990.

12. G. Donovan, *Vermont skiing industry, 1973–74* (Montpelier: Agency for Devel. and Comm. Affairs, Econ. Res. Rept. 7304).

13. Vermont Agency of Env. Cons., *Vermont second home inventory, 1973,* Montpelier, 1974.

14. This section is based on L. C. Irland and T. G. Siccama, *National forest on three sides,* unpub. rept. to The Conservation Foundation, 1975. USDA-FS, 1977, *Final environmental statement, Deerfield River land use plan,* Green Mountain National Forest, Rutland, VT, 1977.

15. Armstrong and Briggs, *Valuation of Vermont forests* (note 8 above).

16. R. L. Baker, Controlling land use and prices by using special gain taxation: the Vermont experiment, Env. Affairs 4 (Summer 1975), provides a full analysis of this tax; see also R. G. Healy and J. S. Rosenberg, Land use and the states (2d ed., Baltimore: Johns Hopkins Univ. Press, 1979), p. 69ff.

17. Armstrong and Briggs, *Valuation of Vermont forests* (note 8 above).

18. J. W. Sewall Co. and Market Decisions, Inc., *Northern Forest Lands Council land conversion study* (Concord, NH: NFLC, Apr. 1993); USDA-FS, *Northern forest lands study* (Rutland: USDA-FS, Apr. 1990, p. 13ff and app. D.)

19. The Irland Group, "Case studies in land use change: overview of the northern forest lands," in USDA-FS, *Northern forest lands study* (Rutland, VT, Apr. 1990, app. D).

20. Anon., *Adirondack Park in the 21st century* (Albany: Comm. on the Adirondacks in the 21st century, Apr. 1990); see also R. M. Whitney, "Forces for change in forest land ownership and use: the large landowner's situation," in C. S. Binkley and P. R. Hagenstein, *Conserving the North Woods* (New Haven: Yale School of For. and Env. Studies Bull. 96, 1989); L. C. Irland, "Wall street in the woods," *Appalachia,* Sept. 1990, pp. 18–22; and Land and Water Assoc. and Market Decisions, Inc., *Summary of the Commission's current land use policies and their net effects after 20 years of development in Maine's Unorganized Areas* (Augusta: Maine Land Use Regulation Commission, 1994).

21. See, e.g., M. I. Bevins, "Seasonal home impact on taxes," *Farm and Home Science* (Winter, 1973); R. Frazer and T. Donovan, *Vacation home*

survey of 8 Vermont towns (Montpelier: Agency for Devel. and Comm. Affairs, 1972); *Downhill in Warren* (Montpelier: Vermont Public Interest Research Group, 1972); E. L. Johnson, *The effect of second home development of Ludlow, VT* (Springfield: South Windsor County Regional Planning Comm., 1973); and B. Payne, R. Gannon, and L. C. Irland, *The second home recreation market in the northeast* (Washington, D.C.: USDI Bureau of Recreation, 1975); A. Klein and D. Phelan, *Second homes and vacation homes: potential impact and issues* (Springfield, IL: Council of Planning Librarians Exchange Bibliography No. 839, 1975). For in-depth analysis of social effects of a proposed ski expansion, see USDA-FS, *Loon Mountain expansion final EIS* (Laconia, NH: White Mountain Nat. For. 1992).

22. Much of this literature is unpublished. See, e.g., C. D. Campbell and T. A. Barthold, *Sources of property taxes in New Hampshire, 1985*, Dartmouth College Dept. of Econ., unpub. paper, esp. pp. 21–22, Rept. to NH office of State Planning.

23. Unpublished data, Vermont Dept. of Employment Security; see also for a detailed analysis, T. L. Brown and N. A. Connelly, *Tourism in the Adirondack region of New York,* New York State College of Agr. and Life Sci., Cornell Univ., Dept. of Nat. Res., Res. and Ext. Series No. 21, 1984.

24. R. L. Ragatz Assoc., *Recreational properties: the market for privately owned recreational lots and leisure homes,* Rept. to the U.S. Council on Env. Quality, 1974; and R. L. Ragatz, "Trends in the market for privately owned seasonal recreational housing," in LaPage, *Proc.,* vol. 1, pp. 179–194 (note 11 above).

25. Armstrong and Briggs, *Valuation of Vermont forests* (note 8 above).

26. R. H. Widmann and T. W. Birch, *Forest-landowners of Vermont, 1983,s USDA-FS NEFES Res. Bull. NE-102.*

27. *J. T. Boros, N. Engalichev, and W. G. Gove,* The timber industry of New Hampshire and Vermont, USDA-FS NEFES Res. Bull. NE-35, 1974.

28. R. R. Nathan Assoc. and Resources Planning Assoc., *Recreation as an industry,* Appalachian Regional Commission, Appalachian Res. Rept. No. 2, 1966, p. 3. For a stronger statement regarding the same issues, see R. R. Gottfried, "Observations on recreation-led growth in Appalachia," *Amer. Econ.* 20 (1967): 54–60; see also P. E. Polzin and D. L. Schweitzer, *Economic importance of tourism in Montana,* USDA-FS INTFRES Res. Paper INT-171, 1975. M. Frederick, *Tourism as a rural economic development tool: an exploration of the literature,* USDA ERS Bibliogr. of Agr. No. 122, 1992, cites 113 references and indexes them carefully with brief abstracts; see also J. F. Ziegler, ed., *Enhancing rural economies through amenity resources: a national policy symposium* (Univ. Park, School of Hotel, Rest., and Inst. Mgmt., Penn. State Univ., 1991), which provides a useful overview. Also, two papers by T. L. Brown, "Forest conservation, forest recreation, and tourism"; and

"The forest products industry: interrelationships and compatibility," in *Technical appendix,* North. For. Lands Council, USDA-FS, SPF, Feb., 1994, Durham, NH.

29. See Healy and Rosenberg, Land use, ch. 3; and F. O. Sargent, *Vermont's Act 250, enabling legislation for environmental planning,* Proc., Conference on rural land use policy in the northeast, Northeastern Center for Rural Devel. Publ. No. 5, 1975. For a critical review see J. McClaughry, "The new feudalism," *Env. Law* 5 (spring 1975); also R. G. Healy, *Land use and the states* (Baltimore: Johns Hopkins Univ. Press, 1979), ch. 4.

30. 10 Vermont Statutes Annotated ch. 151; 24 Vermont Statutes Annotated pt. 2, ch. 91; 24 USA 4302 on Act 200, see J. M. DeGrove, *The new frontier for land policy: planning and growth management in the states* (Cambridge: Lincoln Inst. for Land Policy, 1992), ch. 5.

31. See discussions on Maine, in ch. 4 above; and J. Collins, "The Adirondack Park: how a green line approach works," in C. M. Klyza and S. C. Trombulak, eds., *The future of the northern forest* (Hanover: Middlebury College Press, by Univ. Press of New England, 1994); and see citations in Chapter 2 on the Pinelands.

32. John K. Wright, "The changing geography of New England," in J. K. Wright, ed., *New England's prospect* (New York: Amer. Geo. Soc., 1933), p. 464.

33. For background on the 1980s boom, see, e.g., S. Sass, *How much is that building in the window?* (Boston: Fed. Res. Bank of Boston Regional Rev. 3(3): 6–12, Summer 1993); and L. E. Browne, "Why New England went the way of Texas rather than California," *New England Econ. Rev.,* Jan./Feb. 1992, pp. 23–41.

Notes on Chapter 8

1. The history is ably told in R. Nash, *Wilderness and the American mind* (New Haven: Yale Univ. Press, 1982, 3d ed.). Admirers of this volume will also want to read Nash, *The rights of nature: a history of environmental ethics* (Madison: Univ. of Wisconsin Press, 1989), which contains a vast bibliography. Also, F. Graham, *The Adirondack Park* (New York: Knopf, 1978). The full history of the conservation movement in the northeast has yet to be written. A preliminary version of portions of this chapter was published as "Wildness and wilderness in the northeastern United States — challenge for the coming century, *Int. Jour. Wilderness* 2(3): 27–30;48, 1996.

2. M. B. Davis, ed., *Eastern old-growth forests: prospects for rediscovery and recovery* (Covelo, CA: Island Press, 1996); for New England, an extensive list of special places is described and illustrated in D. B. Bennett, *The forgotten nature of New England: a search for traces of the original wilderness* (Camden: Down East Books, 1996). Also, T. R. Crow, "Old growth and

biological diversity: a basis for sustainable forestry," in Anon., *Old growth forests: what are they? How do they work?*, Proc., Conference on old growth forests (Toronto: Canadian Scholar's Press, 1990), pp. 49–62; J. M. Klopatek, et al., "Land-use conflicts with natural vegetation in the U.S.," Env. Cons. 6(3): 191–200 (1979); their table 11 shows percentages of each state remaining in natural vegetation. A more recent overview is R. F. Noss, E. T. Laroe III, and T. M. Scott, *Endangered ecosystems of the U.S.: a preliminary assessment of loss and degradation,* USDI Nat. Biol. Service, Biol. Rept. 28 (Feb. 1995).

3. The story is briefly told in L. C. Irland and S. M. Levy, "Southern New England water supply lands," *J. New England Water Works Assn.* 91 (March 1977), pp. 12–39.

4. Despite its western flavor, J. K. Agee and D. R. Johnson, eds., *Ecosystem management for parks and wilderness* (Seattle: Univ. of Washington Press, 1988), is useful.

5. Attempts have been made to measure these values through carefully designed interviews: A. Gilbert, R. Glass, and T. More, "Valuation of eastern wilderness: extramarket measures of public support," in C. Payne, J. M. Bowker, and P. C. Reed, comps., *Economic value of wilderness, Proc. of a conference,* USDA-FS SEFES Gen. Tech. Rept. SE-8, 1992; see also papers in same volume by Irland and by Loomis. On a broader plane, current thinking on the ethical issues is well-expressed in the Nash volumes cited above.

6. For detail on the trail, see USDI, NPS, *Comprehensive plan for protection, management, development, and use of the Appalachian National Scenic Trail, Harpers Ferry, WV,* Sept. 1981.

7. M. I. Bevins and D. P. Wilcox, *Outdoor recreation participation — an analysis of national surveys, 1959–78,* Vermont Agr. Exp. Sta. Bull. 686, 1980. See papers by Betz and Cordell and by Lucas and Stanley, in A. H. Watson, comp., *Outdoor recreation, Benchmark, 1988,* Proc., National Outdoor Recreation Forum, USDA-FS SEFES Gen. Tech. Rept. SE-52, 1989 for a detailed national overview; a regional focus is supplied by R. Warnick, "Outdoor recreation trends in the northeast: markets and activities," in W. D. Ostrofsky and W. B. Kroha, eds., Our forest's place in the world, *New England and Atlantic Canada's forests,* Proc. New England SAF and Maine Chapter, TWS, and NE For. Pest Council (Maine Agr. Exp. Sta. Misc. Publ. 738, 1997). For a detailed national analysis, outlook, and summary of policy issues, see H. K. Cordell, et al., *Analysis of the outdoor recreation and wilderness situation in the U.S.: 1989–2040,* Tech. Doc. supporting the 1989 USDA-FS, RPA Assessment, USDA-FS RMFRES Gen. Tech. Rept. RM-189, 1990.

8. B. Wagner and E. Spencer, "Historical trends in back country use," *Appalachia* 42(4) (1979): 129–142.

9. USDI, NPS, 1982–1983 *Nationwide recreation survey* (Washington, D.C.: GPO, 1986), p. 22.

10. For background, see the overview in L. C. Irland, *Wilderness economics and policy* (Lexington, MA: Heath-Lexington, 1979), ch. 9; and the comprehensive treatise by J. C. Hendee, G. H. Stankey, and R. C. Lucas, *Wilderness management,* USDA-FS Misc. Publ. 1365 (Washington, D.C.: GPO, 1978); D. N. Cole, M. E. Petersen, and R. E. Weas, *Managing wilderness recreation use: common problems and potential solutions,* USDA-FS INT FRES Gen. Tech. Rept. INT-230, 1987; W. R. Burch, Jr., ed., *Long distance trails* (New Haven: Yale School of For. and Env. Studies, 1979). See R. E. Manning, *Studies in outdoor recreation* (Corvallis: Oregon State Univ. Press, 1985), for a valuable survey of sociological literature.

11. R. W. Guldin, "Wilderness costs in New England," *J. For.* 78(9) (1980): 548–552; and S. D. Reiling and M. W. Anderson, *Evaluation of the cost of providing publicly-supported outdoor recreational facilities in Maine* (Orono: Univ. of Maine Agr. Exp. Sta. Bull. 793, 1983); B. H. Martin, ed., *Fees for outdoor recreation on lands open to the public* (Gorham, NH: Appal. Mtn. Club, 1984).

12. The quote is from G. P. Marsh, *Man and nature: or, physical geography as modified by human action,* ed. by D. Lowenthal (Cambridge: Harvard Univ. Press, 1964), at pp. 203–204. For early history, a speech by Gov. Charles Evans Hughes at the *1908 White House Conference of Governors on Conservation* (Washington, D.C.: GPO, 1909), pp. 314–325 is of interest; the Clinton quote is at p. 314. A short version of the Park story is told by J. Mitchell in "The man who married the mountains," pp. 182–208, in *Dispatches from the deep woods* (Lincoln: Univ. of Nebraska Press, 1991); and by P. G. Terrie, "Imperishable freshness: culture, conservation, and the Adirondack Park," *For. and Cons. Hist.* 37: 132–141 (1993); W. C. White, *Adirondack country* (Syracuse: Syracuse Univ. Press, 1985), is a readable and nontechnical overview; as is J. E. Keller, *Adirondack wilderness: a story of man and nature* (Syracuse: Syracuse Univ. Press, 1980). Helpful on the Park's origins is P. G. Terrie, "Forever wild forever: the forest preserve debate at the New York State Constitutional Convention of 1915," *New York Hist.* 70(3): 251–275, July 1989. The Adirondack Council annually reports on the Park in its *State of the Park Report.* The Centennial edition was in 1992 (Elizabethtown, NY: The Adirondack Council, 1992). The best introduction is F. Graham, Jr., *The Adirondack Park, a political history* (New York: Knopf, 1976); Commission on the Adirondacks in the 21st Century, *The Adirondack Park in the 21st Century* (Albany: State of New York, 1990), and appendices; see also the plans: *Adirondack Park Land Use and Development Plan, March 6, 1973; and State Land Master Plan, June 1, 1972* (Ray Brook: Adirondack Park Agency). Related studies include D. H. Vrooman, "An empirical analysis of determinants of land values in the Adirondack Park," *Amer. J. Econ. and Sociology* 37 (April 1978): 165–177; P. H. Gore and M. Lapping, "Environmental quality and social equity: wilderness preservation in a depressed region, New York State's Adirondacks," *Amer. J. Econ.* and *Sociology* 35 (October 1976): 349.

13. C. S. Sargent, *The forests of the United States* (Washington, D.C.; dept of the 10th Census 1884) pp. 494–510.

14. Chapter 77 in A. Evers, *The Catskills: from wilderness to Woodstock* (Garden City: Doubleday and Co., 1972).

15. Baxter State Park Authority, *Baxter State Park plan* (Millinocket, ME, Jan. 1994); see also the Agency's annual reports.

16. USDA-FS, *Roadless areas review and evaluation, Final env. statement,* Washington, D.C., 1973.

17. USDA-FS, RARE-II, *Supplement to draft EIS, Northern Appalachian and New England states,* June 1978; and Summary — *Final env. statement,* Jan. 1979.

18. C. H. W. Foster and D. R. Foster, *Managing the greenwealth: the forests of Quabbin* (Cambridge: Harvard Univ., Kennedy School of Govt. Env. and Nat. Res. Program, 1994).

19. L. Dietz, *The Allagash* (Thorndike, ME: Thorndike Press, 1968).

20. *National Wild and Scenic Rivers Act,* PL 90–542.

21. USDI Bureau of Outdoor Recreation, *Penobscot River wild and scenic proposal,* Final Env. Statement, 1976.

22. J. Heinrichs, "Thirteen mile woods," Amer. For. (Sept. 1980): 50–52.

23. New England River Basins Commission, *The river's reach,* Boston, 1976.

24. Massachusetts Dept. of Env. Mgt., Scenic Rivers Programs, *Massachusetts scenic and recreational rivers,* Draft, June 1979.

25. New York State Parks, Recreation, and Historic Preservation Agency, *1994 State comprehensive outdoor recreation plan* (Albany: DEC, 1994).

26. Office of Comprehensive Planning, *Wild, scenic, and recreational rivers for New Hampshire,* Concord, 1977; and Maine Bureau of Parks and Recreation, *Maine rivers assessment,* 1982; see also J. E. Hickey, "Proposed rivers preservation program for Connecticut," *Connecticut Woodlands* 45(3) (1980): 3–7.

27. Memo from Regional Director, HCRS, *Final list of potential wild and scenic rivers* (Philadelphia: USDA HCRS Northeast Region, Mar. 8, 1979). Examples of recent studies include, USDI, NPS, *The Maurice River (NJ) and its tributaries; National wild and scenic river study* (Philadelphia: NPS, Mid-Atlantic Region, 1993); USDI, NPS, *Farmington wild and scenic river study (CT)* (Boston: NPS, North Atlantic Region., 1995).

28. Maine Land Use Reg. Comm., *Action program for management of lakes in Maine's unorganized areas* (Augusta: Maine Dept. of Cons., 1989). R. F. Perkins, "Maintaining the working landscape of the North Woods," *New England Landscape One: 81–90, 1989;* J. A. Bley, *An investment in*

stewardship: a case study of Lowell and Co. Timber Assoc. ownership of Attean Township (Boston: Lowell-Blake & Assoc., 1992).

29. For example, see W. G. Scheller, "Stalking the red herring: sportsmen's rights and wilderness preservation," *Appalachia* No. 171 (1980): 63–71.

30. The literature on the subject of wilderness is immense. A fine place to start is A. Leopold, *A Sand County almanac* (New York: Oxford Univ. Press, 1966). I have summarized the arguments in L. C. Irland, *Wilderness economics and policy* (Lexington, MA: Heath-Lexington, 1979). For the technically minded, J. Krutilla and A. Fisher, *The economics of natural environments* (Baltimore: Johns Hopkins Univ. Press, 1975), is still the classic statement; see also J. Sax, *Parks without handrails* (Ann Arbor: Univ. of Michigan Press, 1980); D. W. Lime, ed., *Managing America's enduring wilderness resource* (St. Paul: Univ. of Minnesota Tourism Center, Ext. Serv., and Agr. Exp. Sta., 1990).

31. A widely used guideline, offered by *The Brundtland report,* is to dedicate 12 percent of the landscape to preservation. Whether this is a realistic long-term goal for the northeast deserves further debate.

32. M. Kellett, *A new Maine woods reserve: options for protecting Maine's northern wildlands* (Boston, MA: The Wilderness Soc., 1989), see map in next chapter; see also, e.g. National Audubon Soc., *The great northern forest* (New York, 1994), with its map listing 10 "proposed wildlands" of New York and northern New England; and B. McKibben, "An explosion of green," *Atlantic Monthly,* April 1995, in which he advocates massive wilderness reservations.

Notes on Chapter 9

1. See, e.g. T. R. Berger, *A long and terrible shadow: white values, native rights in the Americas, 1492–1992* (Vancouver: Douglas & McIntyre, 1991); P. Brodeur, *Restitution: the land claims of the Mashpee, Passamaquoddy, and Penobscot Indians of New England* (Hanover: Univ. Press of New England, 1985). For a readable overview, see T. Morgan, *Wilderness at dawn: the settling of the North American continent* (New York: Simon and Schuster, 1993). A beautifully illustrated volume treating "Norumbega" — New England and the Maritimes — emphasizes cultural and geographic aspects of the European takeover of this region: E. W. Baker, et al., eds., *American beginnings: exploration, culture, and cartography in the land of Norumbega* (Lincoln: Univ. of Nebraska Press, 1994). For a current view, see H. Robtoy, D. Brightstar, T. Obomsawin, and J. Moody, "The Abenaki and the northern forest," in C. M. Klyza and S. C. Trombulak, *The future of the northern forest* (Hanover, Middlebury College Press by Univ. Press of New England, 1994). An excellent regional overview is in M. Williams, *Americans and their forests* (New York: Cambridge Univ. Press, 1989), ch. 2.

2. Useful regional overviews are in P. W. Gates, *History of public land law development* (Washington, D.C.: GPO, 1968), ch. II with numerous citations; and in R. H. Brown, *Historical geography of the U.S.* (New York: Harcourt, Brace, and World, 1948), chs. 1–4 and 6, 7, and 9. C. E. Clark, *The eastern frontier* (New York: Knopf, 1970), is an excellent source for the early history of landownership in Maine. Land use studies summarized in ch. 6, notes 1 and 2, illustrate the local picture. On the Great Proprietors, see Clark, *The eastern frontier.* For a vivid summary of colonial land policy and its social implications, see R. Hofstadter, *America at 1750: a social portrait* (New York: Vintage, 1973), pp. 142–151; also J. F. Martin, *Profits in the wilderness* (Chapel Hill, Univ. of North Carolina, 1991); E. T. Price, *Dividing the land: early American beginnings of our private property mosaic* (Univ. Chicago Geogr. Res. Pap. 238, 1995). Relevant sections of D. W. Meinig, *The shaping of America: a geographical perspective on 500 years of history* (New Haven: Yale Univ. Press, 3 vols., 1986, 1998) provide historic context.

3. C. E. Brooks, "Overrun with bushes — frontier land development and the forest history of the Holland Purchase, 1800–50," *For. and Cons. Hist.* 39(1): 17–26, provides a detailed review of the settlement history of these lands.

4. B. Jordan, P. Liacauras, and W. W. Posvar, *The Atlas of Pennsylvania* (Philadelphia: Temple Univ. Press, 1989), p. 83 ff. On settlement of this region, see T. G. Jordan and M. Kaups, The *American backwoods frontier: an ethnic and ecological interpretation* (Baltimore: Johns Hopkins Univ. Press, 1989), which has detailed maps.

5. See relevant sections of J. T. Schlebeker, *Whereby we thrive: a history of American farming, 1607–1972* (Ames: Iowa State Univ. Press, 1975), esp. ch 1; see also U. P. Hedrick, *A history of agriculture in the State of New York* (Albany: for the New York State Agr. Soc., 1931). Reference to squatters in Pennsylvania is in Schlebeker at p. 13. A colorful history of patroon land ownership (the 1.5 million acre Hardenburgh patent) and squatters in the Catskills is found in A. Evers, *The Catskills: from wilderness to Woodstock* (Garden City: Doubleday and Co., 1972). See Price, note 2, above, part IV.

6. D. S. Powell, et al., *Forest resources of the U.S., 1987,* USDA-FS RMFRES Res. Bull. RM-234 (1993). T. W. Birch, Private forest-land owners of the U.S., 1994, USDA-FS NEFES Res. Bull. NE-136(1996), presents detailed results of a new national ownership survey. For older surveys of New England owners, see notes to ch. 8 in *Wildlands and Woodlots.* Most recent available Forest Service surveys in this region include abundant detail on landowners and their plans and attitudes. They are *Maine,* 1982 data, USDA-FS NEFES Res. Bull. NE-90; *New York,* 1980 data, USDA-FS Res. Bull. NE-78; *Vermont,* 1983 data, USDA-FS NEFES Res. Bull. NE-102; *New Hampshire,* 1983 data, USDA-FS NEFES Res. Bull. NE-108; *Pennsylvania,* 1978 data, USDA-FS NEFES Res. Bull. NE-66 (a new Pennsylvania survey should be available by the time this book is published); New Jersey is also forthcoming.

7. H. Gannett, asst. by J. E. Whelchel, "Lumber," in *12th Census of the U.S., 1900, Manufacturers — Part III — Special reports on selected industries* (Washington, D.C.: U.S. Census Office, 1902), p. 828.

8. For mid 1990's figures, see ch. 7 of Irland, *Land, timber, and recreation in Maine's northwoods* (Orono: Univ. of Maine Agr. and For. Exp. Sta. Misc. Pub. 730, 1996).

9. L. C. Irland, "Wall street in the woods," *Appalachia Bull.*, Sept. 1990, pp. 18–21; and J. W. Sewall Co. and Market Decisions, *Northern Forest Lands Council land conversion study,* April 1993, Rept. to Northern Forest Lands Council.

10. A large and well-publicized set of land sales of the former Diamond International Corporation in the mid 1980's helped trigger public concern about the stability of private forest landownership in northern New England and New York. Contrary to widely publicized fears, little of this land ended in subdivisions, and large tracts in New York, New Hampshire, and Maine were sold to public agencies. The facts are shown in Forum on land sales by Coburn Lands Trust and the former Diamond International Corporation, in *Technical appendix,* Northern Forest Lands Council, USDA-FS, SPF, Durham, NH. For a history of major owners and their acreage owned in Maine in the mid 1990's, see Irland, *Land, timber and recreation,* ch. 7.

11. A useful discussion of common and undivided ownership is in A. H. Wilkins, *Ten million acres of timber* (Woolwich, ME: TBW Books, 1978), pp. 5–23.

12. USDA-FS, *National Plan for American Forestry,* U.S. Congr. Sen. 73d Cong. 1st Sess. Doc. No. 12, 1933, hereafter cited as "Copeland Rept.," p. 967; U.S. General Accounting Office, *Private mineral rights complicate the management of eastern wilderness areas* (Washington, D.C.: Gen. Acctg. Off. GAO/RCED-84-101, July 1984). For a graphic view of how oil development can affect forests, see the aerial photo at p. 117 in F. J. Marschner, *Land use and its patterns in the U.S.* (USDA Agr. Hbk. 153, 1950). The scene is near Bradford, PA.

13. For an overview of this topic, see F. C. Zinkhan, et al., *Timberland investments* (Portland: Timber Press, 1992); R. A. Sedjo, ed., *Investments in forestry: resources, land use and public policy* (Boulder: Westview Press, 1985); and L. C. Irland, "Timber and timberland as an investment option," *The Consultant,* Spring 1994.

14. See L. M. Schepps, "Maine's public lots: the emergence of a public trust," *Maine Law Rev.* 26(2) (1974): 317–372; State Forestry Department, *Report on public reserved lots,* Augusta, 1963. L. C. Irland, "Rufus Putnam's ghost: an essay on Maine's public lands, 1783–1820," *J. For. Hist.* 30(2) (1986): 60–69; R. McCullough, *Landscape of community a history of communal forests in New England* (Hanover: Univ. Press of New England, 1995). The region's nine states also received a total of three million acres of federal

lands outside the region. These were sold, the funds applied to support public eduction — the origin of the land grant colleges. A useful survey of the development of town and county forests is in the chapter "Community Forests," p. 843ff in the *Copeland Report.*

15. *Pennsylvania Game News* 66(1) Jan. 1995.

16. See L. C. Irland, "Outdoor recreation supply in the Maine woods: issues for the future," in C. Murdoch and D. Stone, eds, *First annual Munsungan conference proc.* (Orono: Univ. of Maine Agr. Exp. Sta. Misc. Rept. 378, 1993), pp. 21–34; and other papers in the same volume.

17. Counting ownership in this diverse group of owners is nearly impossible. One reason is their sheer number; another is that many of them hold only partial interests and often pass acquired lands to state or federal agencies for management. The Land Trust Alliance of Washington, D.C. estimates that about 905,000 acres had been acquired by local and regional land trusts by 1990. Of this acreage, some 170,000 had been transferred to other agencies. Easements represented 191,000 acres. Data in table 32 show a larger amount than these estimates would suggest; Land trusts protected 14 million acres as of 1994, *AREI Updates* No. 13 (1995). Pennsylvania accounted for more than one-fourth of this activity. For further information, see Land Trust Exchange, *Report on 1985 national survey of government and nonprofit easement programs,* Dec. 1985. Publications of the owning groups often fail to distinguish between fee and leased land and to give useful acreage summaries. For colleges and land trusts, it is difficult to arrive at a listing of organizations involved. A useful place to begin any review of land ownership for a given state will often be the State Comprehensive Outdoor Recreation Plans issued periodically by park and recreation agencies.

18. G. Abbott, Jr., "Land trusts: innovations on an old New England idea," *New England Landscape,* vol. 1, 1989, pp. 13–24.

19. J. P. DeBraal, 1991, *Foreign ownership of U.S. Agricultural land through Dec. 31, 1990,* USDA ERS, Res. and Tech. Div., ERS Staff Rept. AGES 9120. This survey is updated annually. If a corporation is more than 10 percent foreign-owned, its entire ownership is considered to be foreign-owned for this survey.

20. The "Northern Forest Lands" issue has generated a vast literature: USDA-FS, *Northern forest lands study,* Rutland, VT, 1990. A readable overview of the issue's development is in ch. 21 of D. Dobbs and R. Ober, *The northern forest* (White River Junction: Chelsea Green Publishing Company, 1995); N. Boucher, "Whose woods these are," *Wilderness,* 53(186) Fall 1986, pp. 18–42; Papers in *Habitat, J. Maine Audubon Soc.,* 3(6), June/July 1986; "The northern forest: a special report," *Appalachia,* Sept. 1991; H. E. Echelberger, A. E. Luloff, and F. E. Schmidt, *Northern forest lands: resident attitudes and resource use,* USDA-FS NEFES Res. Pap. NE-653, 1991; C. S. Binkley and P. R. Hagenstein, eds., *Conserving the northwoods* (New Haven:

Yale School of For. and Env. Studies Bull. 96, 1991); C. H. W. Foster, *Of vert and vision: ensuring the legacy of the northern forest of New England and New York* (Harvard Univ. Kennedy School of Govt. Center for Science and International Affairs, Disc. Pap. 92–13, 1992). *Appalachian Mountain Club, et al., Inventory and ranking of the key resources of the Northern Forest Lands of Vermont, New Hampshire, and Maine,* Gorham, NH, Res. Dept., processed rept. (Sept. 1993). Comments on the process used to develop the Council's recommendations are offered in a short paper by the Council's Executive Director, C. Levesque, "Northern Forest Lands Council: a planning model for use of regional forest land," *J. For.* 93(6): 36–38 (1995); and Carl Reidel, "The northern forest, a dead end," Amer. For. (Mar./Apr. 1993).

21. J. A. Michaels, et al., *New York-New Jersey highlands regional study,* USDA-FS, State & Pvt. For., Durham, NH, n.d. (ca. 1992).

22. See, e.g., paper by D. W. Gross, H. Carter, C. Geisler, and G. R. Goff, "The northeastern forest as a new commons," in T. Lepisto, ed., *Sustaining ecosystems, economies, and a way of life in the northern forest* (Washington, D.C.: The Wilderness Soc., 1993), with commentary by Reidel, in same volume; paper by Foster, note 20 supra; and R. G. Healy and P. B. Ristow, "Shared ownership of forest lands: experience and prospects," in Binkley and Hagenstein, *Conserving,* note 20 supra. A historical perspective is provided in R. W. Judd, *Common lands, common people: the origins of conservation in Northern New England* (Cambridge, Mass.: Harvard University Press, 1997). Current land acquisition isues in Maine are reviewed in L. C. Irland, Policies for Maine's Public Lands: a long-term view, in *Maine Choices: 1999,* (Augusta: Maine, Maine Center for Economic Policy, pp. 7–22).

Notes on Chapter 10

1. P. N. Carroll, *Puritanism and the wilderness* (New York: Columbia Univ. Press, 1969), p. 183ff; and C. Bridenbaugh, *Fat, mutton, and liberty of conscience: Society in Rhode Island, 1639–1690* (Providence: Brown Univ. Press, 1974), p. 75. A useful long-term view is in D. W. MacCleery, *American forests: a history of resiliency and recovery* (Durham: For. Hist. Society, 3d printing, 1994).

2. R. Albion, *Forests and sea power* (Cambridge: Harvard Univ. Press, 1926); and J. J. Malone, *Pine trees and politics* (Seattle: Univ. of Washington Press, 1964), are standard works on the mast trade. S. F. Manning, *New England masts and the King's Broad Arrow* (Kennebunk: Thomas Murphy, 1979), is brief and contains vivid illustrations, several of which are reproduced in this volume; see also C. E. Clark, *The eastern frontier* (New York: Knopf, 1970), p. 98ff. Valuable background on the colonial economy and the role played by forest products is found in C. E. Carroll, *The timber economy of Puritan New England* (Providence: Brown Univ. Press, 1973); E. J. Perkins, *The economy of colonial America* (New York: Columbia Univ. Press, 1980);

and D. R. McManis, *Colonial New England: a historical geography* (New York: Oxford Univ. Press, 1975). On the timber trade in Britain, see also B. Latham, *Timber: a historical survey* (London: George G. Harrap, 1957).

3. H. Chapelle, *The American fishing schooner, 1825–1935* (New York: W. W. Norton, 1973), p. 76.

4. USDA-FS, *Timber depletion, lumber prices, lumber exports, and concentration of timber ownership,* Rept. on Sen. Res. 311 (Washington, D.C.: GPO, 1920). Hereafter, "Capper Report," so called after Senator Capper, whose Senate Resolution called for the report. Data from H. B. Steer, *Lumber production in the U.S.: 1799–1946,* USDA Misc. Publ. 669, Oct. 1948. On changing regional competitive advantage, see C. Colgan and L. C. Irland, "The "sustainability dilemma": observations from Maine history," in R. E. Barringer, ed., *Toward a sustainable Maine* (Portland: Muskie Inst., Univ. of So. Maine, 1993), pp. 53–116.

5. G. B. Emerson, *A report on the trees and shrubs growing naturally in the forests of Massachusetts* (Boston: Little, Brown, 1887).

6. G. P. Marsh, *Man and nature* (1864; reprint, Cambridge: Harvard Univ. Press, 1964), p. 258.

7. F. B. Hough, *Cultivation of timber and the preservation of forests,* House Rept. 259, 43d Congress, 1st Sess., March 18, 1874. C. S. Sargent, *The forests of the United States,* (Washington, D.C.: Dept of the 10th Census 1884) pp. 494–510.

8. *Report of the Forestry Commission of New Hampshire,* Concord, 1885.

9. A. Cary, in *Report of the Forest Commissioner,* 1896, Augusta.

10. *Report of the Forest Commissioner,* Augusta, 1894 and 1919, p. XXVIII.

11. E. W. B. Russell, "The 1899 New Jersey state geologists' report," *J. For. Hist.* 32(4): 205–211 (Oct. 1988); see discussion of the Adirondacks in ch. 8 and notes cited there; J. T. Rothrock, *Forest devastation in Pennsylvania* (Harrisburg: Penn. Dept. of Forestry, 1916); and J. S. Illick, *Forest situation in Pennsylvania,* Penn. Dept. of Forestry Bull. 30, 1923.

12. James S. Whipple, "Address," in *Proc. of the Conf. of the Governors,* The White House, May 13–15, 1908, (Washington, D.C.: GPO), p. 99.

13. A. E. Moss, *Annual report,* CT Agr. Exp. Sta., Storrs, 1915, p. 197ff.; and H. O. Cook, "A forest survey of Massachusetts," *J. For.* 27 (1929): 518–522. Other forest surveys and analyses of this period are R. C. Averill, W. B. Averill, and W. I. Stevens, *A statistical forest survey of seven towns in central Massachusetts* (Petersham, MA: Harvard Forest Bull. No. 6, 1923); A. F. Hawes and R. C. Hawley, *Forest survey of Litchfield and New Haven counties,* Connecticut, CT Agr. Exp. Sta. Bull. 162, Forestry Publ. No. 5,

1909; R. C. Bryant, "Rept. of the Committee on the Timber Supply of Connecticut," in *A forest policy for Connecticut,* Rept. of the Connecticut Forestry Assoc., Nov. 24, 1920. A useful summary is A. Wilkins, *The forests of Maine* (Augusta: Maine Forest Service Bull. No. 8, 1932).

14. F. W. Card, *Forests of Rhode Island,* Rhode Island Agr. Exp. Sta. Bull. No. 88, Oct. 1902.

15. *Capper Report* generally.

16. *Capper Report,* p. 28; and J. Illick, *Pennsylvania trees* (Harrisburg: Penna. Dept. of Forestry, Bull. 11, 1923).

17. H. I. Baldwin, *Forestry in New England* (Boston: National Resources Planning Board, Region 1, Bull. No. 70, 1942).

18. USDA-FS, *Forests and national prosperity: a reappraisal of the forest situation in the U.S.* (Washington, D.C.: Misc. Publ. 668, 1948).

19. V. L. Harper, *Timber resources of New England and New York with special reference to pulpwood supplies,* USDA-FS NEFES Sta. Paper No. 5, 1947.

20. C. J. Gadzik, J. H. Blanck and L. E. Caldwell, *Timber supply outlook for Maine: 1995–2045.* (Augusta, ME: Dept. of Conservation, 1998). Maine Department of Conservation, "The state of the forest and recommendations for forest sustainability standards," Draft report to the Joint Standing Committee of the 119th Legislature on Agriculture, Conservation and Forestry, Jan. 1999.

21. T. W. Birch, "Land ownership and harvesting trends in eastern forests," in *Proc., 20th Annual Hardwood Symposium* (Nashville: Hardwood Research Council, Nat. Hardwood Lumber Assn., 1992).

22. Recent bulletins on the region's forests include: D. S. Powell and T. J. Considine, Jr., *Analysis of Pennsylvania's forest resources,* USDA-FS NEFES Res. Bull NE-69, 1982; C. L. Alerich, *Forest statistics for Pennsylvania — 1978 and 1989,* USDA-FS NEFES Res. Bull. NE-126, 1993; R. H. Widmann, *Forest resources of Pennsylvania,* USDA-FS NEFES Res. Bull. NE-131, 1995; R. T. Brooks, D. B. Kittredge, and C. L. Alerich, *Forest resources of southern New England,* USDA-FS NEFES Res. Bull. NE-127, 1993; D. S. Powell and D. R. Dickson, *Forest statistics for Maine, 1971 and 1982,* USDA-FS NEFES Res. Bull. NE-81, 1984; and Anon., *Forest for the future, Report on Maine's forest to the legislature, the Governor, and the people of Maine* (Augusta: Maine Dept. of Cons., 1988). For more recent details, see note 20 supra. Current sources on the region's forest resources include E. Baum, J. Falk, and D. Vail, *Maine's forest economy: crisis or opportunity, vol. 2, Summary of Survey Responses* (Augusta: Mainewatch Inst., 1988); R. Brooks, T. Frieswyk, and A. Ritter, *Forest wildlife habitat statistics for Maine 1982,* USDA-FS NE-RB-96, 1984; J. Selser, K. Hendren, J. Ecker, and H. Hill, *The Maine forest: its future, A perspective and plan of the Maine Forest Service*

(Augusta: Maine Dept. of Cons., 1985); D. M. Griffith and C. L. Alerich, *Forest statistics for Maine, 1995,* USDA-FS NEFES Res. Bull. NE-135, 1996; D. DiGiovanni and C. Scott, *Forest statistics for New Jersey — 1987,* USDA-FS NEFES Res. Bull. NE-112, 1987; T. J. Considine, Jr., *Analysis of New York's timber resources,* USDA-FS NEFES Res. Bull. NE-80, 1984; P. R. Brooks, comp., *The forest resources of New York: a summary assessment* (Albany: Forest Resources Planning, NYSDEC, 1981); T. Considine, Jr. and T. Frieswyk, *Forest statistics for New York, 1980,* USDA-FS Res. Bull. NE-71, 1982; C. L. Alerich and D. A. Drake, *Forest statistics for New York: 1980 and 1993,* USDA-FS NEFES Res. Bull. NE-132 (1995); Rhode Island Office of State Planning and Div. of For. Env., *Rhode Island forest resources management plan,* Providence, 1984; Penn. Bur. For., *Assessment and issues for Pennsylvania forest resources* (Harrisburg: Dept. of Env. Res. Mar. 1985), with update; Bur. For., *Heading into the 21st Century,* Harrisburg, Nov. 1992; Vermont Dept. of For., Parks, and Recr., *Vermont forest resources plan, part 2, Assessment* (Montpelier: Agency of Env. Cons., 1983), which contains excellent maps; Forest Resources Plan Steering Committee, *Draft New Hampshire forest resource plan* (Concord: Div. For. and Lands, 1995); and *Results of an interim forest inventory of New Hampshire's timber resource* (Concord: Div. Forest and Lands, 1995).

23. N. P. Kingsley, *The timber resources of southern New England,* USDA-FS NEFES Res. Bull. NE-36, 1974.

24. R. H. Ferguson and N. P. Kingsley, *The timber resources of Maine,* USDA-FS NEFES Res. Bull. NE-26, 1972. N. P. Kingsley, *The forest resources of Vermont,* USDA-FS NEFES Res. Bull. NE-46, 1977; and N. P. Kingsley, *The forest resources of New Hampshire,* USDA-FS NEFES Res. Bull. NE-43, 1976. R. H. Ferguson and C. E. Magee, *The timber resources of New Jersey,* USDA-FS NEFES Res. Bull. NE-34, 1974; D. M. DeGiovanni and C. T. Scott, *Forest statistics for New Jersey — 1987,* USDA-FS NEFES Res. Bull. NE-112, 1990.

25. These paragraphs excerpted from L. C. Irland and D. I. Maass, "Characterizing the north's hardwood resource," in S. B. Jones and J. A. Stanturf, eds., *Hardwood forest opportunities: creating and expanding businesses* (State College: Penn. State Univ., School of Forest Resources, 1989).

26. Examples include J. C. Finley, S. B. Jones, and J. Pell, "Northern Appalachian forest: how are we doing?" in 1997 Penn State School of Forest Resources Issues Conference — *Forest sustainability: what's it all about?,* Penn. State College of Agr. Sci., School of For. Res., 1997, pp. 34–43; and remarks by state forester Jim Grace in the same volume. Also, *Final report of New Hampshire Forest Liquidation Study Committee* (Concord, NH: Div. of For. and Lands, Feb. 1998).

27. An excellent overviews of the current status of certification programs for forestry and wood products are found in special issues of the J. For., April

1995 and Feb. 1999. More detail on the Seven Islands experience is found in the piece by McNulty and Cashwell in the April 1995 issue. For an earlier view, see L. C. Irland, "Wood producers face green marketing era," *Wood Technology,* Mar./Apr., 1993, pp. 34–36. The literature continues to expand: D. R. Carter and F. D. Merry, "The nature and status of certification in the U.S.," *For. Prod. J.* 48(2): 23–28 (1998); V. M. Viana, et al., eds., *Certification of forest products: issues and perspectives* (Washington, D.C.: Island Press, 1996); and C. Upton and S. Bass, *The forest certification handbook* (Delray Beach, FL: St. Lucie Press, 1996); and C. M. Mater, V. A Sample, J. R. Grace, and G. A. Rose, "Third-party, performance-based certification," *J. For.* 97(2):6–12, 1999.

28. Sargent, p. 489. But see H. Gannett, "Lumber" in *12th Census of the U.S., 1900 — Manufacturing, part III, special repts. on selected industries,* Washington, D.C., 1902.

29. K. E. Skog and I. A. Watterson, "Residential fuelwood use in the U.S.," *J. For.* 82(12): 742–747 (1984); see also fuelwood use surveys in New England including L. Palmer, R. McKusick, and M. Bailey, *Wood and energy in New England,* USDA Econ. Stat. and Coop. Serv., Bibliogr. and Lit. of Agr., No. 7, 1980.

30. The classic analysis of this problem is based on the work of G. E. Likens, F. H. Bormann, and others at the Hubbard Brook Experimental Watershed in the White Mountains. The most useful general report on their work is Bormann and Likens, Pattern and process in a forested ecosystem (New York: Springer Verlag, 1979). For detailed studies, see *Proc., Impact of intensive harvesting on forest nutrient cycling,* State Univ. College of Env. Sci. and For., Syracuse, 1979. For a valuable overview, see C. T. Smith, C. W. Martin, and C. M. Tritton, eds., *Proc. of the 1986 symposium on the productivity of northern forests following biomass harvesting,* USDA-FS NEFES Gen. Tech. Rept. NE-GTR-115.

31. K. E. Skog, "Projected wood energy impact on U.S. forest wood resources," in Proc. *1st biomass conference of the America's* (Golden, Co., Natural Renewable Energy Lab, 1993), vol. I, pp. 18–32). For earlier literature 1978–1980, see note 10 of *Wildlands and Woodlots.* Useful references include: R. Chamberlin and C. High, *Economic impacts of wood energy in the northeast, Summary Report,* Resource Policy Center, Thayer School of Engineering, Dartmouth College, 1986; J. Flumerfelt and K. Mahmood, *Final report of the Commission on Comprehensive Energy Planning* (Augusta: Econ. & Energy Policy Div., Maine State Planning Office, 1992); New York State Energy Office, Dept. of Public Service, and Dept. of Env. Cons., *Draft New York State energy plan, 1991 Biennial Update, vol. I, Summary Report; Vol. II, Assessment Repts.* A valuable overview is found in Northeast Regional Biomass Program, *Mission Accomplishments, Prospects 1992* (Washington, D.C.: CONEG, 1992).

32. D. S. Solomon and T. B. Brann, *Ten-year impact of spruce budworm on spruce-fir forests of Maine,* USDA-FS NEFES Gen. Tech. Rept. NE-165, 1992; also, relevant sections of L. C. Irland, et al., *The spruce budworm outbreak in Maine in the 1970s: Assessment and directions for the future* (Orono: Univ. of Maine Agr. Exp. Sta. Bull. 819, 1988).

33. For developing views on forest health, a series of articles in the July 1994 *J. For.* is useful, also D. H. Smith, W. H. Smith and W. E. Reifsnyder, "Silviculture, pests and disease, forest and wildland fires," in J. G. Griffiths (ed.), *Handbook of agricultural meteorology* (New York: Oxford University Press, 1994), pp. 210-219.

34. P. Niemalä and W. J. Mattson, "Invasion of North American forests by European phytophagous insects: legacy of the European crucible?" *BioScience* 46(10): 741–753 (1996); A. M. Leibhold, et al., "Invasion by exotic forest pests: a threat to forest ecosystems," *For. Sci. Monogr.* 30: 1–49 (1996); and D. A. Orwig and D. R. Foster, "Forest response to the introduced hemlock woolly adelgid in southern New England, USA," *J. Torrey Bot. Soc.* 125(1): 60–73 (1998).

35. USDA-FS, *Gypsy moth management in the U.S.: a cooperative approach. Final EIS* (Radnor: State and Private Forestry, Nov. 1995, 5 vols.); R. W. Campbell, *Gypsy moth: forest influence,* USDA-FS Agr. Info. Bull. 432, 1979; *The gypsy moth: research toward integrated pest management,* USDA-FS and Sci. and Educ. Agency Bull. 1584, 1981; and D. A. Gansner, S. L. Arner, R. R. Hernsley, and S. L. King, *Defoliation potential of gypsy moth,* USDA-FS NEFES NE-INF-117-93, 5 pp. On the ice storm, see L. C. Irland, "Ice Storm 1998 and the forests of the Northeast," *J. For.* 96(9): 32–40, 1998.

36. Valuable recent overviews are M. Miller-Weeks, *Northeastern area forest health report,* USDA-FS, Northeastern Area NA-TP-03-93, 1993, which contains valuable maps and graphics; D. Twardus, et al., 1995, *Forest health assessment for the northeastern area, 1993,* USDA-FS, Northeastern Area NA-TP-01-95; J. J. Stoyenoff, J. J. Witter, and B. Leutscher, *Forest health in the New England States and New York* (Ann Arbor: Univ. of Michigan School of Nat. Res. and Environ., unnum. pub., 1998); and I. Millers, D. S. Shriner, and D. Rizzo, *History of hardwood decline in the eastern U.S.,* USDA-FS NEFES Gen. Tech. Rept. NE-126, 1989. On forest pests, see J. F. Anderson and H. K. Kaya, *Perspectives in forest entomology* (New York: Academic Press, 1976); see also W. A. Sinclair and R. J. Campana, "Dutch Elm disease — perspectives after 10 years,: *Search-Agriculture* 8(5) (1978), Cornell Univ. Agr. Exp. Sta. On spruce budworm, see ch. 4. On fires, see A. Wilkins, *Ten million acres of timber* (Woolwich, ME: TBW Books, 1978).

37. See references in note 6 in ch. 2.

38. Useful perspectives on forests and wood use over time are found in M. Clawson, "Forests in the long sweep of American history," *Science* 204 (1979): 1–7; and R. V. Reynolds, "How much timber has America cut?," *J.*

For. 33(1) (1935): 34–41. On the difficulties of comparing early timber volume estimates, see M. Dietz, "Review of the estimates of the sawtimber stand on the U.S., 1880–1946," *J. For.* 45(12) (1947): 865–874. I have not discussed here the findings of the *1958 Timber Resources Review* (data for 1952), the *Timber Trends Report of 1965* (data for 1963), the *Outlook* (1973), or the *USDA-FS National Assessments.* Recently available is USDA-FS, *RPA assessment of the forest and rangeland situation in the U.S.: 1993 update,* Washington, draft, 1993; which in turn relies heavily on R. Haynes, ed., *Analysis of the timber situation in the U.S.: 1989–2040,* USDA-FS RMFRES Gen. Tech. Rept. RM-199, 1990; USDA SCS, *The second RCA appraisal: soil, water, and related resources on nonfederal land in the U.S.,* Washington, D.C., USDA, 1989, supplies a valuable overview of rural land conditions, including water and forests, with abundant state-by-state data.

Notes on Chapter 11

1. For detailed background on the region's industries, numerous state-level publications are available, for example, Anon., *Vermont forest harvest in summary and perspective* (Waterbury, VT: Agency of Natural Resources, Dept. of Forests, Parks, and Recreation, annual, processed); W. Gove, *The forest resources and wood using industries of New Hampshire* (Durham: Univ. of N.H. Coop. Ext., 1993); R. L. Nevel, Jr., N. Engalichev, and W. C. Gove, *The timber industries of New Hampshire and Vermont,* USDA-FS NEFES Res. Bull. NE-89, 1986. R. L. Nevel, Jr., E. L. Stochia, and T. Wahl, *New York timber output,* USDA-FS NEFES Res. Bull. NE-73, 1982; B. B. Norris, *Locational analysis of Pennsylvania's secondary wood industry,* Penn. State Univ. Hardwoods Devel. Council Arch. Ser. No. 1, 1990; E. H. Wharton and J. L. Bearer, *The timber industries of Pennsylvania,* USDA-FS NEFES Res. Bull. NE-130; Anon, *Forest industry in northern New York* (Albany: Empire State Forest Prods. Assn., 1985); R. L. Nevel, P. R. Lammert, and R. H. Widmann, *Maine timber industries — a periodic assessment of timber output,* USDA-FS NEFES Res. Bull. NE-83, 1985; L. C. Irland and C. W. Murdoch, *Value-added processing in Maine's wood industry* (Orono: Univ. of Maine Agr. Exp. Sta. Misc. Rept. 364, 1992); R. L. Nevel, Jr., and E. H. Wharton, *Timber industries of southern New England,* USDA-FS NEFES Res. Bull. NE-101, 1988; R. S. Bond and A. M. Loud, "Lumber production and marketing changes by sawmills in Massachusetts: 1957–89," *North J. Appl. For.* 9(2):67–69 (1992) For nearby Canada, see Anon., *State of Canada's forests, 1996–97: learning from history* ("Canada's Green Plan"), Ottawa: Forestry Canada, 1997; and Anon., *Quebec's forest resources and industry, statistical information* (Quebec: Ministry of Forests, Annual).

2. For these vanished industries, see, generally, W. H. Rowe, *The maritime history of Maine: three centuries of shipbuilding and seafaring* (Gardiner, ME: Harpswell Press, 1989). J. A. Goldenberg, *Shipbuilding in colonial America* (Charlottesville: Univ. Press of Virginia for the Mariners Museum,

1976). Also, relevant sections of M. Williams, "Industrial impacts on the forests of the U.S., 1860–1920," *J. For. Hist.* 31(3) 1987; and the book, *Americans and their forests: a historical geography* (New York: Cambridge Univ. Press, 1989), from which this article was excerpted; J. Perlin, *A forest journey* (New York: W. W. Norton, 1989), contains abundant detail, esp. chs. 10 and 12; and T. R. Cox, R. S. Maxwell, P. D. Thomas, and J. J. Malone, *This well-wooded land: Americans and their forests from Colonial times to the present* (Lincoln: Univ. Nebraska Press, 1985). A well-illustrated, non-technical treatment is W. G. Youngquist and H. O. Fleischer, *Wood in American life, 1776–2076* (Madison: For. Prods. Res. Soc., 1977); B. McMartin, *Hides, hemlocks, and Adirondack history: how the tanning industry influenced the region's growth* (Utica: North Country Books, 1992). C. S. Sargent *The forests of the United States* (Washington, D.C.: Dept. of the 10th Census, 1884), pp. 494–510 contains an extensive section on Pennsylvania's tanning industry.

3. R. P. T. Coffin, Kennebec, *Cradle of Americans* (New York: Farrar and Rinehart, 1937), p. 176.

4. For more, see D. C. Smith, *History of lumbering in Maine, 1861–1960* (Orono: Univ. of Maine Press, 1972). A well-illustrated overview of Maine sawmills from the 17th to 19th centuries is found in P. Rivard, *Maine sawmills: a history* (Augusta: Maine State Museum, 1990), its depictions would suit the entire region for this period. Also, relevant sections of J. E. Defebaugh, *History of the American lumber industry* (Chicago: Amer. Lumberman, 1906); R. J. Hoyle, "Changes in the wood-using industries of New York State since 1912," *Southern Lumberman*, Jan. 15, 1949, pp. 39–42; and L. C. Irland, *Is timber scarce?* (New Haven: Yale School of For. and Env. Studies Bull. 83, 1974), ch. 4, and notes therein.

5. Counting employment based on wood in a region is difficult. Log exports and imports, and trade in lumber and pulp, complicate matters. In complete counts of small firms, disclosure limitations cause underestimates or gaps. To make matters worse, different agencies in a state may issue different estimates of employment in the same industry.

6. See, e.g., D. C. Smith, *History of lumbering*, chs. 9–16; D. C. Smith, *History of the paper industry of the United States, 1690–1970* (New York: Lockwood's, 1971); and J. A. Guthrie, *An economic analysis of the pulp and paper industry* (Pullman: Washington State Univ. Press, 1972), which includes a good summary of papermaking technology and products. H. Hunter, "Innovation, competition, and locational change in the pulp and paper industry: 1880–1950," *Land Econ.* 31 (1955): 314–327 is a valuable summary; also, L. C. Irland, *Maine timber supply: historical notes and speculation on the future*, Proc., Blaine House conference on forestry (Augusta: Maine Dept. of Cons., 1981); E. Amigo, M. Neuffer, and E. Maunder, *Beyond the Adirondacks: the story of St. Regis Paper Co.* (Westport, CT: Greenwood Press, 1980); W. A. Duerr, "Forestry's upheaval: are advances in western civilization

redefining the profession?," *J. For.* 84(1): 20–26, which contains a graphic overview of U.S. wood use trends.

7. Bureau of Industrial and Labor Statistics, *Twentieth annual report,* Augusta, 1906, p. 140.

8. A useful overview on the region's economy can be found in J. K. Wright, et al., *New England's Prospect: 1933* (New York: Amer. Geo. Soc., Spec. Publ. No. 16, 1933). The standard source is R. W. Eisenmenger, *Dynamics of growth in New England's economy, 1870–1964* (Middletown: Wesleyan Univ. Press, 1967). On Pennsylvania's economy, see T. E. Fuller and S. M. Smith, *Road to Renaissance, IV, update on Pennsylvania's industries,* College of Agr., Penn. State Univ., Dept. of Agr. Econ. and Rural Sociology, n.d. (ca 1994).

9. The literature of economic development is vast; programs are complex, and the issues change. Periodically state governments issue Economic Development Plans or Strategies, which, at a minimum, offer summaries of the issues of the moment and usually contain useful background facts. Economic trends can be followed through publications of the Federal Reserve Banks and state agencies. Particularly valuable for general background are: Anon., *Forest resources in the northeast: contributing to economic development and social well-being,* prepared by the 20 Northeastern state foresters, 1985, no location or publisher shown (presume Michigan DNR); A valuable resource, even if the data are dated, is D. L. Brown et al., eds., *Rural economic development in the 1980s: prospects for the future,* USDA ERS Rural Devel. Res. Rept. No. 69, 1988, esp. ch. 16; also, M. B. Lapping, T. L. Daniels, and J. W. Keller, *Rural planning and development in the U.S.* (New York: Guilford Press, 1989); D. John, S. S. Batie, and K. Norris, *A Brighter future for rural America? Strategies for communities and states* (Washington, D.C.: National Governors Assn., 1988); *Rural development: rural America faces many challenges* (Washington, D.C.: GAO/REED-93-35, 1992); Anon., *Future directions in rural development policy* (Washington, D.C.: USDA, 1990); Anon., *Enhancing rural America: national research program* (Washington, D.C.: USDA-FS, 1991); and W. A. Galsston and K. J. Baehler, *Rural development in the U.S.: connecting theory, practice, and possibilities* (Covelo, CA: Island Press, 1995). A readable overview is D. Osborne, *Laboratories of democracy: a new breed of governor creates models for national growth* (Boston: Harvard Business School Press, 1988).

10. O. W. Herrick, "Structure and change in the northern U.S. forest products industry: a shift-share analysis," *For. Prod. J.* 26 (Aug. 1976): 29–34.

11. L. C. Irland and R. Whaley, "Northern forest outlook: a working sketch of regional forest conditions in 2040," in T. A. Lepisto, ed., *Sustaining ecosystems, economies, and a way of life in the Northern Forest* (Washington, D.C.: The Wilderness Soc., 1993), pp. 77–83. The most recent national overview is in P. Ince, *Recycling and long-range outlook timber outlook,*

USDA-FS For. Prod. Lab. Res. Pap. FPL-RP-534 (1994); for more geographic detail: R. Haynes, coord., *Analysis of the timber situation in the U.S.: 1989–2040,* USDA-FS RMFRES Gen. Tech. Rept. RM-109 (Dec. 1990).

12. S. B. Jones and J. A. Stanturf, eds., *Hardwood forest products opportunities: creating and expanding businesses* (State College: Penn. State School of Forestry, 1989), for Oct. 16–19, 1989 Conference, Pittsburgh, PA, 150 pp.; J. A. Stanturf, *Needs assessment report to the Penn. Hardwoods Devel. Council* (State College: Penn. State School of Forestry, 1989); R. S. Whaley, chair., 1989, Capturing the potential of New York's forests, Rept. of the Governor's Task Force on Forest Industry, Albany, 1989.

13. There is no comprehensive treatment on log exports in the northeast. For a number of perspectives, see S. B. Jones, et al., "Export embargo on publicly owned logs: market implications for Pennsylvania and the northeast," *J. For.* 92(2): 41–46, which gives an unusually careful treatment; J. Dillon, "Log exporting: the view from Vermont," *N. Logger,* June 1995, p. 26ff; and L. C. Irland, "Trade in unprocessed wood: impact on the forest and on Maine's economy," *Maine Business Indicators,* Jan. 1995, pp. 4–5 (publ. by Univ. So. Maine College of Business); L. C. Irland, "Should Vermont restrict log exports? No — the case hasn't been made," *Vermont Woodlands,* Spring 1997, p. 18ff, with response by Eric Palola of the National Wildlife Federation taking the opposite position; L. C. Irland, "Northeast USA as a fiber basket for E. Canada mills," in *2d Annual Eastern Canada Wood Products Industry Conference Proceedings,* Sept. 24–26, 1997, pp. 103–112.

14. J. McNutt, R. Haggblom, and K. Ramo, *The global fiber resource picture, in Anon., Wood product demand and the environment* (Madison: Forest Products Res. Soc., 1992), pp. 39–53. *Pulp and Paper* magazine periodically publishes detailed updates on the status of mill projects and on the progress of recycling in particular. P. Engel and B. Moore, "Recycled mills seek options to turn deinking residuals into resources," *Pulp and Paper,* April 1998.

15. P. Koch, *Wood vs. nonwood materials in U.S. residential construction: some energy-related implications* (Seattle: Univ. of Washington, 1991), CINTRAFOR Work. Pap. 36, 38 pp. (also published in *For. Prod. J.,* May, 1992); H. A. Wells, Jr., "Recycled paper: the unresolved issues," *Ren. Res. J.* 7(1991): 7–12; J. Zerbe, *Wood as an environmentally desirable material,* USDA-FS For. Prods. Lab., unpub. paper, 9 pp. + app., n.d.; S. Alexander and B. Greber, *Environmental ramifications of various materials used in construction and manufacture in the U.S.,* USDA-FS PNWRS Gen. Tech. Rep. PNW-GTR-277, 1991, 21 pp.; Forintek Canada and Wayne Trusty & Assoc., Ltd., *Forintek's western laboratory: building in wood to meet environmental objectives, draft,* Vancouver, 1991, processed, 9 pp.

16. A rigorous analysis using an up-to-date model is given in R. S. Seymour and R. C. Lemin, Jr., *Timber supply projections for Maine, 1980–2080* (Univ. of Maine Coll. For. Res. and Agr. Exp. Sta., Misc. Rept. 337, 1989). A

new timber supply forecast was prepared by the Maine Forest Service in 1998. C. J. Gadzik, J. H. Blanck and L. E. Caldwell, *Timber supply outlook for Maine: 1995–2045*, (Augusta: Maine Dept. of Conservation, 1998). For 1970's citations, see *Wildlands and Woodlots.*

Notes on Chapter 12

1. For an entry into state forest policies, see R. R. Widner, ed., *Forests and forestry in the American states* (Washington, D.C.: National Assoc. of State Foresters, [c. 1968]); and the annual reports of the state forestry agencies. The serious student will find that the superagency boom of the 1970s and legislative restraints on published reports have made finding reports more difficult. For citations from the 1970s, see notes from *Wildlands and Woodlots.* A useful regional history is R. I. Ashman, ed., *A half century of forestry, 1920–1970*, New England Section, Society of American Foresters, 1970 (no place of publ. shown). A recent and thorough overview for Massachusetts is Charles H. W. Foster, ed., *Stepping back to look forward: a history of the Massachusetts forest* (Petersham, MA: Harvard Forest, 1998). A standard overview of the field is P. V. Ellefson, *Forest resources policy: process, participants, and programs* (New York: McGraw-Hill, 1992).

2. For an introduction to federal policies, see S. T. Dana and S. Fairfax, *Forest and range policy* (New York: McGraw-Hill, 1980); H. K. Steen, *The Forest Service (Seattle: Univ. of Washington Press, 1976);* and references cited in both works. On the eastern national forests, *W. E. Shands and R. G. Healy, The lands nobody wanted* (Washington, D.C.: The Conservation Foundation, 1977), is valuable. An annotated listing of USDA programs is given in Anon., *Conservation promoted by various USDA programs,* Agr. Res., Sit. and Outl. Rept. AR-30, May 1993. Also, R. Neil Sampson and L. A. DeCoster, *Public programs for private forestry* (Washington, D.C.: American Forests, 1997).

3. J. P. Kinney, *Forest legislation in America prior to Mar. 4, 1789* (Ithaca: Cornell Univ. Agr. Exp. Sta., New York State College of Agr., Bull. 370, Jan. 1916). In a useful review, Y. Kawashima, "Forest conservation policy in New England," *Hist. J. of Mass.,* 20: 1–15 (Winter 1992), argues that the policies were ineffective.

4. R. L. Bushman, *From Puritan to Yankee: character and social order in Connecticut, 1690–1765* (Cambridge: Harvard Univ. Press, 1967), p. 4; see also O. Handlin and M. F. Handlin, *Commonwealth: a study of the role of government in the American economy: Massachusetts, 1774–1861* (New York: Columbia Univ. Press, 1947); and J. R. T. Hughes, "What difference did the beginning make?," *Amer.* Econ. Rev. 67(1) (1974): 15–20.

5. On the origins of New England's land system, land disposal policies, and their effects, see note 5, ch. 11 in *Wildlands and Woodlots.* An excellent overview of land policy in New York is in D. M. Ellis, et al., *A history of New York State* (Ithaca: Cornell Univ. Press, 1967), chs. 7 and 13. Also references

in ch. 9, note 2 above. For historic perspective and continental context, see D. W. Meinig, "The shaping of America: a geographical perspective on 500 years of history," vol. I, *Atlantic America, 1492–1800* (New Haven: Yale Univ. Press, 1986), pp. 100–109; 231–244.

6. L. M. Schepps, "Maine's public lots: the emergence of a public trust," *Maine Law Rev.* 26(2) (1974): 217–272.

7. This is a simple caricature made necessary by brevity. For a penetrating and full account, see J. Willard Hurst's, *Law and economic growth: the legal history of the lumber industry in Wisconsin, 1830–1915* (Cambridge: Harvard Univ. Press, 1964), the quote is at p. 601. Also, for Maine, see D. C. Smith, *A history of lumbering in Maine, 1861–1960* (Orono: Univ. Of Maine, 1972), chs. 7, 11, 12.

8. The conservation history of this period has yet to be written. See reports of the U.S. Outdoor Recreation Resources Review Commission and the Public Land Law Review Commission; and Dana and Fairfax, *Forest and range policy,* chs. 7–12. Key works setting the tone for the era were Stuart Udall's *The quiet crisis* (New York: Avon Books, 1964); and Rachel Carson's *Silent spring* (Boston: Houghton Mifflin, 1994).

9. Works embodying the green backdrop philosophy are W. H. Whyte, *The last landscape* (New York: Doubleday-Anchor Books, 1970); and Ian McHarg's influential work, *Design with nature.*

10. L. C. Irland and J. E. Connors, "State nonpoint source programs affecting forestry: the 12 northeastern states," *North J. Appl. For.* 11(1): 5–11 (1994); and L. J. Hawks, et al., "Forest water quality protection," *J. For.,* 91: 48–54.

11. See ch. 1, note 7.

12. Dahl, T. E., C. E. Johnson, and W. E. Frayer, *States and trends of wetlands in the coterminous U.S., mid 1970s to mid 1980s* (Washington, D.C.: USDI Fish and Wildlife Serv., 1991), p. 2.

13. National Wetlands Policy Forum, 1988, *Protecting American's wetlands: an action agenda* (Washington, D.C.: The Conservation Foundation), ch. 2.

14. For background, see ch. 6, notes 22–24.

15. See, e.g., Anon., *Final report of the Maine Commission on Outdoor Recreation* (Augusta: Legislative Office of Policy and Legal Analysis, 1988); D. Hudnut and S. Golden, *Public ways in the northern forest lands study area,* Study by NPS for the NFLS (Boston: NPS, n.d. [ca 1990]).

16. A recent volume, while aimed at federal agencies, provides an excellent overview of the problems of setting priorities: National Research Council, *Setting priorities for land conservation* (Washington, D.C., 1993); see also Anon., *A land conservation strategy for the New England Region* (Cam-

bridge: Lincoln Inst. for Land Policy, n.d. Monogr. 86–89); and Anon., *A greenway for the Hudson River Valley: a new strategy for preserving an American treasure* (Albany: Hudson River Valley Greenway Council, 1989, 2d ed.). A recent proposal to reform the Federal Land and Water Conservation Fund, the principal federal funding program for state and local conservation land acquisition, is in *An American network of parks and open space: creating a conservation and recreation legacy* (Washington, D.C.: USDI, Nat. Park Serv., Aug. 1994).

17. National Research Council, *Forestry research: a mandate for change* (Washington, D.C.: Nat. Academy Press, 1990); and American Forest and Paper Assn., *Forestry research for the 1990s: industry's suggested priorities* (Washington, D.C.: AFPA, 1993). For a more technical view, see P. V. Ellefson, ed., *Forest resource economics and policy research: strategic directions for the future* (Boulder: Westview Press, 1989).

18. See, of a vast literature, A. Reiser, "Managing the cumulative impact of coastal land development: can Maine law meet the challenge?," *Maine Law Rev.* 39(2): 322–389 (1987). J. DeGrove, *The new frontier for land policy: planning and growth management in the states* (Cambridge: Lincoln Inst. for Land Policy, 1992); Regional Planning Assoc., *The region tomorrow, Apr. 1991, Issue #1; Where we stand: principles for the regional plan, New York, Apr. 1991*; and also Regional Planning Assoc., *3 working papers on the open space imperative*, 1987; R. C. Einsweiler and D. A. Howe, *Managing "the land between": a rural development paradigm* (Cambridge: Lincoln Inst. for Land Policy, 1993), working paper; and the interesting perspective in T. M. Beckley, "Leftist critique of the quiet revolution in land use control: two cases of agency formation," *J. Planning Ed. & Res.* 12: 55–66 (1992).

19. USDA-FS, *Forests and national prosperity* (Washington, D.C.: GPO, 1948), USDA-FS Misc. Publ. 668, p. 6.

20. Aldo Leopold, *A Sand County almanac* (New York: Oxford Univ. Press, 1966), p. 187.

21. The movement for regulation is well summarized in relevant sections of W. G. Robbins, *American forestry: a history of national, state, and private cooperation* (Lincoln: Univ. of Nebraska Press, 1985).

22. See, e.g., P. V. Ellefson, A. S. Cheng, and R. J. Moulton, *Regulation of private forestry practices by state governments* (St. Paul: Univ. of Minnesota Agr. Exp. Sta. Bull. 605, 1995); and A. Cheng and P. V. Ellefson, *State programs directed at the forestry practices of private forest landowners: program administrators' assessment of effectiveness* (St. Paul: Dept. of For. Res. Staff Paper Series No. 87, March 1993). For a useful perspective, see D. Salazar, "Regulatory politics and environment: state regulation of logging practices," *Res. in Law and Econ.*, 12: 95–117 (1989); and J. L. Greene and W. C. Siegel, *Status and impact of state and local regulation on private timber supply*, USDA-FS RMFRES Gen. Tech. Rept. RM-255 (1994); also, C. E.

Martus, H. L. Haney, Jr., and W. C. Siegel, "Local forest regulatory ordinances: trends in the eastern U.S.," *J. For.* 93(6): 27–31 (1995).

23. G. P. Marsh, *Man and nature* (Cambridge: Harvard Univ. Press, 1864; reprint, 1964), p. 259. For more on this important episode, see Dana and Fairfax, *Forest and range policy,* p. 160ff; and Steen, *The Forest Service,* ch. 13, pp. 222–237, and pp. 259–271.

24. An accessible review of policy options, with references, is State Resource Strategies, "New directions in conservation strategies: a reconnaissance of recent experimentation and experience," in *Technical appendix,* North. For. Lands Council, Feb. 1994, USDA-FS, SPF, Durham, NH.

25. A. Golodetz and D. R. Foster, "History and importance of land use and protection in the North Quabbin region of Massachusetts," *Cons. Biol.* 11(1): 227–235 (1996).

26. For a discussion of practical problems of landscape management in multi-owner areas, see V. A. Sample, "Building partnerships for ecosystem management on mixed ownership landscapes," *J. For.* 92(8): 41–44 (1994); and other papers in the same issue. Also, M. G. Rickenbach, D. B. Kittredge, Don Dennis, and T. Stevens, "Ecosystem management: capturing the concept for woodland owners," *J. For.* 96(4): 18–24 (1998). For another region but relevant for its policy approach, see S. D. Roberts and G. R. Parker, "Ecosystem management: opportunities in the central hardwood region," *North J. Appl. For.* 15(1): 43–48 (1998). A similar proposal, to establish "Legacy Forests," was offered by Charles H. W. Foster in, *Of vert and vision: ensuring the legacy of the northern forest of New England and New York* (Boston: Harvard Univ., Kennedy School of Government, Center for Science and International Affairs, paper 92–13, Dec. 1992). See also recommendations in part V of S. G. Thorne, et al., *A heritage for the 21st century: conserving Pennsylvania's native biological diversity* (Harrisburg: Penn. Fish and Boat Comm., 1995). Excellent reviews on using conservation easements are M. E. Boelhower, *Forests forever: a comprehensive evaluation of conservation easements on working forests in Maine, New Hampshire, and Vermont* (Concord: N.H. Cons. Inst., April 1995); and K. Wiebe and A. Tegene, *Partial interests in land: policy tools for resource use and conservation,* USDA-ERS, Agr. Econ. Rept. 744 (1996). A proposal for a regional wildlife refuge using many of these techniques was developed for the *Connecticut River Watershed: Silvio O. Conte National Fish and Wildlife Refuge: Action Plan and EIS* (USFWS, Turners Falls, MA, May 1995).

27. Leopold, *A Sand County almanac,* p. 101.

28. For full citations on the northern lands, see notes 19–22 in ch. 9.

29. Leopold, *A Sand County almanac,* p. 217ff. Anyone interested in conservation should read the entire book.

30. Catskill Center for Conservation and Development, *Summary guide to the terms of the Watershed Agreement* (Arkville, NY: October, 1997);

Maine Bureau of Public Lands, *Bigelow Preserve management plan* (Augusta: Dept. of Conservation, 1989). "Conservancy acquires 11,300 acre wildlands," *Conserve,* Jan. 1994 (W. Penn. Conservancy); Anon., "Two Vermont groups announce major timberland purchase," *North. Logger,* Mar. 1998, p. 36.

Notes on Chapter 13

1. J. S. Brooks, *Address, Proc. of Conference of Governors, the White House,* May 13–15, 1908 (Washington, D.C.: GPO, 1909, p. 100).

INDEX

Page numbers followed by *f* refer to figures; by *t,* tables, by *p,* photos.